Papers presented at the
Eleventh Technical Conference of
the British Pump Manufacturers' Association

PUMP TECHNOLOGY:
New challenges – where next?

Churchill College, Cambridge, England

18-20 April, 1989

Jointly sponsored by BHRA and BPMA,
co-sponsored by NEL (East Kilbride) and
Process Industries Division of the
Institution of Mechanical Engineers

Published jointly by
BHRA (Information Services), The Fluid Engineering Centre, Cranfield
SPRINGER – VERLAG
Berlin · Heidelberg · New York · London · Paris · Tokyo

First published in 1989 jointly by

BHRA (Information Services), The Fluid Engineering Centre,
Cranfield, Bedford MK43 0AJ, UK
Tel. (0234) 750422 Telex: 825059 Fax: (0234) 750074

and Springer-Verlag (Berlin, Heidelberg, New York)

© BHRA / Springer-Verlag 1989

British Library Cataloguing in Publication Data applied for.

ISBN 0 947711 68 6	**BHRA The Fluid Engineering Centre**
ISBN 3 540 51208 X	**Springer-Verlag Berlin Heidleberg New York**
ISBN 0 387 51208 X	**New York Heidelberg Berlin**

Editor's Note

Papers were reproduced from the Authors' original typescript in order to minimise delay.

The views expressed in these papers are those of the Authors and do not neccessarily represent the views of either the British Pump Manufacturers' Association or BHRA, The Fluid Engineering Centre or Springer-Verlag.

It is recommended that in citing work reported in this collection of papers the following convention be adopted e.g. 11th Technical Conference of the B.P.M.A., pp . . . (April 1989).

Editor: B. Glanfield

This publication is protected by international copyright law. All rights reserved. No part of this publication may be reproduced, stored in a retrieval system, or transmitted in any form or by any means, electronic, mechanical, photocopying, recording or otherwise, without the permission of the publishers.

Eleventh Technical Conference of the
British Pump Manufacturers' Association

PUMP TECHNOLOGY

New challenges – where next?

Eleventh Technical Conference of the
British Pump Manufacturers' Association

STEERING COMMITTEE

Dr A G H Coombs *(Chairman)*	Sterling Fluid Products
Mr M Bicknell	Watson Hawksley Consulting Engineers
Mr H Birchall	Flygt Pumps Limited
Mr A Chiappe	Hayward Tyler Limited
Mr T Fuggle	SSP Pumps Limited
Mr P Gundry	Energy Industries Council
Mr S J A Manning	Sheerway Technology Group Limited
Mr L Mathewson	Mono Pumps Limited
Mr J Peters	National Engineering Laboratories, East Kilbride
Mr D G Perry	Ingersoll Rand Co Limited
Mr P Robinson	KSB Pump Manufacturing Limited
Mr P Rogers	Crane Packing Limited
Mr G Scobie	National Engineering Laboratories, East Kilbride
Mr D W Standish	Girdlestone Pumps Limited
Mr L M Teasdale	Worthington Simpson Division, Dresser UK Limited
Mr K Turton	Loughborough University of Technology
Mr J W Veness	HMD Seal/Less Pumps Limited
Mr N M Wallace	Flexibox Limited
Mr B Glanfield *(Secretary)*	British Pump Manufacturers' Association

British Pump Manufacturers' Association
Artillery House, Artillery Row, Westminster, London SW1P 1RT, UK

Eleventh Technical Conference of the
British Pump Manufacturers' Association

NEW CHALLENGES – WHERE NEXT?

Introduction

The Conference sets out to examine some of the new challenges which face the pump industry arising from, for example, environmental issues or the need to pump multiphase fluids in the extraction of oil and gas from under the sea. These requirements place heavy emphasis on the achievement of high operating reliability in all components of a pump and call for new approaches to design and the use of new technology materials. The papers have been selected to stimulate discussion both on these issues and on the latest state of the art in devising new solutions for more familiar problems.

This Conference is being held in conjunction with the National Engineering Laboratory, and it is also sponsored by the Institution of Mechanical Engineers. The Steering Committee wishes to place on record its appreciation of the support received from both organisations.

A G H Coombs
Chairman, Steering Committee

Papers presented at the
Eleventh Technical Conference of
the British Pump Manufacturers' Association

CONTENTS

Paper		Page
1	On the flow in a centrifugal pump near shut-off head with positive and negative flows A Engeda and M Rautenberg, Institute of Turbomachinery, University of Hannover, F R Germany	1
2	Diagnosing pump problems from their noise emissions signature R Palgrave, Ingersoll-Rand Pumps, UK	9
3	Operating experience with scrubber pumps made from metallic materials in flue gas desulphurisation (FGD) plants using the lime scrubbing process based on limestone P R W Robinson and A Heumann, KSB Pumps Ltd, UK	29
4	Pumps for flue gas desulphurisation (FGD) P A Wheeler and M A Bevan, Warman International Ltd, UK	43
5	A new approach to the design of high capacity storm and foul pumping stations M R Sinclair, Flygt Pumps Ltd, UK and G Bjorkander, Flygt AB, Sweden	69
6	Energy cost management in water supply N Cullen, WRc Engineering Centre, UK	87
7	Retrofitting silicon carbon bearings to a process pump O V Bertele, Consultant, UK	99
8	Seals for hydrofluoric acid applications R Wallis, John Crane (UK) Ltd, UK	105
9	The gas-liquid performance of a centrifugal pump: priming using the shroud reflux method R K Turton, Loughborough University of Technology, UK	127
10	Meeting the challenge of the modern chemical process pump P Fabeck and K H Sierwald, Durco Process Equipment Ltd, UK	137
11	Design considerations for pumping applications of variable frequency drives B P Zell, Aurora Pump, A Unit of General Signal, USA	161
12	On a pilot pump using the Weis-Fogh mechanism M Tsutahara and T Kimura, Kobe University, Japan	177
13	HRC—a new sealing concept for standardized pumps in the chemical industry U Reinfrank and H Nowak, Feodor Burgmann Dichtungswerke GmbH & Co, F R Germany	185
14	The development of special seals for multiphase pumping N M Wallace, Flexibox Ltd, UK	197

15	The development of a seal for subsea pumps—a challenge for the seal manufacturer W Schmoeller, D Zues and G Sivertsen, Feodor Burgmann Dichtungswerke GmbH & Co, F R Germany	225
16	Development in pump condition monitoring B G Murray, G A Ratcliffe and S Palmer, Development Engineering International Ltd, UK	245
17	Thermodynamically-based pump performance monitoring M A Yates, Advanced Energy Monitoring Systems Ltd, UK	259
18	The development of profiled mechanical seal faces for positive liquid transfer P R Rogers, John Crane (UK) Ltd, UK	273

11th International Conference of the
British Pump Manufacturers' Association

New Challenges – Where Next?

18-20 April, 1989
Churchill College, Cambridge

PAPER 1

ON THE FLOW IN A CENTRIFUGAL PUMP NEAR SHUT-OFF HEAD WITH POSITIVE AND NEGATIVE FLOWS

A. Engeda and M. Rautenberg
Institute of Turbomachinery, University of Hannover, Hannover, Germany

Summary

This paper discusses and emphasizes the treatment of shut-off condition as a point in a continious H-Q curve consisting of negative and positive flows. The systematic study of this region on both sides of the shut-off head will not only give better understanding of shutt-off condition but will also greatly contribute to a deeper understanding of the flow mechanism at part load conditions.

Nomenclature

b = Blade height
c = Absolute velocity
d = Diameter
g = Gravitational constant
H = Pump head
n_s = $\omega (Q)^{1/2}/(gH)^{3/4}$, specific speed
P = Power input to pump
Q = Flow rate
R = Radius of curvature
Re = Reynolds number
u = Tangential velocity
w = Relative velocity
Z = Number of blades
BEP = Best efficiency point
β = Blade angle
η = Efficiency

subscripts

1 = At inlet
2 = At outlet
h = Hydraulic
m = Meridional
opt = Optimal

Conference organised and sponsored by the British Pump Manufacturers' Association
in conjunction with NEL and BHRA, The Fluid Engineering Centre.
Co-sponsored by the Process Industries Division of the Institution of Mechanical Engineers.

1. Introduction

One of the difficult tasks in centrifugal pump design is the prediction of the shut-off head and the part load head-flow (H-Q) curve course. This is primarily due to the absence of proper knowledge about the flow mechanism in this area.

In an attempt to contribute to a clearer understanding of the flow near shut-off condition, the flow mechanism is discussed by considering negative and positive flows. The positive flow region consists of a primary forward flow, some complex form of recirculation (reverse flows) and various cross flows (secondary flows) in the pump, the exact behavior of these phenomenon and their relationship to separation in the impeller and to the impeller geometry is still a acontested topic. In the negative flow other than the fact, that energy is dissipated, mainly due to shock losses not much more is known about the flow in the region. There is a tendency to rationalize that in the positive, negative and shut-off regions recirculating flows contribute a major part of the absorbed power and determine flow-head (H-Q) curve course in the region. Recirculatory flows in this region are known to be unsteady but periodic, and pulsate back and forth, thus transferring angular momentum from impeller to the casing and the inducer.

2. The head-flow (H-Q) curve behavior

There has been quite a number attempts in the last few decades to establish a correlation between the basic design parameters of a centrifugal pump (d_1, β, b_1, d_2, β_2 and z) and its H-Q curve behavior, but it seems what has been reached is not yet satisfactory.

The general form of an H-Q curve is best associated with the designer's yardstick, the specific speed n_s. The specific speed is generally known as:

$$n_s = \frac{n \cdot (Q_{opt})^{1/2}}{(H_{opt})^{3/4}} \qquad (1)$$

or by using the continuity and Euler's equations it could be expressed as:

$$n_s = k \cdot \frac{(d_1/d_2)((b_1/d_2)\tan \beta_1)^{1/2}}{\left[1-(d_1/d_2)^{1/2}(b_1/b_2)(\tan \beta_1/\tan \beta_2)\right]^{3/4}} \qquad (2)$$

which shows the direct dependence of design Head and flow rate (H_{opt} and Q_{opt}) on the basic design data, but in addition to this these optimal combination of design data affect to a large extent the whole course of the H-Q curve. However, the exact determination of the complete H-Q curve purely on the basis of geometrical datas is still impossible.

For a centrifugal pump the H-Q curve in the normal working range is based on one dimensional flow analysis and normally expressed by the well known equation

$$H = \eta_h \frac{u_2^2}{2}\left(h_o - \frac{c_{m2}}{u_2}\tan \beta_2\right) \qquad (3)$$

where h_o is the Busemann's slip factor and η_h the hydraulic efficiency. The greatest drawback in this one dimensional approach is primarily the absence of knowledge about the two parameters h_o and η_h over the whole H-Q range and in the second place the assumption when using velocity triangles that the relative velocity w_1 approaches at inlet the impeller velocity u_1 and at outlet w_2 approaches zero, as flow is reduced to zero. Meaning the absolute velocities c_1 and c_2 also approach zero at shut-off head, but in reality shut-off head is closed valve condition NOT zero flow in the impeller, therefore all w_1, w_2, c_1 and c_2 behave differently other than in the above simple assumption.

If it is assumed that optimal design parameters have been achieved and yet this produces unsatisfactory H-Q form (usually at part load), then the most likely corrections are the positions and magnitudes of the inlet and outlet blade edges

(b_1 and b_2), which are believed to be critically related to the velocity profiles at inlet and outlet particularly at part load. Other than few experimental works showing the effect of changes in the positions or magnitudes of b_1 or b_2 on the H-Q course, there is no clear understanding of the mechanism involved. This is also true for all the other design parameters, where the relationships between them and the H-Q course have been extensively studied experimentally.

3. Part load head-flow (H-Q) curve

The general slope and trend of the head-flow (H-Q) curve's course between design flow and shut-off head is more or less dependent on the specific speed, n_s. But local discontinuities (curve knicks) and local slope changes (droping curves) are difficult to account for.

The primary misconception in this H-Q area is the assumption that the inlet flow to the impeller is based on an incidence flow variation, where the inlet flow based on cascade experience is often described by the variation of the velocity triangle as flow is reduced or increased. Recent experiences hint at a different inlet flow model other than incidence variation. There are strong evidences to suggest that the actual inlet flow at part load consists of <u>a blocked area</u> at the outer diameter and an <u>axisymmetrical flow</u> at the centre. The blocked area is due to back flow out of the impeller, where the relative magnitude of the ratio of the blocked are to the area with axisymmetrical flow depends on the position in the H-Q curve.

As flow is continuously reduced to shut-off head centrifugal impellers inevitably show a jet-wake flow behavior followed by back flow (recirculation), the development of wake area and its extent has been the topic of numerous researchs but as yet there is no conclusive quantitative explanation. One of the simplest means available to the designer to judge the flow behavior in this H-Q range is the particle equilibrium analysis shown in figure (1), which are used to estimate the relative magnitudes of the,

curvature effects ----- $\dfrac{w^2 d_m}{R_c}$ and $\dfrac{w^2 d_m}{R_b}$

centrifugal effect ----- $r \omega^2 d_m$

Coriolis effects ----- $2 \omega w d_m$,

which together with skin friction and diffusion effect are then used to account for H-Q curve course at part load. The effects of local curvature in turbulent boundary layer and the effects of rotation could be expressed by the Richardson number

$$Ri.c. = \frac{\text{centrifugal force due to curvature (streamline)}}{\text{inertia forces}}$$

$$Ri.r. = \frac{\text{local forces due to Coriolis}}{\text{inertia forces}}$$

or by the Rosby number

$$R_o = \frac{\text{curvature effect}}{\text{rotation effect}}$$

This two parameter seem suited for judging irregularities in the H-Q curve course at part load, but they have not yet been sufficiently tested and correlated.

The classical approach to explain the H-Q course at part load is based on the one dimensional Bernoulli equation of a centrifugal impeller

$$H = \underbrace{\frac{c_2^2 - c_1^2}{2g}}_{\substack{\text{dynamic} \\ \text{rise}}} + \underbrace{\frac{u_2^2 - u_1^2}{2g} + \frac{w_1^2 - w_2^2}{2g}}_{\substack{\text{centrifugal} \quad \text{diffusion} \\ \text{rise} \qquad\quad \text{rise} \\ \text{static rise}}}$$

where the balance between static and dynamic pressure rise play a dictating role in inducing back flow (recirculation) at part load, and hence affecting the H-Q curve course. At a part-load flow stage where the jet-wake structure is already established, the fluid in the wake zone does not have the pressure rise due to the diffusion of the relative velocity, therefore at that spanwise section the balance between the dynamic pressure and the centrifugal pressure rise determine the consequently following flow structure in the impeller or at its exit, in addition to the already existing jet-wake structure.

Figure (2) shows typical design data for centrifugal impellers, but such data are optimized for the design point and do not have much influence on the details of the H-Q curve course between design flow and shut-off head. Unwanted irregularities in the H-Q curve are usually corrected by changes in the leading edges at inlet or outlet, or by changes in inducer or recuperator devices. Typical examples are
- extension of the leading edge into the suction zone
- increases in the blade height at outlet (b_2)
- inclining the blade edge at exit
- modification of the suction pipe or recuperator (volute, diffusor etc.).

Such modifications have been shown in a number of cases to have significant influence on the H-Q curve course at part load, but their exact influence on the flow mechanism: distribution of primary and secondary flow, separation, and development of the jet-wake structure are only suppositions and difficult quantitatively to account.

It is well known that if a closed container filled with water is rotated about an axis outside of itself (figure (3)), the fluid will tend to rotate in the opposite direction relative to the container because of its inertia. A case, /1/, is known of an investigation where the impeller exit of a fully shrouded impeller has been closed, leaving only the inlet opening to each blade chamber. Tests on this impeller have shown that flow into and out of the impeller at the inlet have detected, and one wonders about the similarity between the rotating container and the impeller. The question is, in the real impeller, if the boundary between the rotating fluid in the impeller and the stationary fluid in the volute could be seen as equivalent to the above impeller closure, and how much this analogy holds at shut-off and near it.

4. Shut-off condition

Shut-off condition consists of a complex flow pattern where unsteady recirculation flow takes place in the impeller, at inlet and in the recuperator. The shut-off power and head are dependent on many factors and effects, of which only very few have been considered to date. Even though the conditions at shut-off differ considerably from the ideal assumption of no flow, it is surprising that the head developed is actually nearly $u_2^2/2g$ based on no-flow assumption where the static pressure devleoped in the impeller is taken to be $(u_2^2-u_1^2)/2g$.

The recirculating flows are unsteady but periodic and travel simultaneously in the forward and reverse direction, thus transferring angular momentum to the recuperator and the inducer. Several experimental works mainly based on flow visualization technique have been carried out in order to establish the exact flow mechanism at shut-off condition, and the effect of the various parameters (d, b, b_2, d_2, β_2, Re, BEP parameters, etc.) on the flow mechanism. But there is no known attempt where the shut-off flow has been related to the positive and negative flow conditions near it (figure (4)), which are easier to visualize, and could also enable the treatment of the shut-off condition as a point in a continious curve and not as an end point.

5. Head-flow (H-Q) curve in the negative flow region

Figure (4) presents a head-flow (H-Q) curve during negative flow in the second quadrant, which arises from a pump running in the normal pump mode but the flow being reversed in it, due to another pump or system downstream producing a head higher than its shut-off head.

The present outhors experience agree with that of Rautenberg et al. /2/ and that of Wachter et al. /3/, who based on centrifugal compressor tests showed that the H-Q curve near shut-off condition with negative and positive flows has similar surge and pulsations behavior, strongly hinting at a related flow mechanism across the shut-off head point. Rautenberg et al. /2/ showed that the H-Q curve during negative flow could be resonably represented as a third-order polynomial throttle curve and that its course is highly affected by the impeller outlet blade angle β_2, but variations in the inlet angle β_1 had no effect on it.

Eventhough the two works /2 and 3/ reported in detail about the H-Q course in the negative flow region, they didn't give any information about the details of the flow mechanisms other than to simply assume that the flow consisted of a high degree of shock losses.

The positive flow region near shut-off condition, consisting of complex recirculation and forward flow, is therefore believed to posses analogus flow mechanism to the negative flow region which consists of some form of recirculation and reverse flow.

6. Conclusions

It has been attempted to treat the head-flow (H-Q) curve around shut-off conditions with positive and negative flows as a continious curve, where the flow mechanism are interrelated. The likely parameters that influence the flow mechanism in this region and determine the H-Q curve course have been discussed.

7. References

1. Stoffel, B. and Krieger, P.: "Experimental Investigations on The Energy Ballance of Radial Centrifugal Pump Impellers at Part Load condition", IAHR 1982, Amsterdam.
2. Rautenberg, M., Kämmer, N. and Mobarak, A.: "Performance Measurements of small Centrifugal Compressors Including Negative Mass Flow Rates", 4 th. International Conference for Mechanical Power Eng., Cairo 1982.
3. Wachter, I., Rohne, K.H. and Löhle, M.: "Der Radialverdichter bei Rückwärtsdurchströmung", VDI-Berichte Nr. 487, 1983.

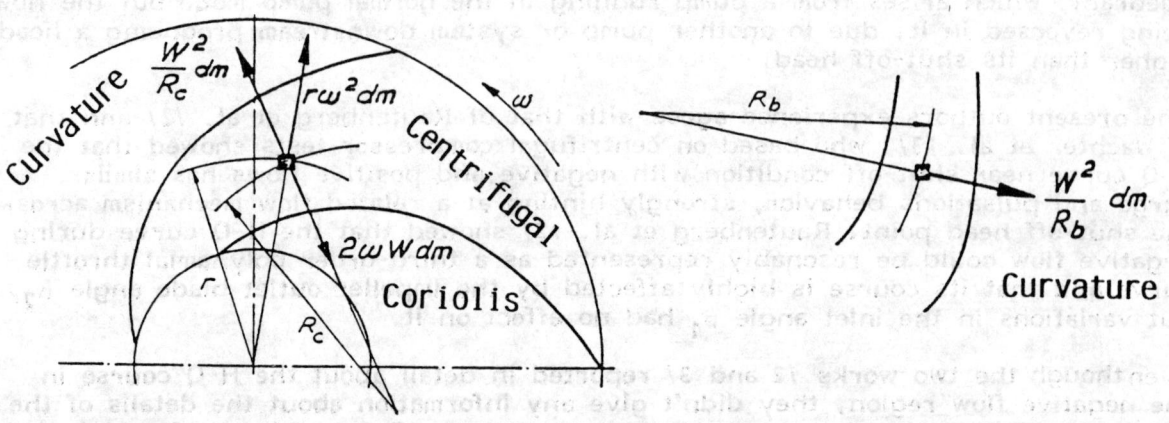

Fig. 1: Particle dynamic-equilibrium analisis for a centrifugal impeller

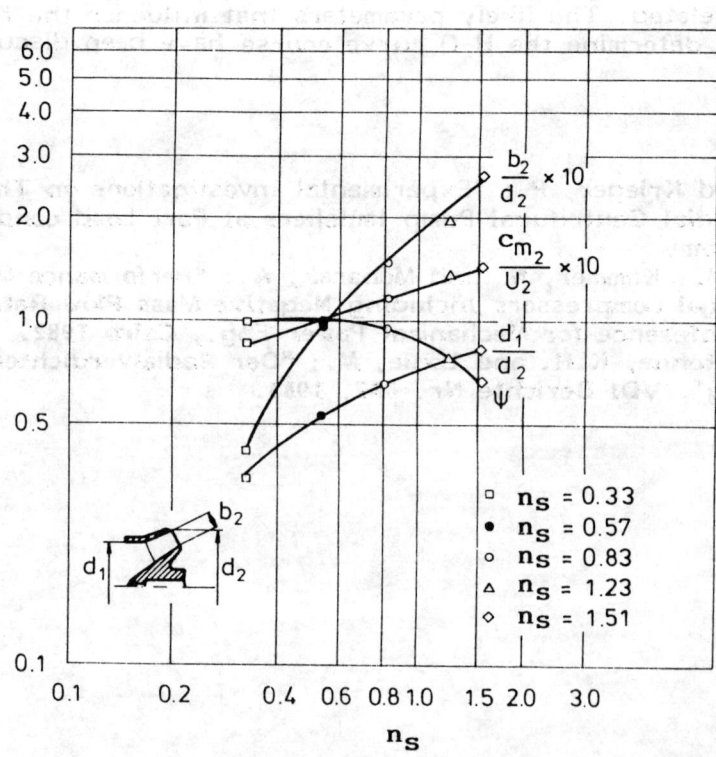

Fig. 2: Typical design data for centrifugal impellers

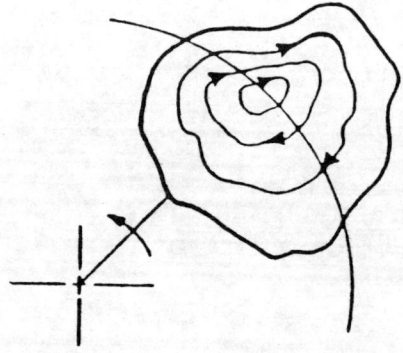

Fig. 3 : Circulatory flow in a rotating container about an axis outside of itself

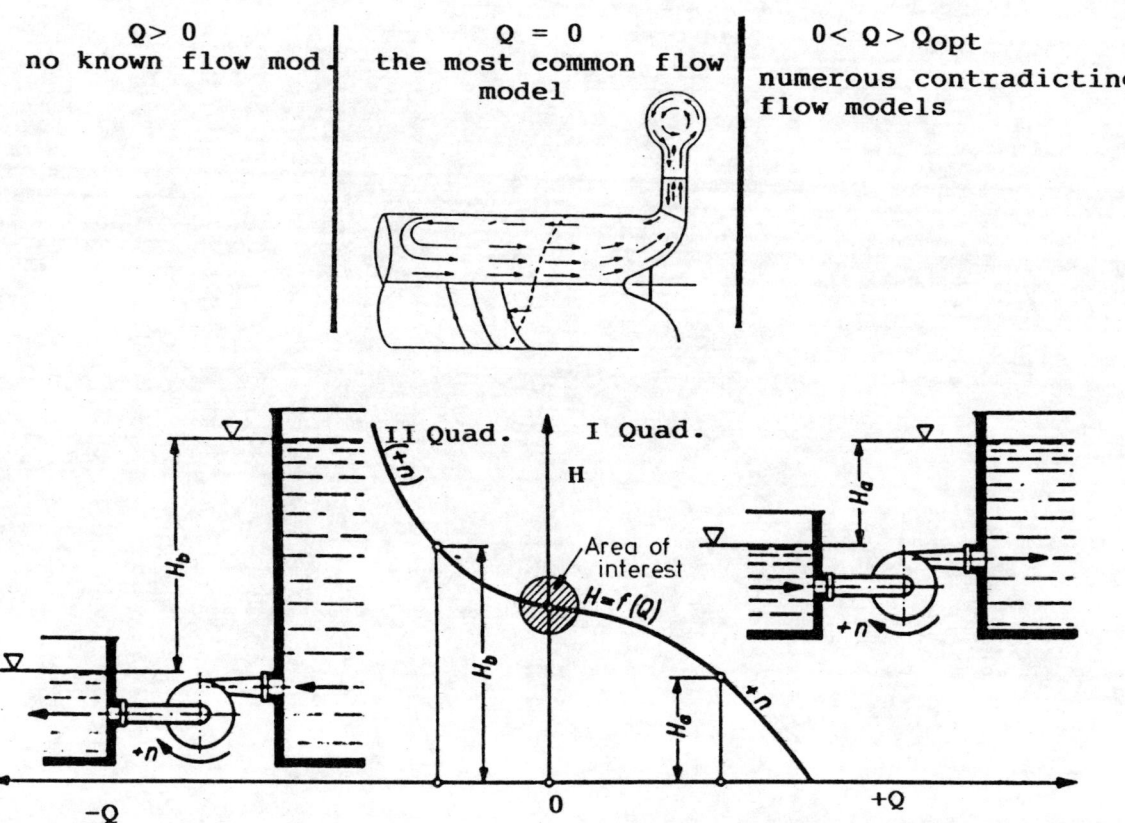

Fig. 4 : A continuous H-Q curve in the first and second quadrant

11th International Conference of the
British Pump Manufacturers' Association

New Challenges – Where Next?

18-20 April, 1989
Churchill College, Cambridge

PAPER 2

DIAGNOSING PUMP PROBLEMS FROM THEIR NOISE EMISSIONS SIGNATURE

R. PALGRAVE, B.A.

INGERSOLL-RAND PUMPS
GATESHEAD, ENGLAND

S U M M A R Y

Noise emission levels from a well designed properly installed water pump can be forecast with some certainty. The noise levels from cavitating pumps that are poorly designed, (or deliberately distorted), Poorly installed, or operating on non-aqueous fluids are less readily predicted.

This paper sets out to show how an idealized noise spectrum may be calculated for any pump. This may then be used as a reference framework against which the noise characteristics of the actual pump may be judged. Different pump problems induce different departures from this ideal so it is possible to use this comparison to help identify the source more uniquely. In particular, this technique may be used to diagnose pumps that, for one reason or another, can not be stopped or dismantled.

1.0 INTRODUCTION

Viewed as an independent noise generator, ie ignoring the driver, a centrifugal pump has two major noise categories.

1.1 Mechanical Noise

Bearings
Mechanical seals
Coupling windage
Bedplate/coupling panel resonance
Gears

These are normally only of consequence in pumps of very low, or very high speed (ie where the outlet top speed is less than 10 metres/second or shaft speeds much above 6000 rpm).

Conference organised and sponsored by the British Pump Manufacturers' Association
in conjunction with NEL and BHRA, The Fluid Engineering Centre.
Co-sponsored by the Process Industries Division of the Institution of Mechanical Engineers.

1.2 Hydraulic Noise

General turbulence within the machine
Stator/rotor interaciton
Acoustic resonance of hydraulic passages
Cavitation
Localised turbulence (stall, backflow, etc)
Leakage through fine clearances
Multi phase fluids effects

In the vast majority of industrial pumps, hydraulic noise is easily the most dominant source, although to the ear, this may well be swamped by other components of the pumpset (ie driver, gearbox, lube systems, orifice devices or part of the plant, control valves, resonance in the pump pipework). This paper is concerned only with the driven machine and its contribution.

It is worth examining each of these elements in some detail in order to see how they can affect the overall acoustic performance of a pump.

1.2.1 Turbulence

In any pump there are certain hydraulic losses which are embraced by the term hydraulic efficiency. These losses result from both fluid friction and separation within the passages of pump. They are also a noise source and so a pump with poor hydraulic efficiency will be noisier in this respect than an efficient one.

The noise from fluid friction will obviously be related to velocity and Reynolds number. It will increase in some way with flow rate and pump speed.

The noise from gross separation will be related to pump flow and will nominally be a minimum at its design point. However, since the process of head generation in an impeller may also result in large scale separation (at least in low to medium specific speed machines) this needs to be considered and will be discussed in the next section.

Both friction and separation are essentially broad band in nature and produce 'white' noise of no special frequency.

1.2.2 Stator/Rotor Interaction

It is widely imagined that the flow leaving an impeller passage and entering a volute/diffuser, is relatively uniform. Even if this were true, and it is not, there would still be a noise source resulting from flow interruptions every time an impeller blade passed close by a stationary part of the casing. This is because the blade has a finite thickness and results in a flow 'shadow' or 'wake' (figure 1.1). These interactions are very discrete with a frequency related to both shaft speed and vane number.

The simple rate of interaction per shaft revolution (vane rate) may be express by;

$$Zr(1) = N \times Zi \qquad \ldots\ldots\ldots (1)$$

Where N is determined from;

$$N \times \frac{Zi}{Zd} = \text{Integer} \qquad \ldots\ldots\ldots (2)$$

Here Z_i = number of impeller vanes
Z_d = number of stator vanes

The rate vane rate has been calculated from a range of impeller and diffuser vane numbers and is shown in Table 1 (single and double volutes are considered as 1 and 2 vaned diffusers).

However, as hinted earlier, the actual picture is more complex since the flow leaving an impeller is very rarely uniform. In impellers of low specific speed (particularly below 1500 in usgpm, ft, and rpm units) the flow within the vane space is permanently separated (figure 1.2). Conceptually this can be visualized as a 'jet/wake' leading to a stepped velocity distribution. In reality it is seldom so clear cut, but this simplification helps understand the noise generation process.

The interaction resulting from the blade 'shadow' combined with the 'jet/wake', is shown in figure 1.3. In this respect the impeller now behaves as if it has twice as many blades so equation 2 should be modified to give a 'pseudo vane rate' as follows:-

$$N \times \frac{(2 \times Z_i)}{Z_d} = \text{Integer} = Z_{r2} \qquad \ldots \ldots (3)$$

Table 1 may still be used by taking twice the impeller vane number. This gives the pseudo vane rate.

The acoustic frequencies (F_i) at which this primary noise source occurs is expressed by;

$$(Z_{r2} \times \frac{RPM}{60}) = F_i \text{ (Hz)} \qquad \ldots \ldots (4)$$

As we shall see later the noise strength at this frequency is dependent upon impeller tip speed though this is obviously attenuated by increasing distance between stator and rotor vanes. Distances normally chosen to reduce discharge pressure pulsations are adequate to reduce this.

1.2.3 Acoustic Resonance

The long narrow hydraulic passages in impellers, diffusers and volute crossovers can form excellent resonators when excited at the appropriate frequency. The resonant frequency (F_r) can be modelled by simple 'organ pipe' theory where;

for 1/2 wave resonance;

$$F_r = n \times (S/(2 \times L_a))$$

for 1/4 wave resonance;

$$F_r = (n-1) \times (S/(4 \times L_a))$$

The most likely excitation source would be at the once and twice vane rate discussed earlier. This noise source is largely confined to machines of high head per stage and is less serious than the resulting fluid borne pressure pulsation (reference 3).

1.2.4 Cavitation

Three types of cavitation can coexist at any time within a pump.

1.2.4.1 Sheet Cavitation

This is a stable type of cavity forming across the blade surface when pumps operate close to their design flow with low suction pressure. It is RELATIVELY quiet and makes its presence felt as broad band noise - generally at frequencies between 2 kHz and 40 kHz.

1.2.4.2 Cloud Cavitation

This unstable cavity forms downstream of the cavity sheet when pumps operate away from their design flow at low suction pressure.

Since it is a 'bubble collapse' phenomena it is very damaging and is relatively the noisiest of the three types. It generally appears at higher frequencies - say 20 kHz to 40 kHz, and gives the familiar sound of 'pumping gravel'.

1.2.4.3 Vortex Cavitation

This highly unstable from of cavity exists when pumps operate at very low flows and in the inlet backflow regime. Although also a bubble collapse phenomena, it can be less damaging because the collapse usually takes place well away from solid surfaces. This type is characterised by random bursts of noise and gives the characteristic cavitation 'crackle'.

When simultaneously the NPSHA gets very close to the 3% head decay line and the pump operates in the backflow regimes, then vortex cavitation organises itself into a very low frequency beat (say 1 Hz to 4 Hz). This is known as cavitating surge.

1.2.5 Leakage Through Fine Clearances

Little information exists on this source of noise but it is likely to be high frequency broad band if data from valve noise is anything to go by.

2.0 NOISE PERCEPTION

Irrespective of its source, hydraulic noise is transmitted via two routes; through the air to create 'local' annoyance, or through the pumped fluid to create more remote annoyance (fluid properteis permitting).

Since the object of this paper is to suggest the use of noise signatures as a means of diagnosis, and since microphones are more frequently available than hydrophones, only airborne noise will be considered here.

The most widely used presentation method for noise data is the so called 'Noise Rating' Charts (figure 2), these portray lines of constant loudness in octave centre band frequencies typically from 32 Hz to 16 kHz. A designers first reaction might be to try and concentrate noise emission in the lower frequency bands since this would reduce the pumps perceived noise rating. However, this might also conflict with the requirement to have the stator/rotor interaction rate (equation numbers 2 and 3) as high as possible, in order to reduce pressure pulsation levels (Reference 4).

Segregating noise into frequency bands of one ocatve width can be legitimately challenged by the purist as a slightly crude process since it will capture data between $\sqrt{2}$ and $1/\sqrt{2}$ times the centre band value. For example, for a centre band frequncy of 1 kHz, data between 707 Hz and 1414 Hz will be captured).

In defence it can be said that an industrial level, this standard is still the most wedely used. Whilst it is eqully true that noise measurement devices of 1/3 octave band width are available, instruments of one octave band width are easily the most common amongst pump vendors and pum owners. For this reason alone, Noise Rating Charts based on a one ocatve band width will be used here.

3.0 DIAGNOSTIC CONCEPT

How can noise data be used as a diagnostic tool?

Basically the site noise signature of a pump is compared to some idealised signature, and any significant depatures identified. The location and extent of these depatures can then suggest certain categories of pump problems. Whilst this comparison can be done in the noise rating curve reference frame, it is more convenient to take a different approach.

Figure 2 shows how a pump noise signature can be disassembled into a number of components. The peak of the signature normally occurs in an octave band containing the 'pseudo vane rate'. The noise strength at this frequncy is, as stated above, dictated by the impeller tip speed to a large extent (figure 3). The way that the signature decays in the octave bands to either side of the pseudo vane rate tends to follow a trend amongst normal commercial machines. This trend can form a useful yardstick for comparative purposes. (Appendix 1).

In order to demonstrate this noise, profiles for a number of 'better' shop tests were examined. In this instance, better meant;

3.1 Pump was not cavitating and the ratio between NPSHA and NPSHR was greater than 3.

3.2 Pump had large enough impeller tip clearance (D3/D2 1.1) to suppress cutwater 'squeeze' vortex.

3.3 Pump was operating close to its nominal design point.

3.4 Modest suction perforamnce (Nss).

3.5 Pump was tested in area remote from shop noise, and in a loop having a low noise control valve.

Within the scope of a manufacturers test bed, this permitted reasonably coherent data sets to be established. The octave band decay on either side of the vane rate octave band was then normalized for each pump and is shown in figure 4. This shows a remarkable similarity, considering the wide range of machines and the use of crude whole octave band resolution.

From this data, an idealized octave decay curve was drawn in order to establish a Standard Signature. Standard Signatures were also generated for a number of known 'problem' pumps and compared with decay curves based upon test data. A pattern began to emerge.

4.0 CASE STUDIES

Figure 5.1 Barrel Case Pump

This shows data from a large 2 MW multistage barrel case pump. Agreement exists at and above the vane interaction rate. Below this there is a very severe depature with a peak that corresponds to a 'once per rev' event. This was subsequently traced to a cracked motor shaft.

Figure 5.2 Ring Section Pump

This shows data from a large ring section pump throttled by a simple gate valve. The turbulence generated by such valves is well known and appears as a depature in the high frequency zone.

Figure 5.3 End Suction Pump

This shows data for a end suction process pump having 3 impeller vanes. Designers normally avoid such low numbers (figure 1.2 shows why), but often this is the only route to very low NPSH requirements and can be acceptable on hydrocarbon service. Such impellers can be very prone to inlet backflow and this phenomena reveals itself as a spike in the low frequency bands.

Because this is an improvised site test there is also some evidence of throttle valve noise.

Figure 5.4 Double Entry Process Pump

This displays data for a large double entry process pump operating under cavitation conditions. It is widely stated that cavitation noise maximises in the 40 kHz - 90 kHz region (reference 2). Whilst this is true, it is less well known that it can influence the pump noise signature even down to frequencies as low as 4 kHz. This data set demonstrates the point. Further examples of this and backflow phenomena are given in reference 1.

Figure 5.5 Self Priming Pump

Here data from a small self priming pump is shown. Even after the pump had gone through its priming cycle it was still found to be noisy. Departure from the idealized signature occur at above vane interaction rate and suggested inlet cavitation. However, this was ruled out be substantially increasing the NPSH during testing - to no avail.

Self prime pumps rely upon a very close stator/rotor clearance fro priming and this can generate tip cavitation. In this case it was found that the machining error had reduced this clearance dramatically, resulting in severe tip vortex cavitation.

Figure 5.6 Split Case Pump

A large cooling water pump demonstrated objectionable noise once installed on site. Comparison of the actual noise siganture with the ideal indicated progressive depatures at the low frequency. This suggested a system related problem. Further review showed probable resonance in the discharge spool piece. This was modified and successfully reduced the noise levels.

5.0 INTERPRETATION GUIDELINES

In many cases of diagnosis, simply narrowing down the field of review is a significant step towards a solution. This helps eliminate sources of low probability.

The perceived noise emissions from a pump are notoriously difficult to classify so far as design/performance origins are concerned. With experience, certain categories emerge; the classic example is a pump that is cavitating heavily and gives out a sound like pumping gravel. Even so, experience pump engineers still have difficulty in discriminating between (say) simple cavitation and vortex cavitation caused by inlet backflow. (The former is caused simply by insufficient suction pressure, whilst the latter is largely independent of suction pressure).

On a manufacturers test bed, the origins of pump noise can often be quickly identified by changing the test parameters and seeing what happens. For example inlet backflow noise will vanish rather quickly as flow is increased, but will not respond greatly to increases in suction pressure. Classic cavitation, on the other hand, will increase progressively as flow increases and will vanish quite suddenly as suction pressure rises.

Once a pump has been installed however, this sort of elimination process can seldon be conducted, at least not to the same extent. It is usually difficult to vary the flow over a wide range and suction pressure is often constrained by particalities of the installation. If the pump forms part of some continuous process, very few adjustments are possible.

Yet this is the classic situation in which a noise issue arises. Unfortunately, with little change of significant performance variation, resolution of the problem becomes difficult. Faced with such a very real situation, the technique described here can help narrow down the analysis and eliminate some possibilities.

The idealized depatures for a number of noise origins are shown in figure 6.

In practice, the octave decay curves have been found the most useful reference frame for diagnostic purposes, and the examples in this paper are shown on such a basis. However, the process can easily be reversed in order to synthesize a Noise Rating type forecast for new, or untested machines. (See appendix 1).

6.1 Cavitation

Inlet cavitation usually shows as a progressive increase in the octave bands 2 kHz and above.

Outlet cavitation, typical in wet recirculation self priming pumps, shows as an increase at and above the pseudo vane rate level. Generally this appears first on the higher octave bands, say 8 kHz and above.

High specific speed machines can also display this effect - due to separation vortices at outlet and/or discharge recirculation.

6.2 Backflow

Inlet backflow typically appears as a spike in the low frequency octave bands. The backflow swirling out from the impeller eye reacts against stationary elements. Because most pumps have only one stationary inlet rib, or because the appraoch is asymetric, the interaction rate corresponds to Zr1. (Table 1 can still be used.

6.3 Speed

For interest, the effect that changing speed has on the overall frequency profile is shown for two different machines.

6.4 System Effects

Noise generated by the installation (ie piping, bends, etc) is generally smeared across the low frequency end of the signature. There is a tendency for it to appear most in the lower frequency octave bands.

CONCLUSION

Certain classic pump hydraulic problems can be differentiated by the way in whcih they force the noise/frequency signature to be modified. This is more readily seen by decomposing the signature into decay rates either side of the vane rate octave band. This technique can be used to diagnose noise problems on site.

ACKNOWLEDGEMENTS

The author would like to thank the management of Ingersoll-Rand for permission to publish this work.

REFERENCES

1) Palgrave R., 'Using Noise Data to Analyse Pump Hydraulic Problems On Site'. ASME Pumping Symposium, San Diego, July 1989.

2) Deeprose W M, King N w, McNutly P J, Pearsall I S. 'Cavitation Noise, Flow, and Erosion'. Conference on cavitation (Edinburgh) IMechE, 1974 pages 373-381.

3) Schwartz R E, Nelson R M. 'Acoustic Resonance Phenomena in High Energy Centrifugal Pumps'. Procedure, First International Pump Symposium, Texas A & M, May 1984, pages 23-28.

4) Makay E, Barrett J A. '10 Ways to Improve High Energy Pump Performance'. Power Magazine, January 1988, pages 37-40.

NOTATION

F_i = Stator/rotor interaction frequency .... Hz.

F_r = Organ pipe resonance frequency ..... Hz.

L_a = Acoustic length of passage .... consistent unit

n = Harmonics 1, 2, 3, 4, etc.

N = Shaft speed .... RPM

S = Speed of sound in fluid .... consistent unit

Z_d = Number of stator vanes.

Z_i = Number of impeller vanes.

Z_{r1} = Simple vane interaction rate/rev.

Z_{r2} = Pseudo vane interaction rate/rev.

TABLE 1

NO OF IMPELLER VANES (Z_i)	S VOL	D VOL	NO OF STATOR VANES (Z_d)								
	1	2	3	4	5	6	7	8	9	10	11	12
1	1	2	3	4	5	6	7	8	9	10	11	12
2	2	2	6	4	10	6	14	8	18	10	22	12
3	3	6	3	12	15	6	21	24	9	30	33	12
4	4	4	12	4	20	12	28	8	36	20	44	12
5	5	10	15	20	5	30	35	40	45	10	55	60
6	6	6	6	12	30	6	42	24	18	30	66	12
7	7	14	21	28	35	42	7	56	63	70	77	84
8	8	8	24	8	40	24	56	8	72	40	88	24
9	9	18	9	36	45	18	63	72	9	90	99	36
10	10	10	30	10	10	30	70	40	90	10	110	60
11	11	22	33	44	55	66	77	88	99	110	11	132
12	12	12	12	12	60	12	84	24	36	60	132	12
13	13	26	39	52	65	78	91	104	117	130	143	156
14	14	14	42	28	70	42	14	56	126	70	154	84
15	15	30	15	60	15	30	105	120	135	30	165	60
16	16	16	48	16	80	48	112	16	144	80	176	96
17	17	34	51	68	85	102	119	136	153	170	187	204
18	18	18	18	36	90	18	126	72	18	90	198	36
19	19	38	57	76	95	114	133	152	171	190	209	228
20	20	20	60	20	20	60	140	40	180	20	220	60

EXAMPLE 1

ANALYSIS

TEST DATA

ⓒ = OCTAVE CENTRE BAND FREQ. Hz	32	63	125	250 Ⓜ	500	1000	2000	4000	8000
INCLUDES FREQUENCIES:									
BETWEEN	22	44	88	176	353	707	1414	2828	5656
AND	44	89	176	353	707	1414	2828	5656	11313
Ⓓ = NOISE TEST DATA dB	77	85	90	89 Ⓟ	85	79	72		

1) Impeller vane number, $Z_i = 6$
   (but treat as if $Z_i = 12$, see equation (3))
   Casing, double volute $Z_d = 2$

2) From Table 1, vane interaction rate (Z_{r_2})
   For this combination = 12 per/rev

3) Consequently, interaction frequency = 12 × revs/sec
   = 12 × (1450/60)
   F_i = 290 Hz

4) 290 Hz lies between 176 and 353 Hz so the noise peak (P) lies in the 250 Hz octave band (M), see above.

ANALYTIC WORKSHEET

Ⓔ = LOG $\frac{C}{M}$	-.89	-.6	-.3	0	.3	.602	.903	1.204
DECAY F = (P - D)	-12	-4	1	0	-4	-7	-10	-17

EXAMPLE 1

Continued.......

E and F are plotted in figure 7.1 and compared with the standard profile. The principal depatures are at the low frequency end. Because this is well distributed, rather than concentrated into a spike, it suggests a significant contribution from the pipe/valve system adjacent to the pump. Since the profile is otherwise normal, it suggests that the pump is normal and that the excess noise is related to the installation.

EXAMPLE 2

SYNTHESIS

Hot oil pump, 7 impeller vanes, double volute, impeller diameter 400 m/m, shaft speed 3560 RPM.

1) Impeller tip speed = $3.142 \times \frac{400}{1000} \times 3560 \times \frac{1}{60}$

    = 74.5 m/s

2) From figure 3, noise in the octave band containing the vane rate is estimated as 88 dB $\pm$ 6 dB = D.

3) Stator rotor vane rate, (Zr_2) from Table 1, for $Zi = 7$ and $Zd = 2$, is 14. (remembering to multiply Zi by 2 before entering Table 1).

    Frequency of interaction (Fi) is therefore;

    $Zr_2 \times$ rev/sec $= 14 \times \frac{3560}{60}$

    $Fi = 830$ Hz

    This lies in the octave band centred upon 1 kHz.

Now using figure 4, the noise strengths in the side octave bands can be estimated.

(C) =	32	64	125	250	500	1kHz	2	4	8
(E)= LOG [C/M]	1.49	1.19	.90	.60	.304	0	.30	.60	.9
dB DEDUCT FROM FIG 4 = (d)	-40	-28	-18	-10	-2	0	-1	-3	-11
RESULTANT (D - d)	48	60	70	78	86	88	87	85	77

This is plotted in 7.2 giving a Noise Rating of 88.

This data, when sumnated, gives an overall noise level forecast of 93 dB at 1 metre from the machine.

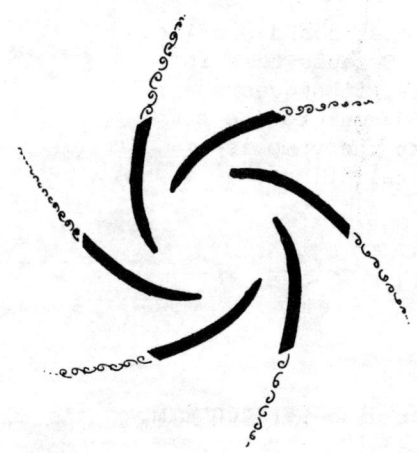

1.1   Blade wake 'shadow'.

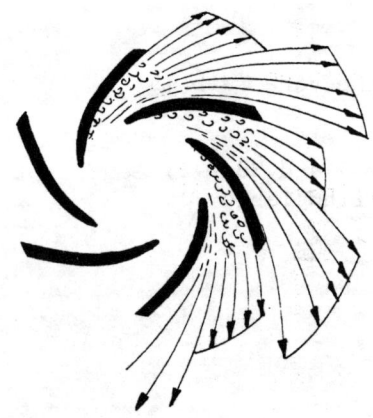

1.2   Jet/wake profile leaving vane to vane space.

1.3   Combined outlet profile creating 'pseudo' vane rate.

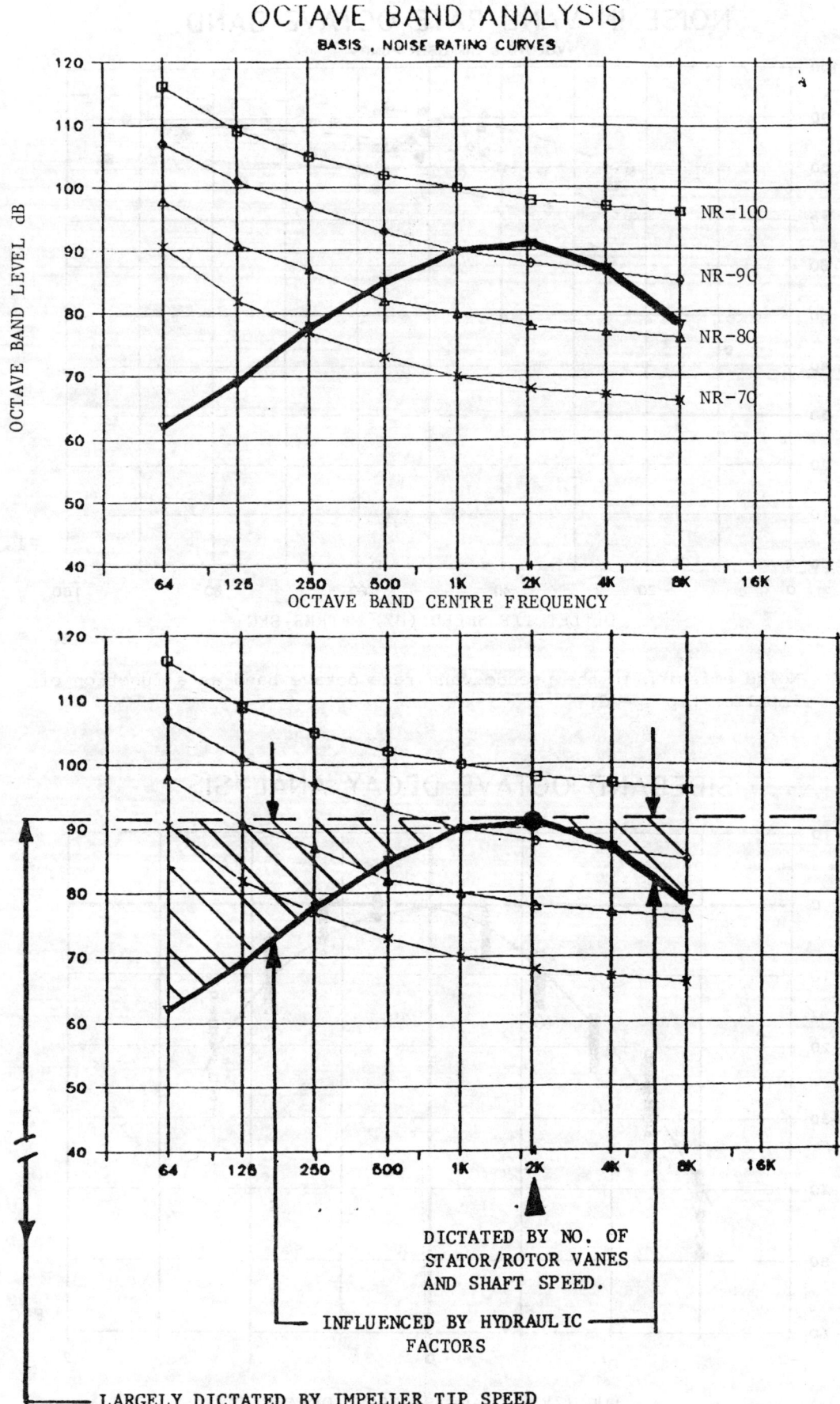

2.0 Noise rating charts and simplified pump siganture structure.

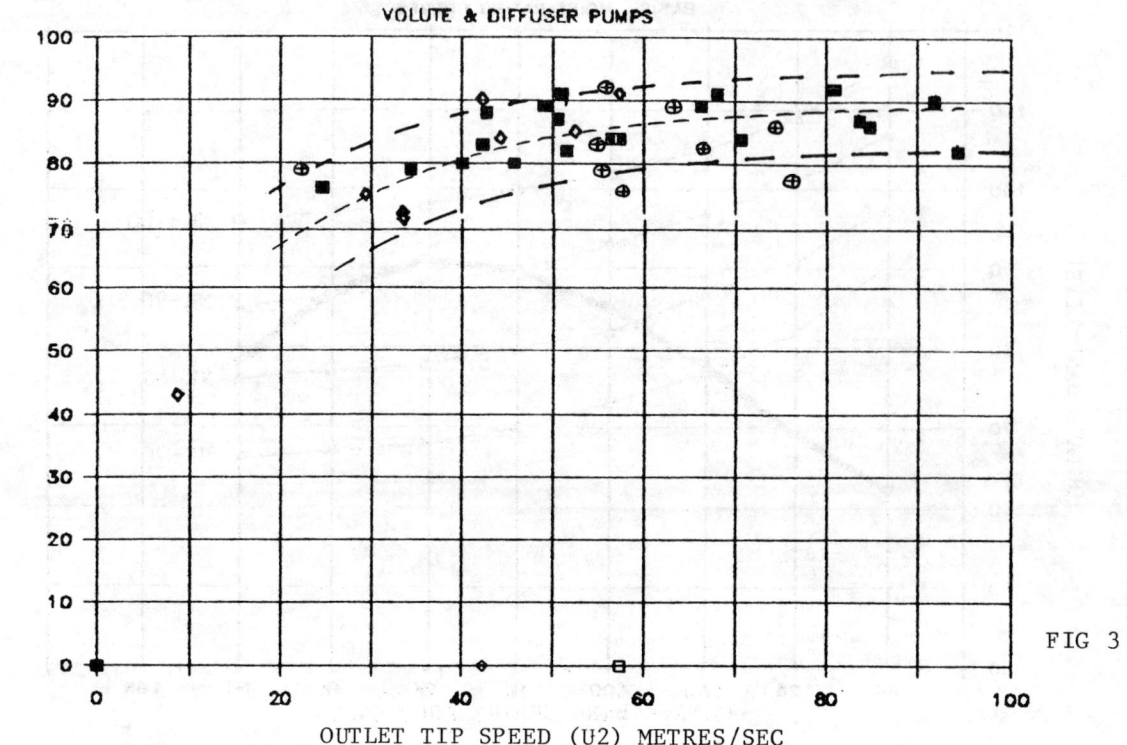

3.0 Noise emission in the pseudo vane rate octave band as a function of impeller tip speed.

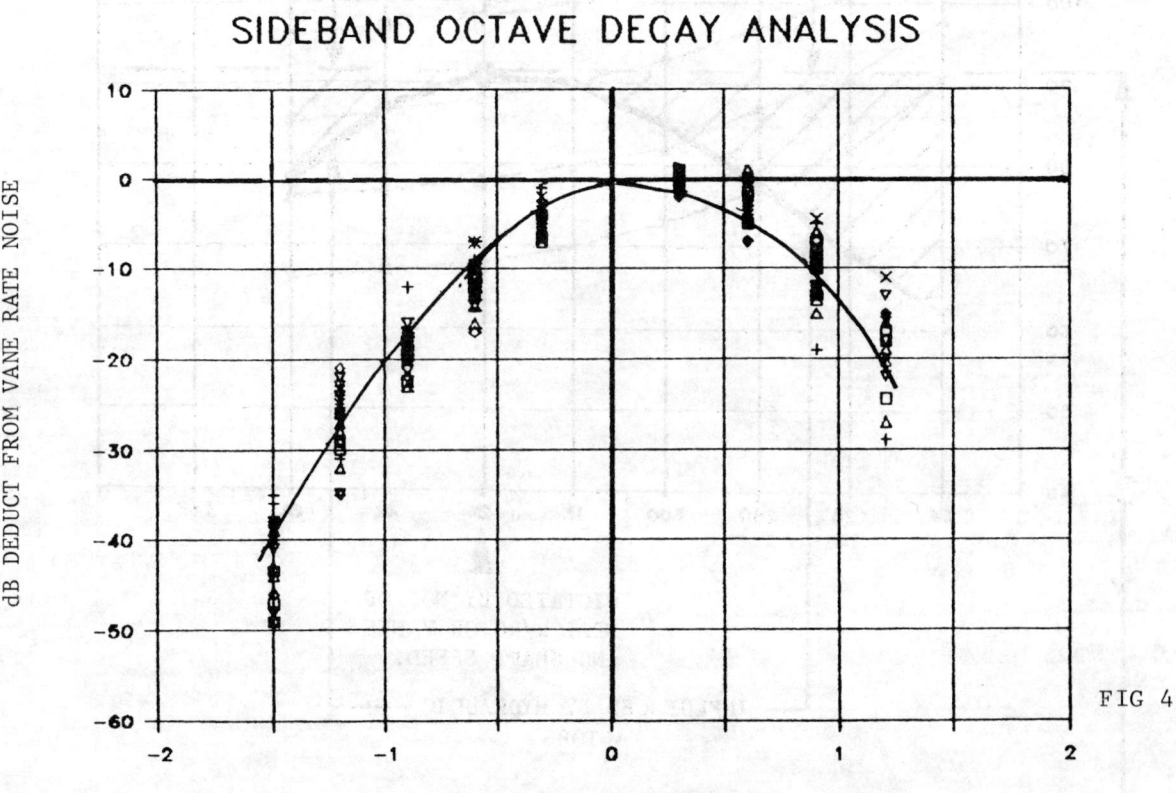

4.0 Noise decay rated either side of pseudo vane rate octave band.

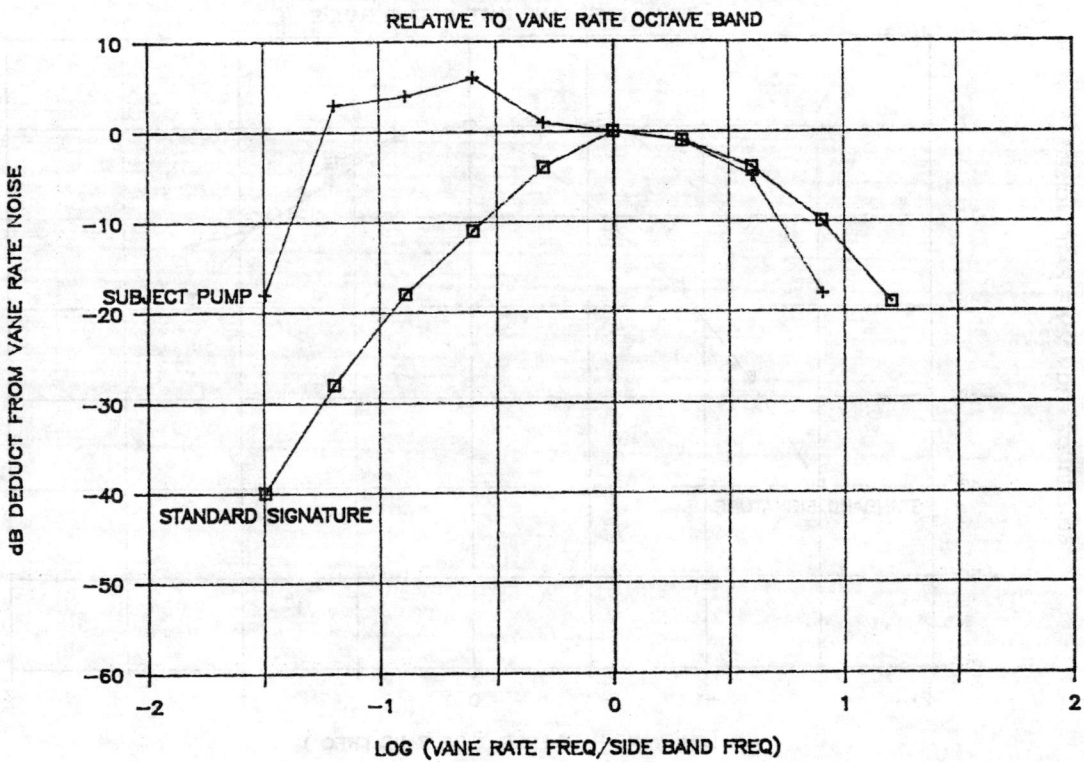

5.0 Case Studies;

    5.1 Large barrel case pump.

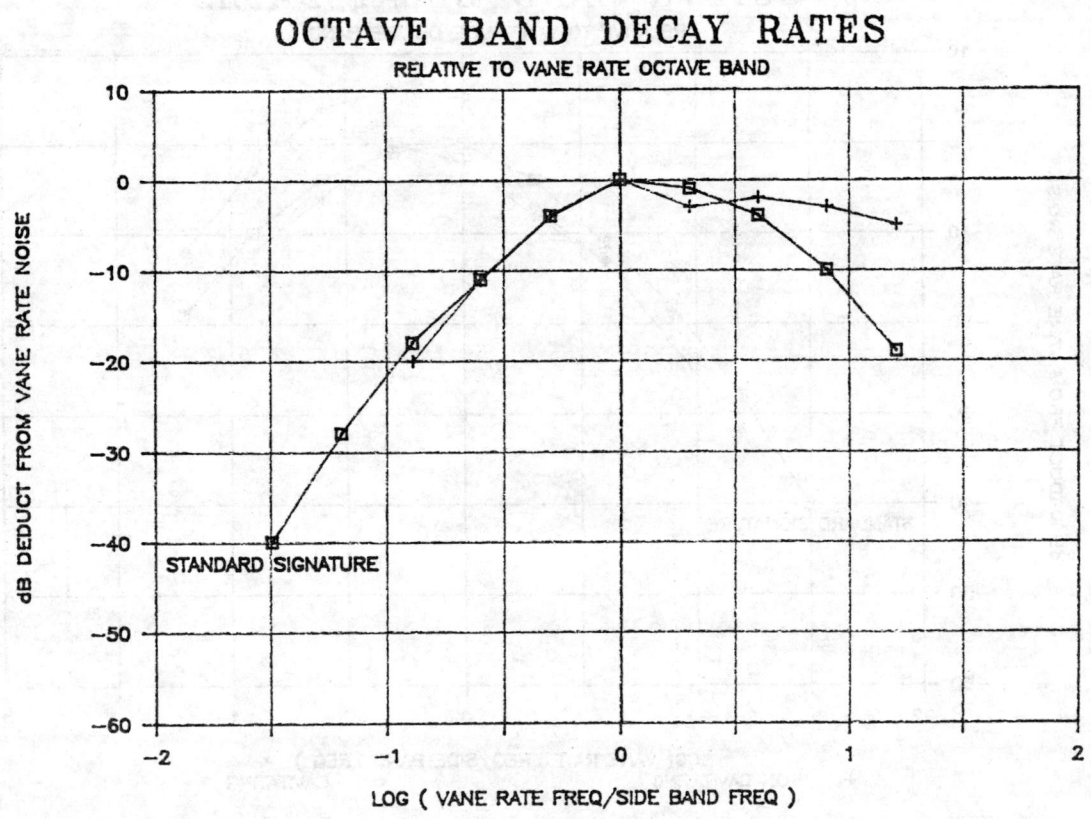

    5.2 Ring section pump.

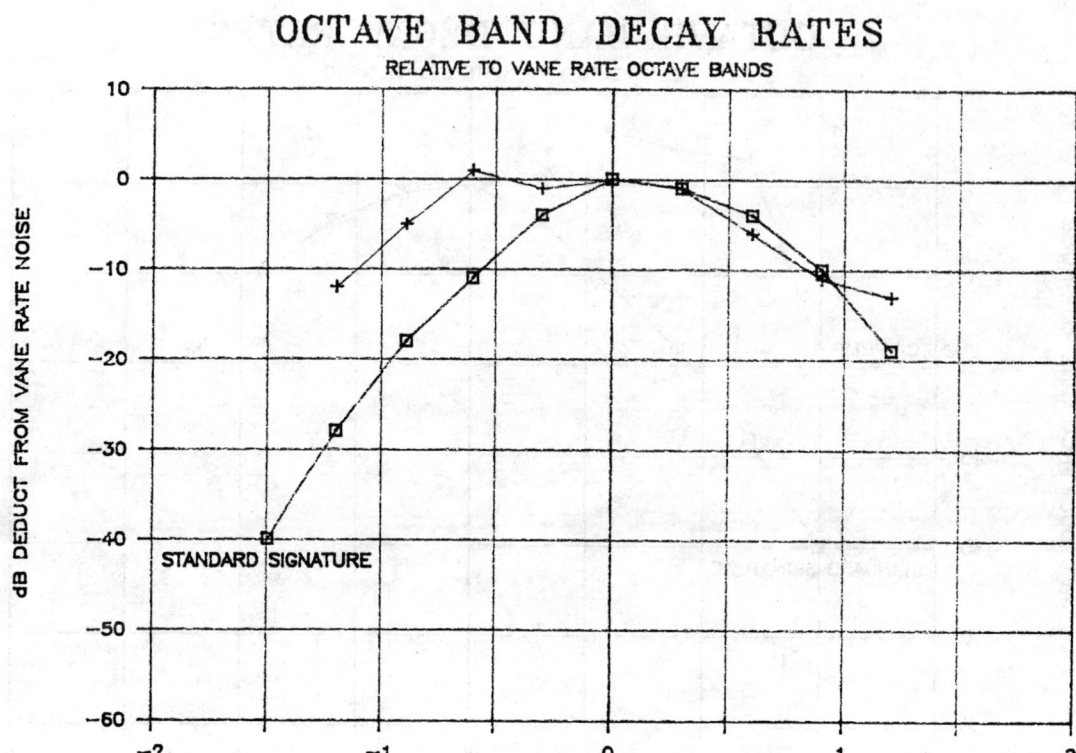

5.3 End suction pump.

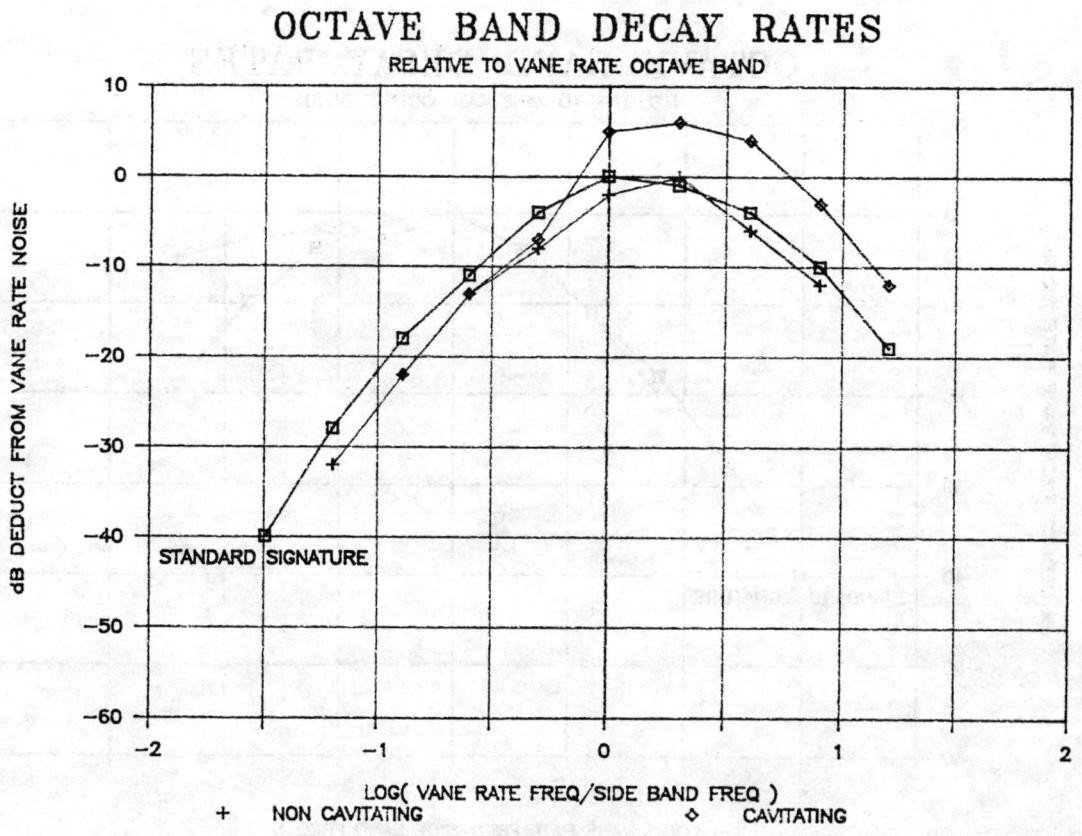

5.4 Large double entry hot oil process pump.

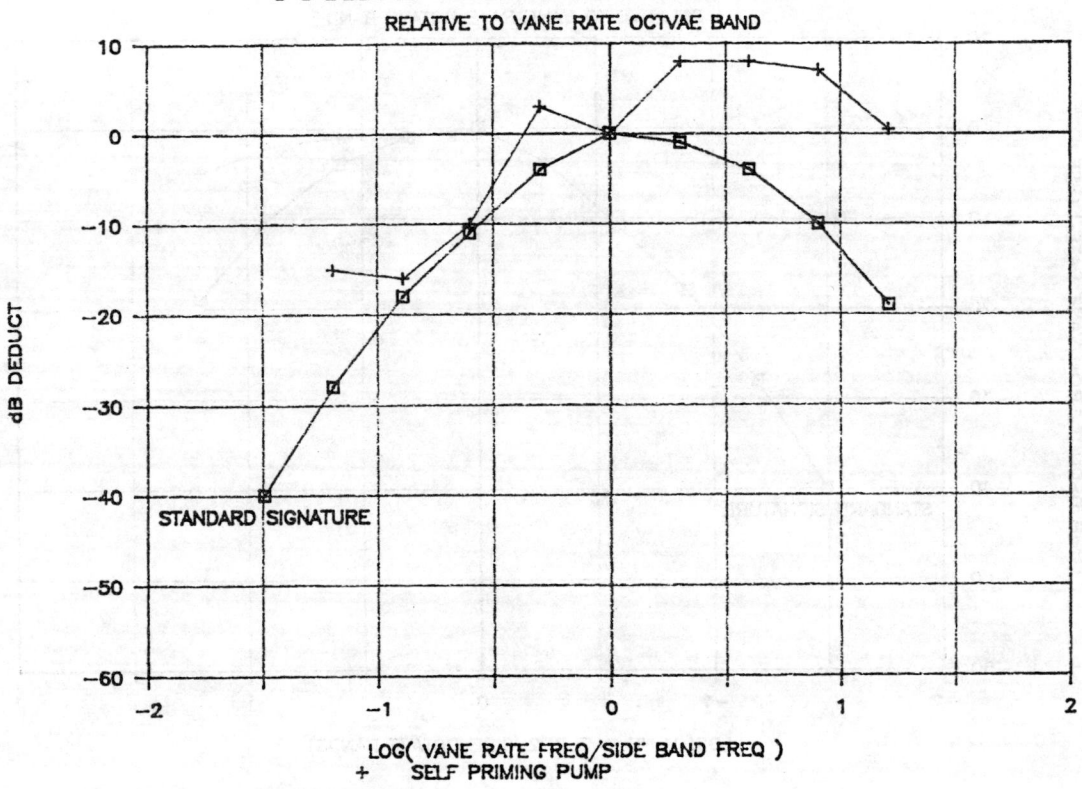

5.5 Small self priming pump.

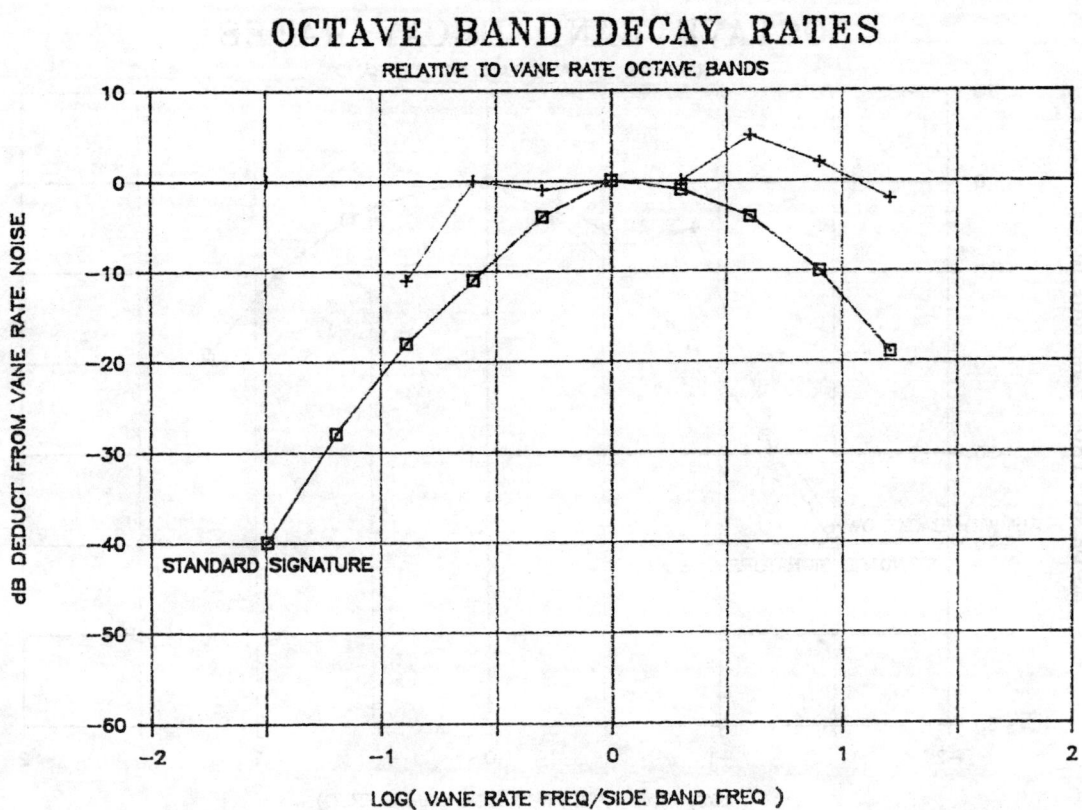

5.6 Large split case cooling water pump.

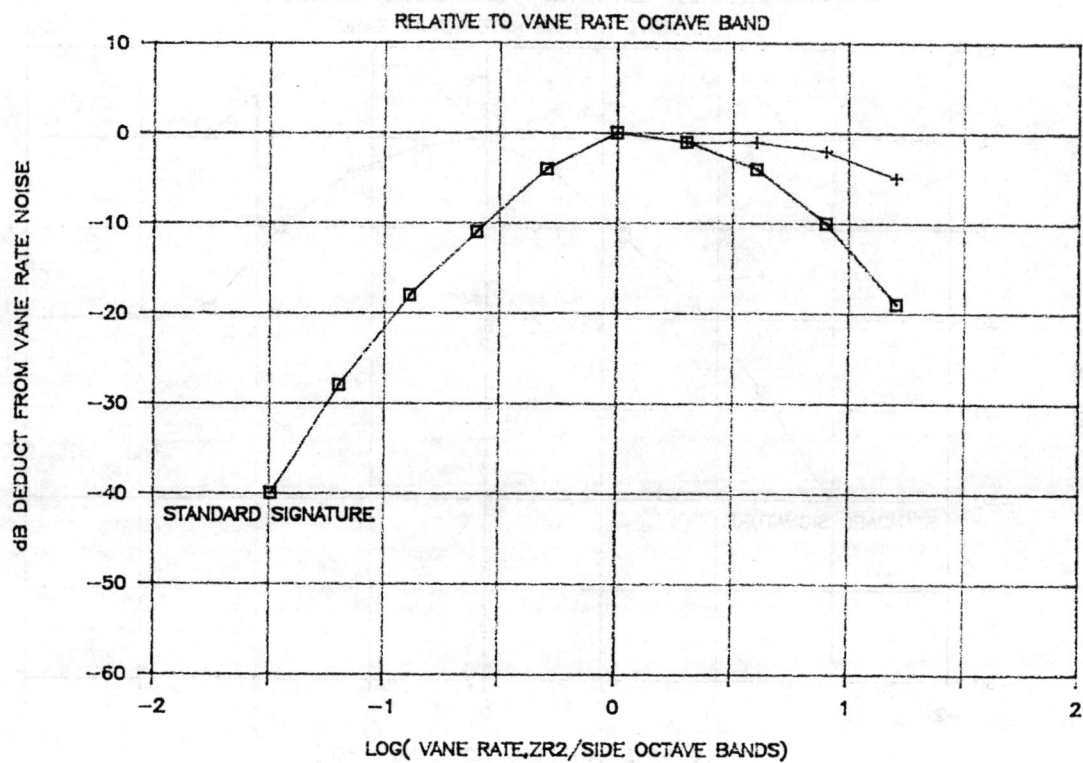

6.0 Typical depatures from the idealized signature and the causes;

6.1 Cavitation.

6.2 Inlet backflow

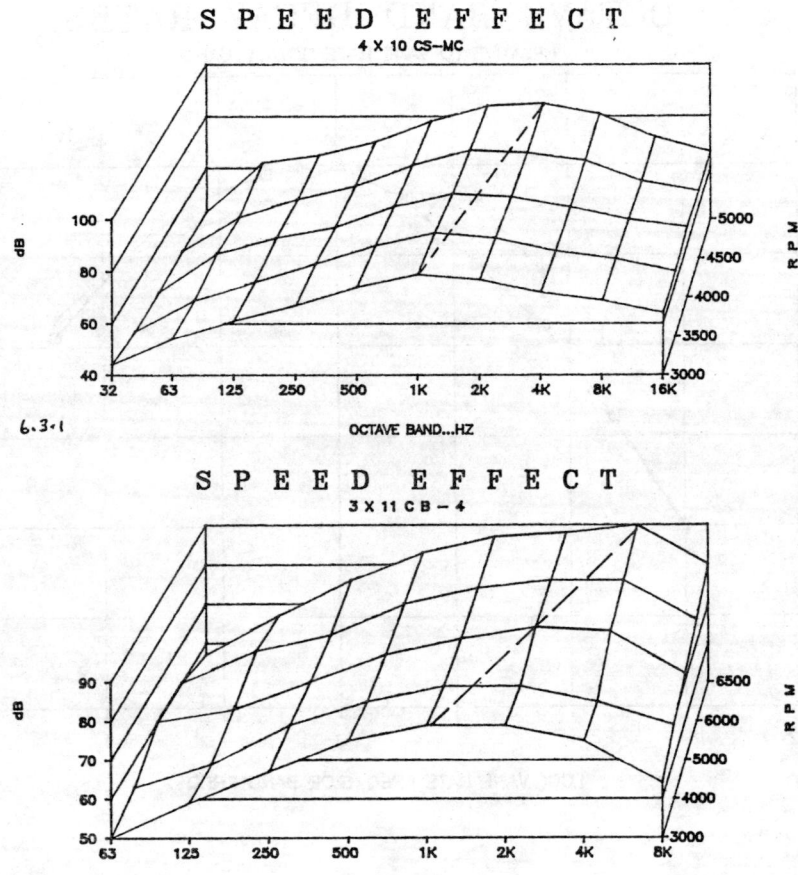

6.3 Effect of speed change.

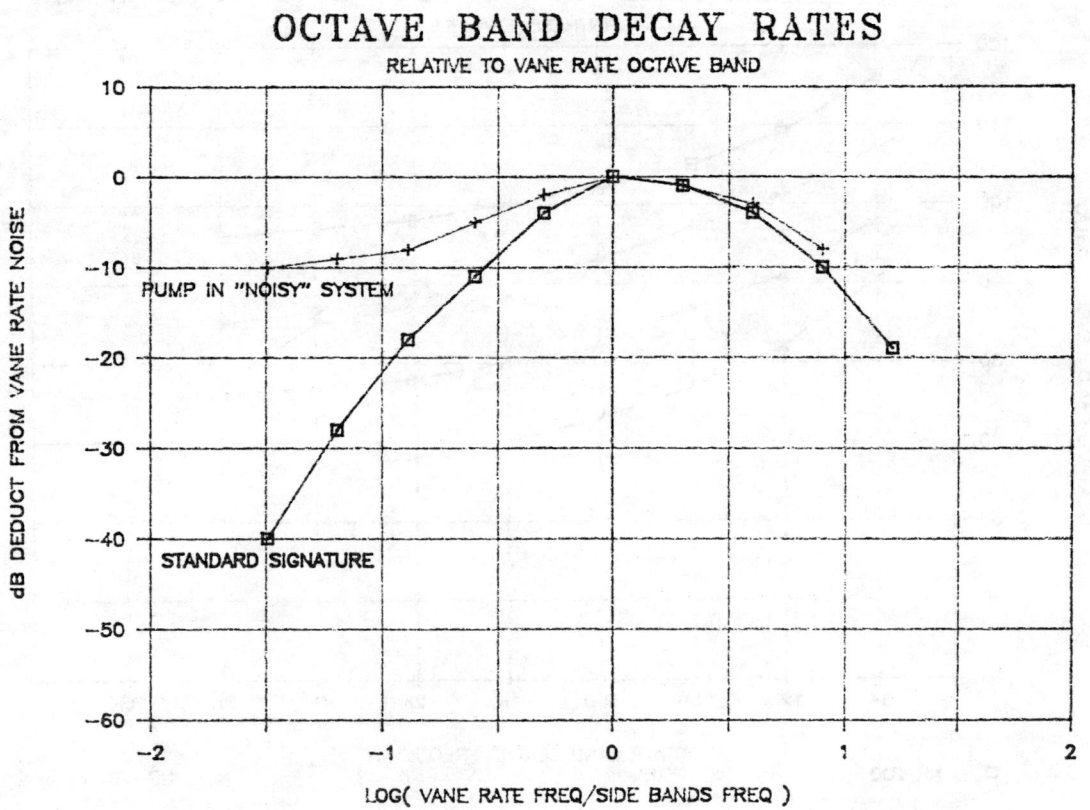

6.4 System effects.

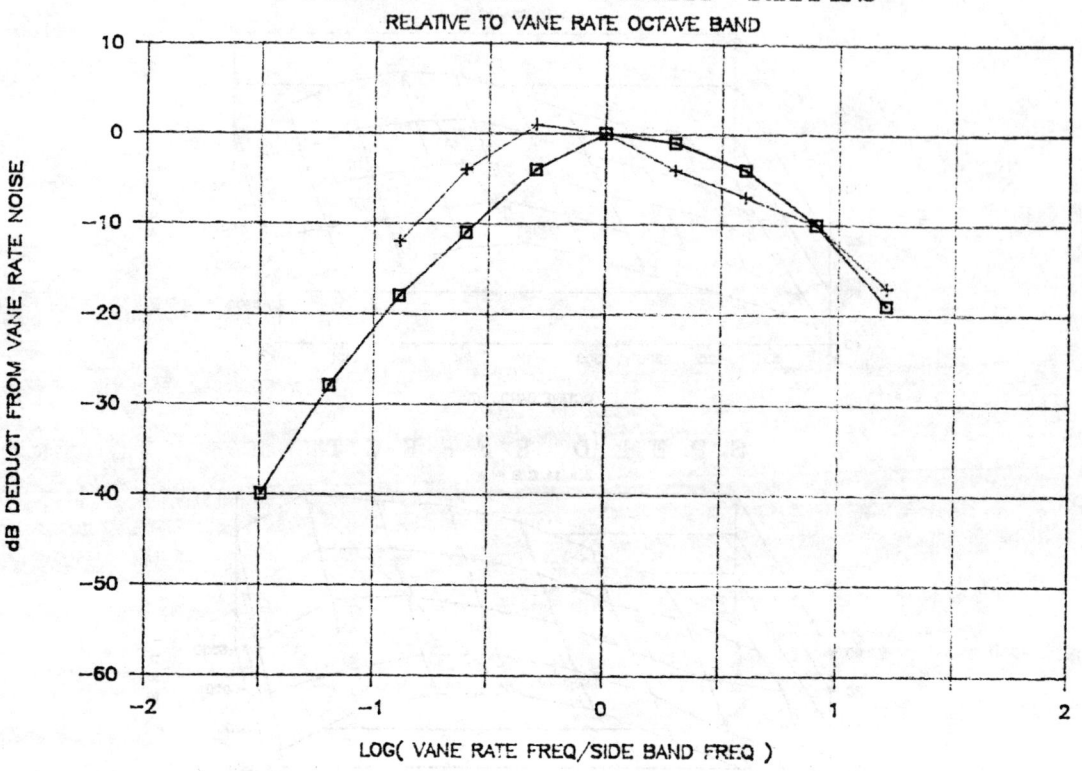

7.0

7.1 Worked example - Analysis

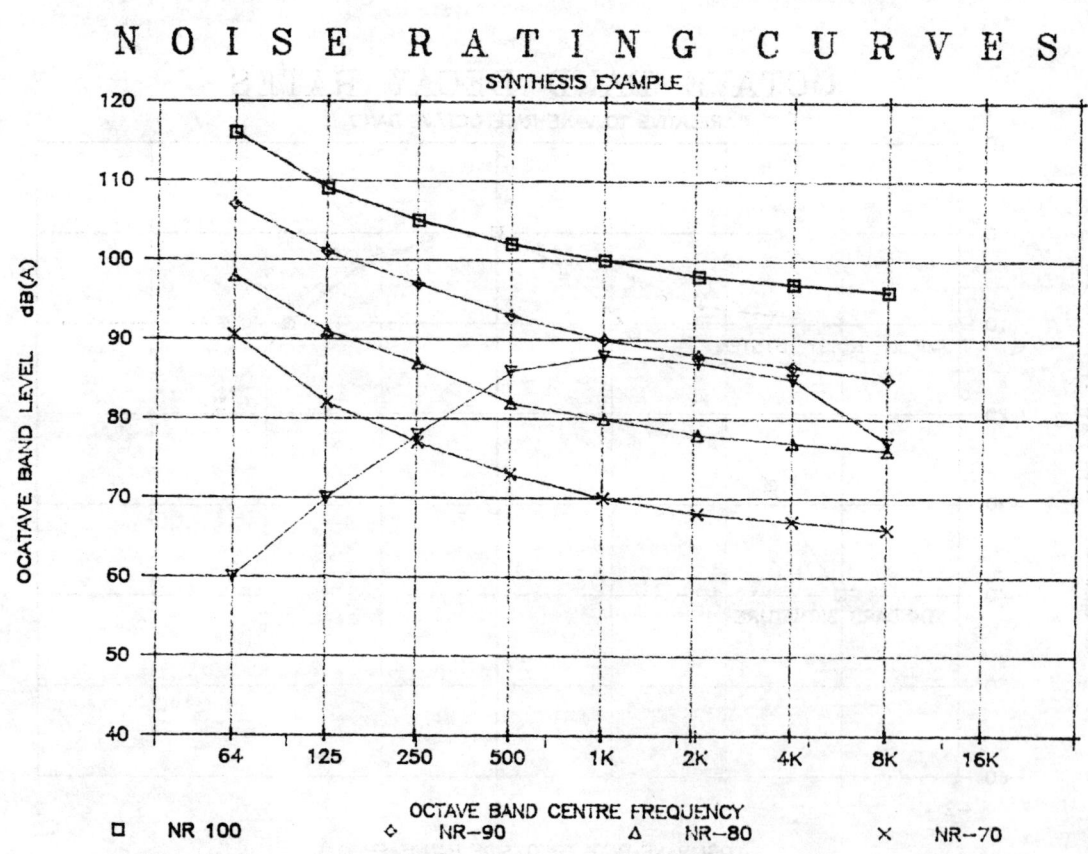

7.2 Worked example - Synthesis

11th International Conference of the
British Pump Manufacturers' Association

New Challenges – Where Next?

18-20 April, 1989
Churchill College, Cambridge

PAPER 3

OPERATING EXPERIENCE WITH SCRUBBER PUMPS MADE FROM

METALLIC MATERIALS IN FLUE GAS DESULPHURISATION (FGD) PLANTS

USING THE LIME SCRUBBING PROCESS BASED ON LIMESTONE

P R W Robinson and A Heumann, KSB Pumps Ltd., UK

SUMMARY

With few exceptions, the absorbent medium used in the lime scrubbing process for the desulphurisation of flue gases has been accepted as limestone $-CaCO_3-$ due to its low cost as compared with the alternative medium unhydrated lime – Ca O. However, the powdered $CaCO_3$ has a far greater hardness than Ca O.

Depending upon the source of the limestone, various amounts of silicate impurities will be present, which further increases the abrasiveness of the solids-laden washing suspension. The first FGD Plants in the Federal Republic of Germany to use limestone as the absorbent medium were commissioned at the end of 1983 so that they have now been operating for between 25,000 and 30,000 hrs. Early limestone plants operated with relatively high speed scrubber pumps, in ignorance of the higher wear stresses, but to-date none of the solution annealed austenitic-ferritic cast steel pump casings has required to be repaired by welding or replaced.

The resistance of solution annealed duplex steels to hydro abrasive wear is no longer sufficient to achieve comparable operating periods with the extremely high stressed internals – i.e. the impeller and wear plate. However, by manufacturing these components from wear resistant chilled cast iron grades, it is possible to achieve the 16,000 hours between service intervals as specified by the operators.

There has been no evidence as yet of major cavitation damage in pumps made from metallic materials, which frequently leads to the failure of Elastomer-lined pumps.

Conference organised and sponsored by the British Pump Manufacturers' Association
in conjunction with NEL and BHRA, The Fluid Engineering Centre.
Co-sponsored by the Process Industries Division of the Institution of Mechanical Engineers.

Modern mechanical seal technology exists for single seals without external flushing for sealing the shaft passage of scrubber pumps. This type of seal has been in use in scrubber pumps on limestone suspensions in the Federal Republic of Germany for approximately 2 years and has achieved operating periods of up to 10,000 hours.

FLUE GAS DESULPHURISATION IN THE FEDERAL REPUBLIC OF GERMANY (FRG)

Commercial flue-gas desulphurisation of coal and oil-fired power stations started in the Federal Republic of Germany at the end of the seventies. The regulation restricting emissions from large-scale furnace plants (GFAVO) which came into force on July 1st 1983 in conjunction with the clean-air act (TA Luft) which became law on March 1st 1986, compels the operators of large-scale furnace installations to retrofit existing plants and fit new ones with flue-gas cleaning equipment to ensure that their sulphur dioxide and nitrogen oxide emissions do not exceed the statutory limits applicable from July 1st 1988.

Of the FGD processes, which were developed mainly in the USA and Japan, it is the lime scrubbing processes which have captured the market with the unhydrated lime - CaO - initially used as the absorbent medium being replaced by limestone - $CaCO_3$ - for reasons of economy. Powdered limestone is harder and, depending on the source of the deposits, often contains impurities. On account of the slower reaction rates as compared with CaO or $Ca(OH)_2$, limestone plants require greater flow rates of the washing suspension with a higher solids content. The exposure of the components to hydroabrasive wear in dynamic washing suspensions with $CaCO_3$ as the absorbent medium is considerably greater than with a $Ca(OH)_2$ suspension of CaO unhydrated lime.

The significance of this was not known in the first FGD plants to be designed for a limestone suspension nor could the consequences thereof be anticipated.

The situation proved to be particularly critical in those plants where - based on the stressing known to be caused by a $Ca(OH)_2$ suspension - scrubber pumps selected for the upper speed and capacity limits were used. The necessary increase in the quantity of the washing suspension to be circulated in limestone plants in order to achieve the guaranteed degree of desulphurisation could only be achieved by allowing the pumps to be operated at 25% overload.

The adverse effects on the operating behaviour and the service life of the wetted parts should become evident as the period of operation increases. Of particularly serious consequence was the speed, which was inadmissibly high for products containing abrasive solids in conjunction with the high flow velocities resulting from operation in the overload range.

Despite these operating conditions, which are extremely unfavourable for centrifugal pumps, to date we do not know of any severe cavitation problems resulting in failure of impeller and casing on pumps made of metallic materials of the kind occurring in Elastomer lined scrubber pumps in FGD plants in the USA, - not even on larger pumps with capacities of over 7000 m^3/h. Fig.1.

The same applies to the shaft seal of scrubber pumps which in the course of time has been changed from the water-flushed packed gland via double-acting, water-sealed mechanical seals to single, self flushed mechanical seals with a rotating seal ring and stationary spring section. Single mechanical seals operating without external flushing water are nowadays the norm in scrubber pumps. This type of mechanical seal, made of high-grade materials, is today no longer regarded as a wearing component of scrubber pumps affecting the maintenance intervals, since the manufacturers of these seals can guarantee 16,000 operating hours between maintenance intervals provided certain conditions are fulfilled. Fig.2.

The first FGD plants built in the FRG for commercial application using the lime scrubbing process on a limestone basis came on stream during 1983. They have now achieved operating periods of 25000 hours or more, so that sufficient operating experience is now available with the scrubber pumps used in these applications to allow well-founded statements to be made on the suitability and success of the technology used. By using information on experience from comparable Elastomer-lined pumps in the USA, it is possible to compare the resistance of scrubber pumps in metallic materials to those with Elastomer linings. The guarantees for periods between maintenance intervals demanded by electricity utilities in the F.R.G are unique for the operation of centrifugal pumps handling solids-laden, abrasive and corrosive products in FGD plants or even in other applications with similar arduous operating conditions.

Generally, operating guarantees of 16,000 hours are standard practice in the F.R.G., calculated from the date of provisional take-over of operation of the FGD plant by the ESU. During this time which, with continuous operation of the plant, corresponds to a period of at least 2 years and to which the period of start-up and trial operation has to be added, it is not permissible for any component to be repaired or replaced.

This extremely stringent requirement had to be taken into account in the hydraulic design configuration of scrubber pumps, particularly in the selection of the drive speed, the construction of the pump and shaft seal and last but not least in the selection of suitable corrosion and wear-resistant materials for the wetted parts.

Additional requirements relating to the availability, efficiency and minimum positive suction head as well as noise emission levels also had to be considered. Fig. 3 and 4

We have succeeded in meeting all these demands to an almost optimum extent, by producing a pump with a hydraulic configuration designed for the duty point, which is of back pull-out design with direct drive via a detachable coupling. The pump features wetted components in metallic materials with adequate resistance to corrosion and abrasion and is sealed by a single mechanical seal which can be operated without external liquid flush Fig. 5.

The impeller and casing in metallic materials which are insensitive to damage from entrained particles, guarantee risk-free operation of the scrubber pumps without the need for a suction strainer, so that even with the largest pumps with flow rates of up to 8000 m³/h and NPSH $_{av}$ values of less than 10m, trouble-free operation is possible without damaging cavitation.

This point appears to be particularly significant when it is remembered that in the FGD plants in power stations in the USA which mainly feature elastomer-lined scrubber pumps, damage by foreign bodies is by far the most common cause of failure. This has meant that in the course of time in many plants the original elastomer-lined impellers have been replaced by metallic impellers.

In addition to the impeller, it is the suction-side wear plate and the shaft seal which are the components which determine the service life and therefore the time between maintenance intervals. The impeller and wear plate are the two wetted components exposed to the greatest amount of hydroadbrasive wear with certain areas being in particular danger due to the localised high flow velocities and turbulence. Fig. 6.

Areas subject to particular localised wear are the suction-side clearance gap between the wear plate and the impeller, the impeller inlet in the area of the suction-side coverplate and the blade outlet edges at the point of transition to the two impeller coverplates.

Silicon carbide (SiC) with its high resistance to wear has been used with great success in the area of the clearance gap in the suction side wear plate in the form of an L-shaped casing wear ring. Even after operating periods of several tens of thousands of hours this material shows no signs of wear. By using an appropriate design configuration these components, manufactured from a very brittle material, can be extensively protected against mechanical damage from foreign bodies or assembly errors. However, as SiC rings are only available at present in sizes up to a maximum diameter of 500mm, this arrangement can only be used in pumps in sizes up to DN 500. Fig. 7.

The results of model wear tests in the laboratory, together with initial experience from FGD plants using a limestone suspension, have shown that the resistance of solution annealed austenitic ferritic cast steels to hydroabrasive wear was not sufficient to achieve the 16,000 operating hours between maintenance intervals specified by the electricity utilities with the most highly stressed components, i.e. the impeller and wear plate.

Therefore, instead of the material NORIDUR 9.4460, materials offering better resistance to hydroabrasive wear but with a slightly lower resistance to corrosion were used for these components in scrubber pumps in plants using the limestone suspension. Depending on the lowest pH value and the highest chloride content of the washing suspension either NORIDUR 9.4460-DA – a development of the austenitic-ferritic cast steel NORIDUR 9.4460 with special heat treatment or NORILOY NL 25 2, a corrosion-resistant chilled cast iron with a ferrite-carbide structure was used.

Criteria for the Selection of Scrubber Pumps based on Operating Experience

The limit for the corrosion-resistance of NORILOY NL 25 2, the corrosion-resistant chilled cast iron used for the internals of scrubber pumps for FGD plants using a limestone suspension, in acidic aqueous solutions is pH = 3.5 and Cl' contents of $\leq$ 60,000 ppm. If, during normal scrubber operation, these limit values are not attained, then there is a danger that on start-up of the FGD plant the levels will either fall below or exceed these limit values when the flue gas is fed into the water-filled scrubber and if the limestone suspension required for neutralisation is not added in the correct amount. Fig. 8.

The material, which is damaged by selective corrosion attack of the ferrite in the structure, is then abraded due to hydroabrasive wear in the area of the corrosion pits. Since the hydroabrasive wear increases considerably as the damage to the component advances, the useful life of NORILOY NL 25 2, components can be adversely affected by brief corrosive overstressing. Fig. 9.

Of decisive importance for the operating behaviour of scrubber pumps and the service life of its wetted components is the optimum adaptation of their performance to the operating data of the scrubber. A very narrow operating range around the best efficiency point is recommended: $0.8\ Q_{opt} \leq Q_{operation} \leq 1.1\ Q_{opt}$.

Operation in this relatively narrow range in the characteristic field of a scrubber can only be achieved if the heads on which the selection of the pump is based correspond as closely as possible to the actual values in the plant. The plant values are determined by the static heads between the individual spray levels and the liquid level in the scrubber sump and the pressure loss in the spray level which is dependant on the flow rate.

If the pressure loss calculation is too low the pump will leave the afore-mentioned recommended operating range and if the flow rate is too small will operate in the partial-load range of the curve. In the case of over calculation of the pressure loss, the pump duty point will move into the overload range.

Such variations in the flow rate of circulating pumps can occur during operation if spray nozzles become worn, which was more common in the first FGD plants where the spray nozzles were generally made from metallic materials. In the ceramic spray nozzles used almost exclusively today, which offer greater resistance to wear, this is no longer a problem. However, the pressure loss in the spray system can increase if individual nozzles become blocked by loosened gypsum deposits or other foreign bodies.

Apart from a few exceptions today, as already mentioned, almost all the scrubber pumps used in FGD plants operating with the limestone scrubbing process are equipped with single mechanical seals which can be operated without external flushing water. To date operating periods of 10,000 to 12,000 hours have been achieved with cartridge type rotating and stationary seal rings of high-grade silicon carbide in scrubber pumps of FGD plants of this type. Extremely positive experience has been obtained with these seals. Occasional problems during start-up operation or in plants operating intermittently were due to the fact it was not possible to observe the limits specified by the manufacturers for pressure to be sealed off by such seals on pump shutdown. After pump shutdown, as a result of back flow into the scrubber of the liquid contained in the discharge pipe, reverse rotational speeds may be reached during pump turbining, depending on the liquid column which acts as the pressure head, at which the impeller back vanes reduce the pressure in the seal chamber to such an extent that it falls below atmospheric pressure; the seal then opens with air being drawn into the seal chamber and until the rotor reaches standstill the contact faces can run dry. By limiting the reverse rotational speed and adapting the back vane geometry accordingly, it is possible to prevent the pressure in the seal chamber from falling below atmospheric.

Conclusions:

Now that the first FGD plants commissioned in the Federal Republic of Germany using the lime scrubbing process on a limestone basis have been in operation for over 25,000 hours it is possible to make well-founded statements on the suitability and the performance of the equipment and materials used in these plants.

The engineering of the scrubber pumps used, their design, construction and the shaft seal as well as the materials of construction have proven to be successful. Despite the, to some extent unfavourable, conditions under which the pumps have to operate - at high speeds and in the overload range - they have achieved without exception the service periods guaranteed to the operator and in some cases such guaranteed service periods have been exceeded by far. Even after 25,000 operating hours it has not been necessary to carry out repair weld on, or renew, any of the pump casings made from solution annealed austenitic ferritic CrNiMo cast steel NORIDUR 9.4460 as a result of wear or corrosion.

On the assumption that the operating conditions in these plants and the stress to which scrubber pumps are exposed can be regarded as typical of FGD plants using the limestone scrubbing process, the conclusion may be drawn that scrubber pumps in suitable corrosion and wear-resistant materials which are equipped with single, self-flushing mechanical seals fulfil the extreme demands of the electricity utilities for a guaranteed period between service intervals of 16,000 hours and therefore provide a very high degree of reliability.

The high availability of FGD plants is guaranteed - without the need for standby pumps - by the back pull-out design of the scrubber pumps which allows rotor replacement to be carried out in only a few hours. This means that as far as the scrubber pumps are concerned all the conditions have been fulfilled to extend the period between FGD plant maintenance intervals to a two-year cycle.

The anticipated periods between maintenance of the wetted components of scrubber pumps in limestone process FGD plants can - according to the experience available to date with casings and ceramic-reinforced wear plates - be assumed to be 25,000 to 30,000 operating hours or with non-reinforced wear plates and impellers 16,000 to 20,000 hours.

If use is made of the possibility available with metallic materials of trimming the most severely wear damaged area at the impeller outlet in order to adapt the head of this impeller to the next lower spray level, then with this impeller a useful operating life of 24,000 hours can be achieved.

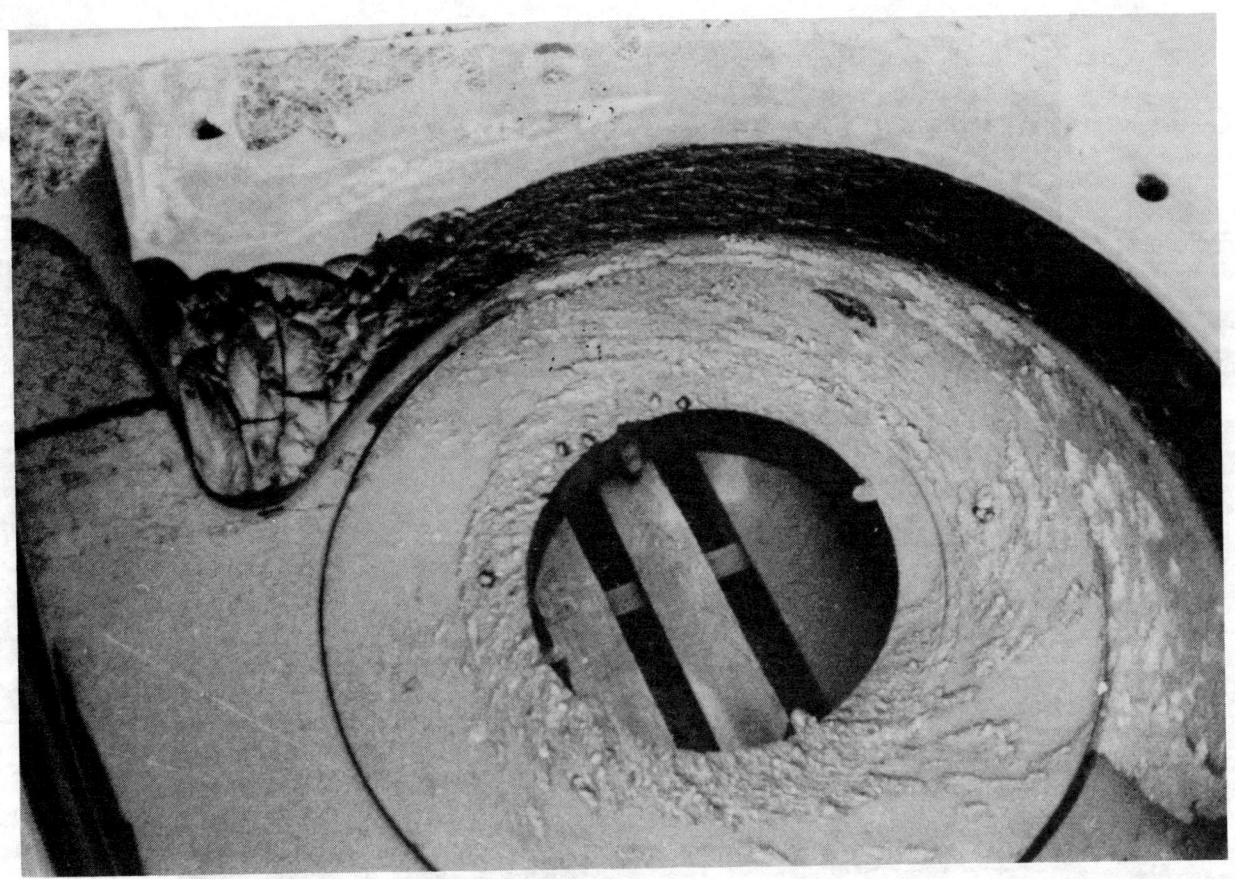

Figure 1.

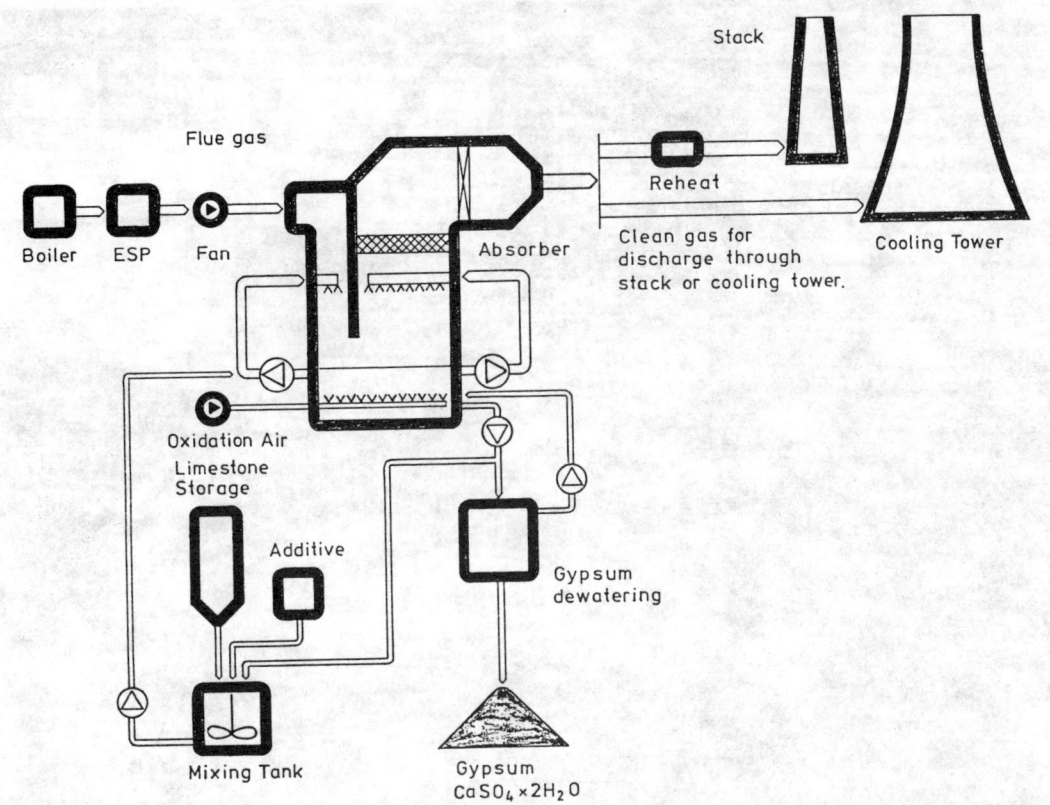

Flow scheme of a wet scrubbing process utilizing limestone ($CaCO_3$)

Figure 2.

Figure 3.

Figure 4.

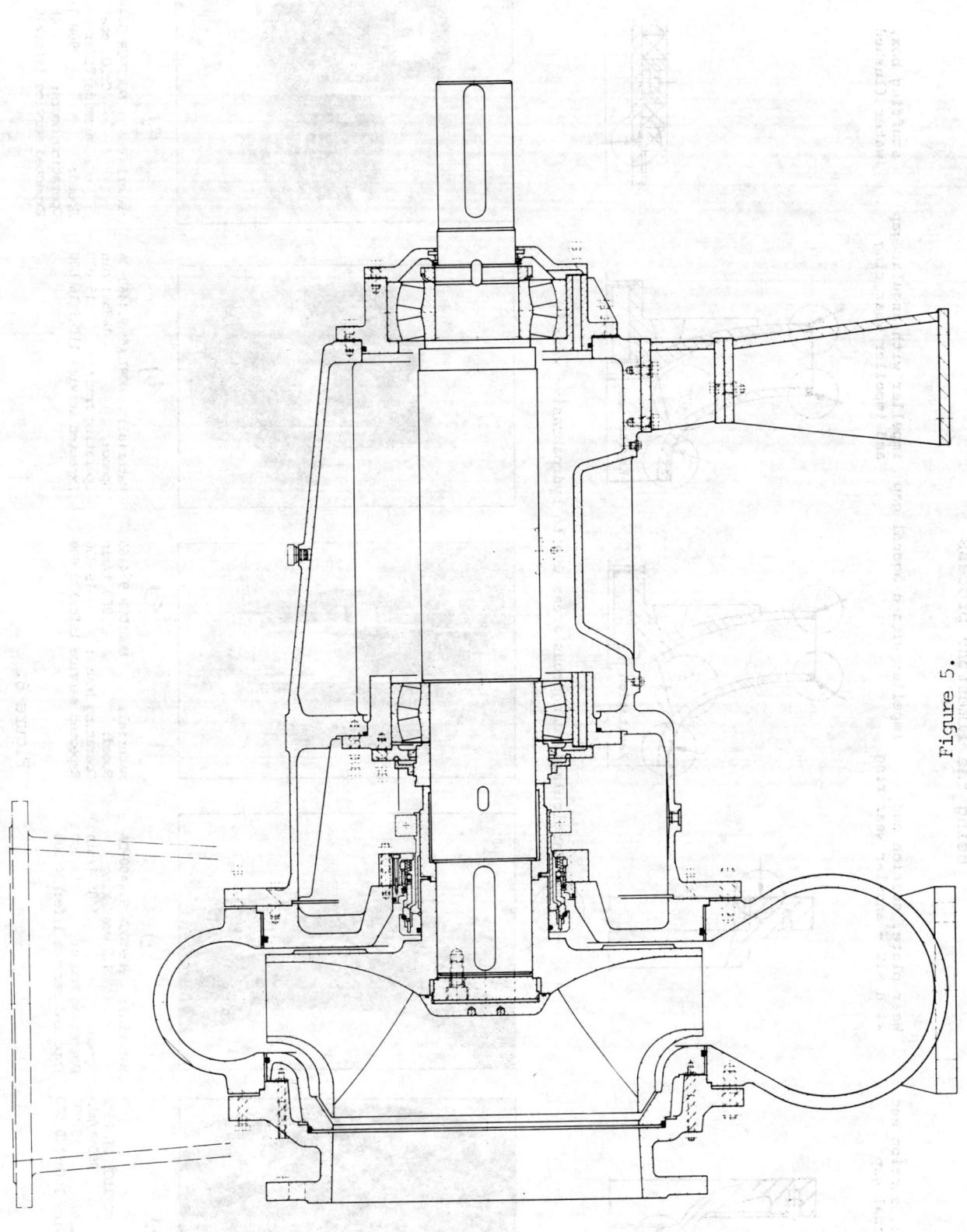

Figure 5.

Wearing parts of scrubber pumps from FGD plants using the limestone process

Wear plate, suction end with diagonal gap	Wear plate, suction end with SiC - angular wear ring	impeller with diagonal gap	impeller with angular gap and impeller wear ring	stuffing box, water flushed
Material: NORILOY NL 25 2 Speed: 960 1/min Operating hours: 15 000 Expected service life:>20 000	Material: NORIDUR 9.4460-DA SiC wear ring Speed: 1485 1/min Operating hours: 15 000 Expected service life:>24 000	Material: NORIDUR 9.4460 Speed: 960 1/min Operating hours: 19 300 Expected service life:>20 000	Material: NORIDUR 9.4460-DA Speed: 1485 1/min Operating hours: 15 000 Expected service life: 16 000	Material: NORIDUR 9.4460/ Cr_2O_3 coating aramide fiber packg. Speed: 960 1/min Operating hours: 8 000 Expected service life:>10 000
a.)	b.)	c.)	d.)	e.)

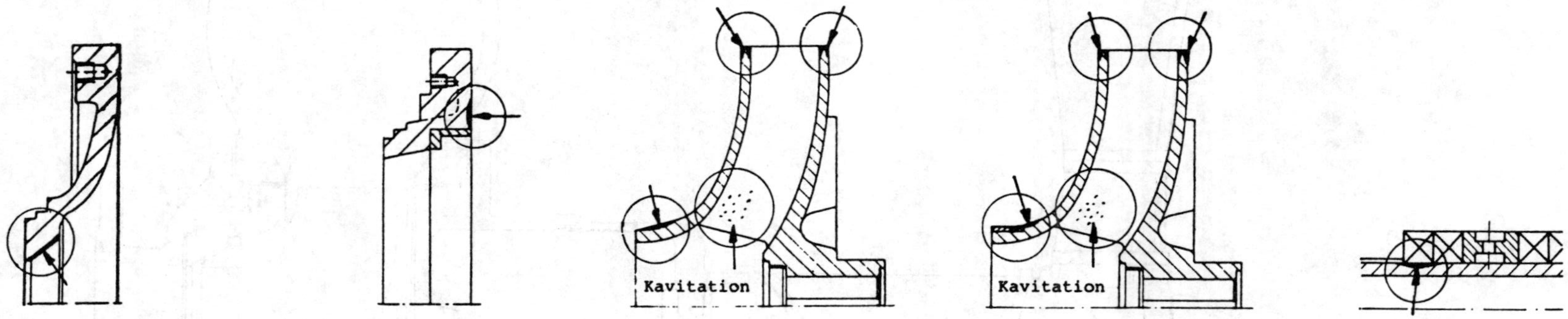

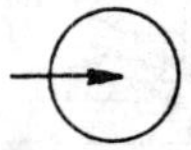

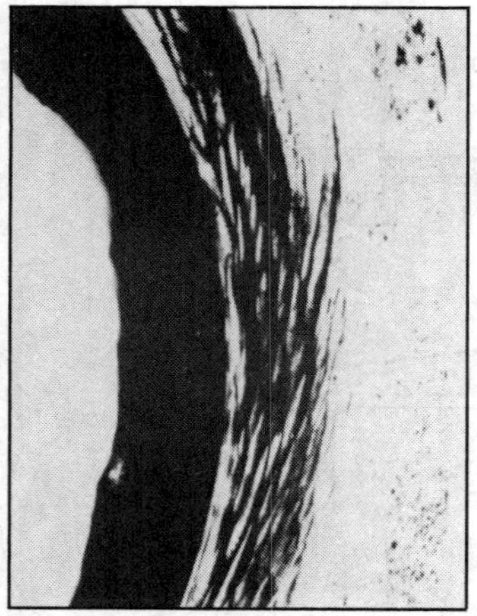

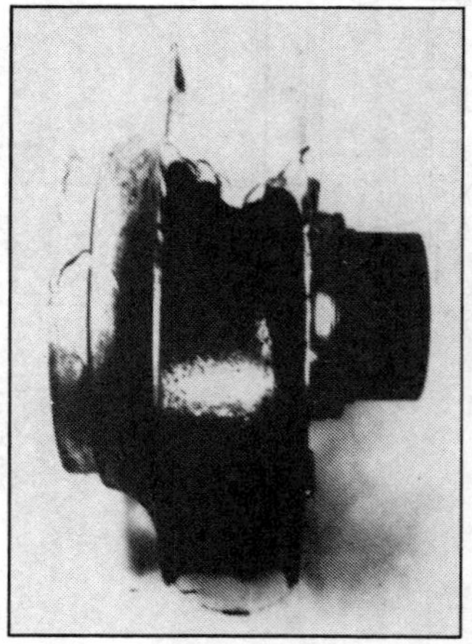

↑ Particularly dangerous areas due to hydroabrasive wear

Figure 6.

KSB-materials in flue-gas desulphurisation plants
using the lime-scrubbing process with limestone-$CaCO_3$ - as the absorbent medium

NORIDUR R 9.4460 a.)

NORIDUR R 9.4460-DA b.)

NORILOY R NL 25 2 c.)

Designation acc. to DIN 17006: G-X3CrNiMoCuN 24 6

Chemical Composition (Mass. %):
C		0,04
Si	max.	1,5
Mn	max.	1,5
Cr		25
Ni		6,0
Mo		2,5
Cu		3,0
N		0,15

Mechanical Properties (Amb. Temp.):
Yield point (N/mm²) $R_{p\,0,2}$	450
Tensile strength (N/mm²) R_m	700
Elongation at rupture (%) A_5	25
Notch impact strength (J) DVM	85
Hardness MB 30	200–260

Heat treatment: 1060 °C / 3 h / Water

Structure: Ferrite/Austenite

Pitting resistance equivalent: 35
(% Cr + 3,3 % Mo + 15 % N)

Limitations of applic. on FGD-Scrubbing liquors:
at 65 °C and continuous operation
pH ≧ 2,0
Cl' ≦ 100 000 ppm

Material loss rate of model test under abrasive conditions:
(Silica sand - water mixture 1:1) 76 (g/m²h)

Designation acc. to DIN 17006: G-X3CrNiMoCuN 24 6

Chemical Composition (Mass. %):
C		0,04
Si	max.	1,5
Mn	max.	1,5
Cr		25
Ni		6,0
Mo		2,5
Cu		3,0
N		0,15

Mechanical Properties (Amb. Temp.):
Yield point (N/mm²) $R_{p\,0,2}$	550
Tensile strength (N/mm²) R_m	700
Elongation at rupture (%) A_5	10
Notch impact strength (J) DVM	10
Hardness MB 30	250–320

Heat treatment: 1060 °C / 3 h / Water + dispersion hardening

Structure: Ferrite/Austenite/intermetallic phases

Limitations of applic. on FGD-Scrubbing liquors:
at 65 °C and continuous operation
pH ≧ 2,5
Cl' ≦ 80 000 ppm

Material loss rate of model test under abrasive conditions:
(Silica sand - water mixture 1:1) 24 (g/m²h)

Designation acc. to DIN 17006: G-S170CrMo 25 2

Chemical Composition (Mass. %):
C	1,7
Si	1,0
Mn	1,0
Cr	25
Mo	2,0

Mechanical Properties (Amb. Temp.):
Tensile strength (N/mm²)	400
Hardness MV 50	400

Heat treatment: Precipitation hardening

Structure: Primary carbides in ferritic matrix with secondary carbides

Limitations of applic. on FGD-Scrubbing liquors:
at 65 °C and continuous operation
pH ≧ 3,5
Cl' ≦ 60 000 ppm

Material loss rate of model test under abrasive conditions:
(Silica sand - water mixture 1:1) 4,9 (g/m²h)

Figure 7.

Figure 8.

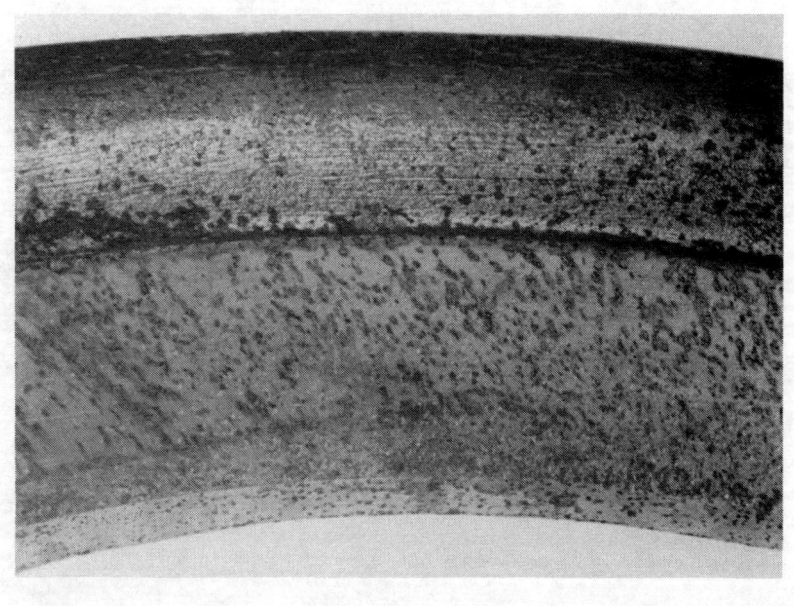

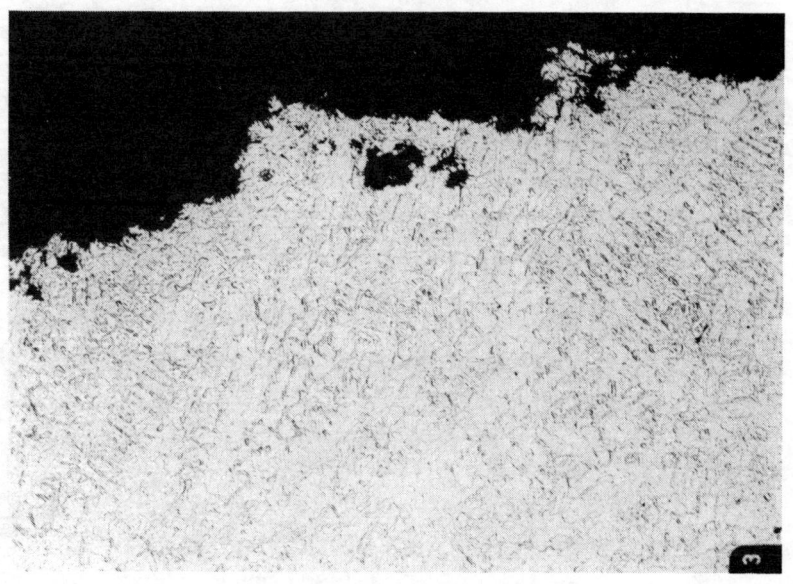

Figure 9.

11th International Conference of the
British Pump Manufacturers' Association

New Challenges – Where Next?

18-20 April, 1989
Churchill College, Cambridge

PAPER 4

PUMPS FOR FLUE GAS DESULPHURISATION - (FGD)

P A Wheeler and M V Bevan, Warman International Ltd, UK

The use of Flue Gas Desulphurisation (FGD) systems in electrical generating plants to control sulphur dioxide (SO_2) emmission from fossil fuel power stations has had a significant growth in the 1970's and 1980's. This is due to the growing concern over the the extent of environmental damage caused by acid rain as a result of atmospheric moisture and SO_2 from burning coal.

Although many different FGD processes are available, only about 10 have been commercially applied. The most common system presently in use is the Wet Limestone Process, because of its low reagent cost and its successful application in a wide variety of high sulphur coals.

WET LIMESTONE PROCESS

This process involves recirculation of limestone slurry to an absorber where the sulphur dioxide laden flue gas is treated. It consists of three major process sections namely;

(a) SO_2 absorbtion
(b) Solid-liquid Separation
(c) Alkaline Additive System

Figure 1 a shows a typical flowsheet of the Wet Limestone Process.

The Alkaline Additive System consists of a Limestone Ball Mill which converts pebble limestone to powdered limestone. Limestone ground to approximately 90%

Conference organised and sponsored by the British Pump Manufacturers' Association
in conjunction with NEL and BHRA, The Fluid Engineering Centre.
Co-sponsored by the Process Industries Division of the Institution of Mechanical Engineers.

less than 200 mesh is mixed with water in an Additive Feed Tank and then fed to the Reaction Tank through the Additive Feed Pump as required to replenish the alkalinity lost in the process slurry during sulphur dioxide absorption. The fresh slurry in the Reaction Tank is then pumped to the Spray Tower by the Recycle pump. This absorbent slurry is sprayed against and contacted with the flue gas in the tower. It reacts with the sulphur dioxide in the gas stream to form calcium sulphite and sulphate. The process slurry is then returned and regenerated in the Reaction Tank by the addition of limestone before recycling to the Spray Tower.

The solid calcium sulphite and sulphate are removed continuously from the Reaction Tank by bleeding a portion of the process slurry to a Thickener and Centrifuge via a Bleed Pump. The thickened calcium salts precipitated at the lower part of the Thickener are fed to the Centrifuge by the Underflow Pump.

The clean filtrate from the Centrifuge is returned to the Thickener for recycling. The clear overflow from Thickener with make-up water are added to the Additive Feed Tank for water replenishment.

LIMESTONE PROCESS WITH PRESCRUBBER

For the above process, the Prescrubber may sometimes be needed to remove some soluable gases (in particular chlorides and fluorides) and fly ash in the flue gas before entering the Sulphur Dioxide Absorber. This is shown in Figure 2 which is derived from the flowsheet for Figure 1. A prescrubber usually operates at low pH and wash water spray is used to purge chlorides and fluorides out of the system as a dilute hydrochloric and hydrofluoric acids. After passing through a waste water treatment system, it is discharged to the sea or local water courses. However, FGD systems with Prescrubber require higher constructions and operating costs. Also more sludge is produced from the waste water treatment system. Therefore, Prescrubbers are seldom used in the USA and West Germany. However, they are used extensively in Japan for removing chlorides and keeping impurities out of the FGD system and the gypsum product. In the UK, Prescrubbers are required due to the very high level of chlorides in the local coals.

SLURRY CHARACTERISTICS

The physical and chemical properties of SO_2 absorber recycle-slurry vary with the type of FGD process and the absorber design selected. Although slurry properties may also show considerable variation in a given system, the residence time of the slurry reaction tank is sufficient to isolate the pumps from sudden transients. The slurry solids content may range from 5-20% by weight; 10-15% is typical. Specific gravities range from 1.05-1.15. The slurry temperature approaches the saturation temperature of the flue gas leaving the absorber module, typically 50-65°C.

For wet scrubber systems utilizing lime or limestone as the primary reagent, primary chemical species include calcium sulphate ($CaSO_4 \cdot 2H_2O$), calcium suphite ($CaSO_3 \cdot 1/2H_2O$) and unreacted reagent consisting of calcium carbonate ($CaCO_3$) for limestone systems and calcium hydroxide ($Ca(OH)_2$) for lime systems. Lesser but significant amounts of calcium chloride ($CaCl_2$), magnesium chloride ($MgCl_2$), magnesium carbonate ($MgCO_3$), and magnesium hydroxide ($Mg(OH)_2$) are also present. For lime systems, the pH at the pumps normally ranges from 6-8. For single-loop limestone systems, the pH typically ranges from 5-6 at the pumps, while for double-loop systems, the pH is typically 4.5-6.5 in the upper loop and 0-2.0 in the lower loop.

Particle sizes range from 1-100 microns or larger. Calcium sulphite crystals form thin plates and plate clusters, while calcium sulphate forms blocklike crystals. Larger particles tend to form in wet-lime FGD systems than in wet-limestone FGD systems.

Abrasive particulate matter can present severe conditions for slurry pump operation. Although calcium sulphate/sulphite particles do not present abrasion problems for lined pumps, large fragments of detached scale from the absorber tank can inflict impact damage. Fly ash and lime grit can result in abrasive wear problems. Fly ash entrainment is generally a problem only during unit start-up or shut-down, when the particulate control device may be by-passed or de-energized, and can be minimized through proper start-up procedures. However, some FGD systems are designed to utilize the inherent alkalinity of fly ash to reduce lime or limestone requirements. Also, unburned fuel oil that may be entrained in the flue gas during unit start-up can accumulate in the recycled slurry and damage some elastomer materials. Entrained or dissolved gas in the slurry can reduce net positive head available (NPSHA) to the pumps and cause cavitiation.

Sodium-based dual-alkali FGD systems use recirculation pumps in a manner similar to that of wet lime or limestone systems, although the recirculation flow rates are lower. Because of the high solubility of the sodium reagent, the recirculated flow is a solution rather than a slurry. However, the potential for abrasion due to entrained fly ash and for corrosion due to high chloride concentrations may require the use of abrasion resistant and corrosion resistant slurry pumps, as in the lime and limestone scrubber systems.

PUMP DESIGN

There are two basic philosophies on the design of pumps for use in FGD circuits, one being a Heavy Duty Slurry Pump, double cased, and the other, a back pull-out, one piece casing, relatively high specific speed machine.

The author's company manufacture the double casing type pump and the advantages are:-

Large impeller diameter - which leads to a low speed of operation.

The advantages are:-

(1) low NPSH requirement;
(2) low eye periferal velocity, which leads to a lower wear rate
(3) low degradation of gypsum

The double cased construction gives the advantages of:-

(1) Maximum safety of operation, the wearing components can wear right through and there is no chance of a pump casing failing in a disastrous manner;

(2) Availability to mix and match materials of construction, the pump can be lined with a variety of materials (See Figure 3);

(3) Point (2) also means that there is a maximum of interchangeability between pumps handling different liquids, the mechanical components including the casing, will be identical.

SHAFT SEALING

This is probably the most severe problem in a slurry pump and there are three common options:-

(a) flushed gland
(b) expeller sealed, dry gland
(c) mechanical seal

FLUSHED GLAND

The full flow flushed gland is the most positive and widely accepted design for most applications where dilution of the product is permitted (Figure 4).

The flushing provides a two fold function:-

 (i) it provides liquid lubrication of the packing
 (ii) it helps flush away particles from entering the stuffing box area

EXPELLER SEAL

The purpose of this design is to keep the product away from the stuffing box whilst the pump is operating. The principle of operation is as follows. The back vanes on the impeller generates a pressure which partially offsets the pressure generated by the main vanes in the impeller. An auxillary impeller, commonly referred to as an expeller, operates in the expeller box cavity generating additional pressure. The pressure in the pump casing is dependant upon the pump generated head plus the pump suction pressure. For this sealing system to operate satisfactorily, the pressure generatedby the main impeller back vanes plus the expeller will balance the pump casing pressure. See Figure 5.

From the above, it will be apparent that no sealing water is required and that there is no dilution of the product.

MECHANICAL SEAL

Whilst there are several major manufacturers, the trend is for a single mechanical seal, product lubricated with the contact faces of silicon carbide. These feature an in board rotating face with stationary springs which are not in contact with the product. These are 'cartridge' units which operate without any external flushing. This configuration means that the seal chamber is not suitable for conversion to a packed gland, without considerable expense. (Our own experience to date in Europe is extremely good as we have not yet had a seal failure. Major manufactures are happy to guarantee the required 16,000 hour life).

Figure 6 shows a seal in which the springs are not in contact with the product.

Figure 7 shows a seal in which the pressure maintaining the contact between the stationary and rotating faces is obtained by, what is in effect, a rubber spring (rubber in shear).

MAINTAINING PUMP EFFICIENCY

The Heavy Duty Slurry Pump design does not have close clearance wear rings at the eye of the impeller but incorporates auxillary sealing vanes on the front of the shroud of the impeller. Wear will occur over the front face of the impeller and throatbush and this can be periodicallly checked and re-adjusted without stripping the pump. This is done by moving the bearing cartridge forward. A pump which incorporates wear rings typically does not have the sealing vanes on the front shroud, will then have a relatively high pressure across the wear rings, this in turn leads to a high velocity and therefore high wear. We would, therefore, suggest that this efficiency drop off in this style of pump will be faster. Also in order to replace wearing a complete **stripdown is required, as opposed to the few minutes required** to adjust the cartridge of the Heavy Duty Slurry Pump. See Figure 8.

INSTALLATION

Problems can and do occur with pumps in FGD installation and in most instances they could have been avoided had enough thought been put into the system design. The following are possible problem areas:-

1. AIR PURGING

    It is recognised that air is used to help complete the chemical reaction in the tank. However, this can cause problems in the pump

    (a) It can build up behind the impeller, which can thus lead to dry running, and subsequent failure of this seal.

    (b) Severe vibration due to bubbles of air/gas going through the impeller.

2. SCALE

    In the absorber tank scale can build up on the sides of the tank. This can then break off in chunks and find its way into the pump. We have instances where strainers are fitted onto the inlet pipe in the sump itself. These can get blocked and then the pump will cavitate. When the pump is stopped and strainer investigated, typically the back flow through the pump will have cleared the screen, and the operator believes that all is well. It should be appreciated that the noise of cavitation is effectively muffled in a rubber lined pump.

If strainers are not fitted then obviously the large lumps will go through the pump and this can cause damage, particularly to rubber components.

3. TRAMP

Inevitably welding rods and all sorts of bits and pieces are present at start up and nozzles can get detatched as wear and tear takes its toll. A strainer is needed to stop this getting into the pump.

SOLUTION

To overcome the above problem, we would refer to Figure 9 and 10. We also recommend that pressure gauges be placed both sides of the strainer so that the operator knows that there is a build up on the screen. Whilst the station is on a low load condition the pump can be shut down, the strainer removed from the pipework, cleared and replaced in a few hours. There is no shut down of the plant required.

We would recommend the reducer be offset as shown, an air release valve could also be incorporated to reduce the amount of gas getting into the pump.

PARALLEL OPERATION

Genuine slurry pumps are designed with wide passages with low velocities, this leads to a flat performance curve. If a system is designed with say four pumps in parrellel then, if at a low load condition, two or three pumps are operating, this leads to these pumps operating at beyond their design capacity. This can cause:-

(a) cavitation problems
(b) higher wear rates due to higher velocities and turbulent flow

We would recomend that each pump should have its own pipeline and designated bank of sprays. See Figures 11 and 12.

PUMP DUTIES

The pump duties would be approximately as shown in Table One.

TABLE ONE

RANGE OF SLURRY PUMP PERFORMANCE REQUIREMENTS FOR LIMESTONE PROCESS

PUMP CONDITIONS	(1) SPRAY RECYCLE PUMP	(2) SLURRY BLEED PUMP	(3) THICKENER UNDERFLOW PUMP	(4) ADDITIVE FEED PUMP	(5) PRESCRUBBER RECYCLE PUMP
Head H (m)	20–35	20–35	10–30	30–40	20–35
Flow Q(l/s)	400–2000	50–3000	15–30	10–30	150–2000
Specific Gravity (Sm)	1.05–1.15	1.05–1.15	1.15–1.30	1.15–1.30	1.00–1.05
Solids by Wt (Cw)	8–15	10–15	20–40	15–40	2–5
Chloride Content (ppm)	300–500	300–500	500–1000	1000–10000	10000–40000
Fluoride Content (ppm)					1000–4000
pH	4.5–6.5	4.5–6.5	4.0–7.0	6.5–14.0	0–2.0
Relative Abrasion	Moderate	Moderate	Moderate	Moderate	Moderate
Relative Corrosion	Moderate	Moderate	Moderate	Moderate	Moderate

MATERIALS SELECTION

In developing a range of materials for Flue Gas Desulphurisation use we had to draw extensively on our experience in general slurry handling.

For Pre-scrubber Applications normally rubber lined pumps are used and so the technology was well known and tried. For the Absorber Spray Recycle applications, however, in some instances rubber was used but on many applications a mixture of metal and rubber gave the required wear life.

Quite often the materials selected were rubber volute and back liner with metal impeller and metal throatbush requiring new technology if a guaranteed life of 16,000 hours was to be achieved.

Our task was to develop an errosion/corrision alloy. The basic approach was to develop a white iron with a supporting matrix similar in composition and chemistry to stainless steels, and with sufficient carbides to provide wear resistance.

The majority of corrosion applications are satisfied by using stainless steels. A similar approach to the chemistry behind stainless steel was used with white irons. The composition of the alloy had to be altered so that the matrix contained at least 12% chromium in solid solution. To obtain this level of chromium, several approaches were used. The chemistry would be altered, by lowering the carbon level, and having an excess of chromium. Only limited carbide formation would occur, resulting in more chromium in the matrix. Alternatively, alloying elements could be added to a basic Fe-Cr-C iron to form complex carbides, and also leave more chromium in the matrix.

As the amount of carbides are reduced, the wear resistance of the alloy will be reduced. A compromise between wear resistance and corrosion resistance was eventually reached.

The phenomenon of wear is complex and still not fully understood. The majority of people associate hardness with toughness but hardness is only one of the many factors working to combat wear.

Any test rig must give the minimum degredation and otherwise approximate to the site conditions taking into account not only the nature of the slurry but also its pH value and the amount of suspended solids.

One particular material may be good for one type of application where a different material may be better suited to another.

The controlling factor is the microstructure of the material used and experience has shown that the martensitic carbidic structure of white irons is superior to other structures having equal or higher hardness values. While microexamination is the most reliable method of checking the quality of white iron castings, hardness level is the most frequently measured property. The Brinell and Rockwell 'C' tests are most commonly employed in quality control. Due to the high hardness level of chrome irons it is necessary to use a tungsten carbide ball in making the Brinell Test. A finish grind with a No.180 grit paper or finer is recommended. (Fig 9)

For convenience in converting difference hardness measurements on chrome irons, a conversion chart is shown below. As the structure becomes coarser (in heavier sections), the Brinell test is comparatively the most consistent and reliable hardness test.

HARDNESS MEASUREMENT EQUIVALENTS

BRINELL HARDNESS TUNGSTEN CARBIDE BALL	ROCKWELL C HARDNESS	VICKERS HARDNESS	SHORE SCLEROSCOPE HARDNESS
750	63-65	830-860	90-93
700	61-63	740-777	84-87
650	59-61	690-720	79-82
600	56-58	630-660	75-78
550	53-55	570-610	60-73
500	50-52	510-540	67-70

Abrasion and corrosion are formidable and very expensive problems in Mining and Industry and can caused crippling costs through loss of production and very high maintenance expenditure.

When a new plant is designed, it is most important that materials of construction be carefully selected to minimise these problems. This requires detailed knowledge of the properties of available materials, their ease of fabrication, their cost and anticipated useful life in the plant. Product contamination and safety aspects must also be considered, as well as the cost of holding large stocks of spare parts.

WEAR

The progressive loss of substance from the operating surface of a body occurring as a result of relative motion of the surface.

The mechanical and chemical causes can be basically classified as follows:

(a) Mechanical causes of wear:
- Abrasion
- Impact
- Pressure
- Erosion
- Cavitation
- Friction

(b) Chemical causes of wear:
>   Corrosion
>   Heat

However, these not only combine and interact in many service conditions, but each of them can operate with many variations.

Abrasion

The degree of abrasion is influenced greatly by the nature of the abrasive such as the size, shape, hardness, texture and whether it is wet or dry. The severity of abrasisve wear is also influenced by physical conditions of operation such as the pressure of impact exerted on the abrasive particle, the concentration of the abrasive, the angle of impingement to the surface and the relative speed of the particles.

There are basically three kinds of abrasion, namely gouging, grinding and scratching. In addition, impact must always be considered when selecting materials for abrasion resistance.

Gouging

This occurs when particles cut into the wearing surfaces with considerable force, tearing offer relatively large particles from that wear surface.

Gouging can occur under low velocity eg when a shovel dipper digs into a pile of rock, or at high velocity, eg with hammers or breaker bars in impact-type pulverisers.

Grinding

This occurs when two wearing surfaces rub together with sufficient force to produce a crushing action on abrasive particles trapped between the two surfaces. This type of wear occurs in roll and jaw crushers, and on track shoes.

Scratching

This occurs as a result of contact between relatively freely moving abrasive particles and wearing surfaces. There is not much crushing of the abrasive particles and impact is normally negligible. Chutes, truck bodies and grizzly screens are subjected to this type of abrasion.

COMBATING ABRASION

Abrasion can be combated to some extent by changes in equipment design and operating conditions, but the simplest and most effective method is usually to use materials with greater abrasion resistance to the product being pumped.

When selecting an abrasion-resistant material, the first step is to determine the type of abrasion involved. The second step requires consideration of impact, which may necessitate toughness in the alloy, possibly at the cost of maximum abrasion resistance.

ABRASION-RESITANT ALLOYS

The large number of available abrasion-resistant irons and steels can most conveniently be classified into seven groups.

HIGH-CHROMIUM MARTENSITIC WHITE IRONS

These are based on 8-32% chromium and an abbreviated list is given in Table 2.

Table 2

DESCRIPTION	TYPICAL COMPOSITION (%)					
	C	MN	Si	Cr	Mo	Other
15 Cr - 3 Mo	2.4-3.6	0.7	0.6	15	3	-
15 Cr - 2 Mo - 1 Cu	3.4	0.7	0.6	15	2	1.0 Cu
20 Cr - 2 Mo - 1 Cu	3.0	0.7	0.6	20	1.7	1.0 Cu
12 Cr - Mo	3.2	0.7	0.6	12	0.5	-
25 - 32 Cr	2.3-3.0	1.0	1.0	27	0.5	-
8 Cr - 6 Ni	3.3	0.6	1.5	8	-	6.0 Ni
10 - 15 Cr	1.2-2.5	1.5	0.6	12	-	-
Warman A05	2.7	1.0	0.5	27	-	-
Warman A49						

The high-chromium irons possess a microstructure of carbides, surrounded by a matrix of martensite or austenite.

The alloys have excellent abrasion resistance in low-stress environments, such as occurs in slurry pumps. They are also outstanding under gouging and high-stress grinding conditions. In general, these alloys have better toughness than the white irons.

High-chromium irons have the advantage of responding to heat treatment and can therefore be annealed before machining, then hardened to a martensitic structure.

COMPOSITION OF THE NIHARD ALLOYS FOR COMPARISON PURPOSES

COMPOSITIONS (%)	TYPE 1 (A03)	TYPE 4 (A06)
Total Carbon	3.2 - 3.6	3.0 - 3.3
Nickel	3.5 - 4.5	4.5 - 5.5
Chromium	1.5 - 2.5	8.0 - 9.5
Silicon	0.4 - 0.7	1.5 - 2.0
Manganese	0.4 - 0.7	0.4 - 0.7
Sulphur (max)	0.12	0.12
Phosphorous (max)	0.12	0.12

All Nihard castings should be heat treated before use to ensure austenite transformation. This treatment increase strength and impact resistance without loss of hardness or abrasion resistance.

Type 1 is the original nickel-chromium white cast iron. Type 4 is a higher alloy composition with superior abrasion, impact and corrosion resistant properties. Performance of the Nihard alloys is inferior to the high chromium white irons in the majority of slurry pump applications.

The nature of the surface has also to be considered. The mechanism of abrasion usually consists of grooves being cut in the base metal followed by plastic deformation of the metal at the edges of the grooves. Leading to fragmentation and detachment from the surface. Then the cycle starts again. Therefore, theoretical studies have to include surface energy, deformation energy and deformation by volume units and rates of work hardening etc. Additional studies have to include the hardness and microstructure of the surfaces to resist abrasive wear.

IMPACT

Wear by impact is the result of a succession of local shock loads on the material surface. If the material is brittle, it may readily fracture. On the other hand, if the material is tough, it will yield by elastic or plastic deformation and fracture will be avoided or postponed until the limit of plastic deformation has been exceeded.

The classical theories concerning shock loading have to be used with caution especially when evaluating the stresses of the impacts, because the estimation of the point of contact with the surface at the moment of impact. The speed of impact can give wear by low, medium and high impact.

The speed of impact can give some precise information. For example, a high speed impact will cause permanent deformation by high stress and the destruction of the surface, by concentrated thermic energy due to the short time of application.

PRESSURE

Wear due to pressure is in relation to the compressive load that the metal is capable of withstanding. Such loads can be either static or low speed cycling. The maximum compressive stress that a material is capable of withstanding prior to deformation, is dependent on the yield strength and the microstructure.

A metal having high yield stress is generally capable of withstanding high compressive loads, but where complex microstructures exist, cracking tends to occur along the carbide boundaries. A metal having a low yield strength will plastically deform under compressive loads, until the compressive strengths of the metal is exceeded and fracture results. Under repeated cycle loads a fatigue failure can also occur if the fatigue endurance limit of the metal is exceeded.

EROSION

Erosion is caused by the impact of small abrasive particles generally conveyed in a liquid or gas. Wear is generally more severe where a turbulent media exists or where the flow changes direction. A knowledge of fluid mechanics is necessary in order to determine the speed and angle of attack of the abrasive particles; these two parameters being the basis of the study of erosion. In effect erosion can be classified into two mechanisms:

(a) Erosion by impact deformation at high angles of impingement by the particles.

(b) Erosion by an abrasive cutting effect a low angles of impingement by the particles.

Numerous studies relating to this problem have been conducted by scientists in many countries. Interesting theories have been put forward which are valid but difficult to implement in industrial situations.

CAVITATION

This wear phenomena results from the rapid formation and dispersion of gas bubbles in a liquid. This results in high speed pressure changes or explosions which cause shock waves to impinge on the containing metal surface, resulting in local deformation. The damage to the surface arises from a similar mechanism to that of erosion by impact deformation, except that with cavitation the solid abrasive particles are replaced by microwaves which produce pitting fatigue, subsequent micro-fissures and metal removal.

WEAR BY FRICTION

Wear by friction, or adhesive wear, can be defined as being that process which gives rise to loss of metal between two interacting surfaces as a result of adhesion of asperities in sliding or rolling contact. Wear by this mechanism is severe and can happen in the kinematic chain of a machine whenever there is an absence of lubrication at the interface of two bodies in contact.

The surface topography on a microscope scale shows that two interacting surfaces make contact at a few isolated points, resulting in high stresses in these areas. This gives rise to plastic flow at the interface and forms microwelds. The resistance to sliding is equivalent to the sum of the shearing forces necessary to break up all these junctions. The wear debris may then act as abrasive particle which accelerate the wear rate. A general law of wear which can be used by design engineers is not yet complete, possibly because of the complexity of the subject.

WEAR BY CORROSION

Wear by corrosion is the most complex of the wear phenomena discussed so far and is a science of its own. There are numerous forms of corrosion mechanisms and corrosive media. Corrosion mechanisms include chemical, electro-chemical corrosion, stress corrosion, corrosion fatigue and elevated temperature corrosion. Corrosion media can vary from acids and alkalines to atmospheric conditions, generally acids (with varying concentrations) constitute the most severe of all corrosive environments, but alkaline and atmospheric conditions can also be corrosive and should, therefore never be overlooked.

The concentration and temperature of a corrosive media is of considerable importance, as this can change the corrosion resistance characteristics of the weld or base metal. It is therefore important that the concentration and temperature are determined accurately.

MICROSTRUCTURE

Figure 13 shows a typical example of A03 and A05 and the difference microstructurally between the two.

Accepting the carbide is the most brittle phase it is probably obvious to see which mircostructure is likely to offer the greater toughness. There is much less obvious path along which a crack could propagate in the A05.

RUBBERS

Natural and synthetic rubbers used in abrasion resistent applications within the power industry range in hardness from 40 to 75A; typical application conditions for various hardness levels within this range are summarised in Table 3

Table 3

APPLICATIONS FOR ABRASION RESISTANT RUBBERS

HARDNESS (Durometer A)	APPLICATIONS
45–55	Wet or dry fine abrasives at relatively high throughput rates (eg slurries)
55–65	General applications involving moderate impact (eg chutes, truck bodies)
65–75	Large abrasive rocks with heavy impact

Apart from hardness, the material factors which influence the abrasion/erosion resistance of rubbers are:

> resilience
> elastic modulus and elongation
> tearing/cutting resistance
> tensile strength

All of the above parameters relate to the ability of the rubber to behave in an elastic manner under the action of the abrading particles, and to resist cutting or tearing under high tensile stresses. In addition, the size and shape of the abrading particles, and the impact conditions (force, angle and velocity) will have a signifiant effect. In general, all rubbers are unsuited for use with impact angles in the range 10–40°, as this gives rise to accelerated wear rates. Rubber wear components are normally designed so as to avoid this range of impact angles; this may involve the use of serrated or 'saw-tooth' profiles on the wear surface.

The abrasion resistance of rubbers is particularly sensitive to the velocity of abrasive particles, velocities which exceed a critical value will give rise to accelerated wear by scouring or cutting mechanisms, as the rubber is unable to behave elastically. A similar trend is observed under high frequency (greater then 10 Hz) impact abrasion conditions. The value of this critical velocity, or impact frequency, for particular rubbers is related to temperature, and abrasive size, density and shape. Maximum recommended abrasive velocities, for slurry applications, are typcially 10m/sec for the softer rubbers (eg Linatex). However eddy currents and localised turbulence can give rise to substantial (up to thirty times) increases in particle velocities, and hence accelerated wear rates.

Rubber wear materials can be formulated to provide a range of properties other than hardness. For example, under impact abrasion conditions, a combination of high modulus and reduced tear resistance may be required; the high modulus will limit the extent of elastic deformation under impact. Conversely, abrasive wear conditions involving fine slurries will involve reduced impact levels, and hence softer, higher tear strength rubbers are preferred. In both cases, appropriate lining thickness are required in order to absorb the impact energy. For soft rubbers (eg Linatex) in slurry abrasion, or under mild conditions (less than one metre fall), minimum rubber thicknessess are shown in Figure 14 as a function of maximum particle size.

Under more severe impact conditions, a harder, higher modulus will be required; Figure 15 shows minimum lining thicknesses for these materials (hardness 60-75A) under normal (90°C) impact conditions.

Rubber has energy-absorbing properties, rubber wear linings also enable substantial noise reductions (up to 10db(A)) to be obtained in FGD and mineral processing equipment.

ABRASION RESISTANT RUBBERS

TYPE	CHARACTERISTICS
Skega	Synthetic rubber, available in various hardness levels. applications include bin and truck linings, mill liners, screen clothes.
Trellex	Synthetic rubbers, hardness range 40-70A. Applications as listed for Skega.
Linatex	Natural rubber, hardness 38-40A. Applications include pumps, hydrocyclones, pipelines.
Linard	Natural rubber, hardness 60A. Suitable for heavy duty abrasion conditions.

CONCLUSION

During the 15 years that Warman International Limited have been providing pumps for the Flue Gas Desulphurisation industry, we have seen many changes and experienced many problems. It is a little disappointing that some of the design errors made 15 years ago are still being incorporated by contractors in their design. In particular, we would cite pipe work arrangements and suction conditions in the absorber.

We have found that the heavier designed slurry pump gives overall reliable performance, in particular providing a more congenial environment for the ancillary equipment, such as mechanical seals, bearings and lubrication systems.

The trend in both single and double loop systems is for better monitoring and control of the final flue gas emission and the need for improved efficiency to give lower power consumption costs. These requirements inevitably impinge on the pump design criteria, and we see much advantage in having pumps that can have a variety of wear materials, and a good degree of material interchangeability.

There is pressure to increase the guarantee period of 16,000 hours for all items of plant and for the rotating machinery. This brings us near the limit of design life for ancillaries such as lip and mechanical seals.

We are now involved in an extensive development programme covering elastomers and white irons. With our slurry handling experience, we are now able to actually pre-determine the carbide and matrix structure of irons, and tailor-make them to suit the particular application, bearing in mind, the corrosive and abrasive properties of the slurry.

The emphasis over the next decade in pump design will therefore be dedicated to lower user costs, specifically:-

1) **Lower power cost, through higher sustained efficiency maintained through the life of the wearing parts;**

2) **Lower maintenance costs, through superior materials of construction, specifically of the wearing parts.**

WET LIMESTONE PROCESS WITHOUT PRESCRUBBER

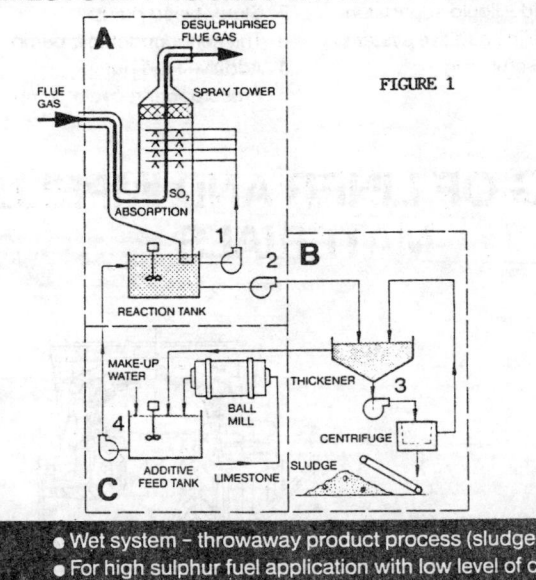

FIGURE 1

Main Features
- Wet system – throwaway product process (sludge)
- For high sulphur fuel application with low level of chlorides and fluorides
- Low reagent cost
- Economical for large power plant

Legends

Process Section:
A SO_2
B Solid - liquid separation
C Alkaline additive system

Pump Duty:
1 Spray recycle pump
2 Slurry bleed pump
3 Thickener underflow pump
4 Additive feed pump

WET LIMESTONE PROCESS WITH PRESCRUBBER

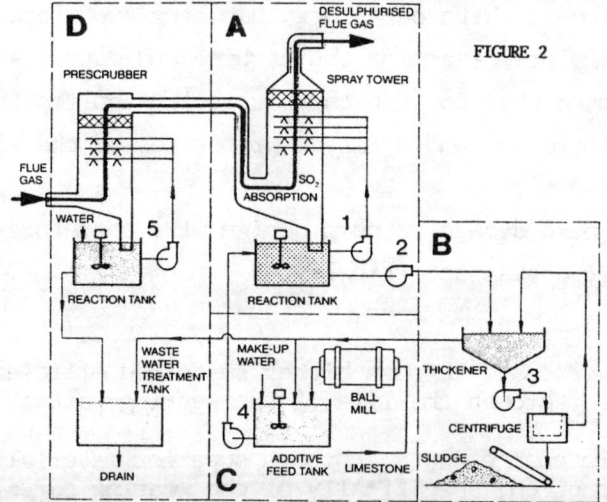

FIGURE 2

Main Features
- Wet system – throwaway product process (sludge)
- For high sulphur fuel application
- Prescrubber is used for fuel with high level of chlorides and fluorides
- Low reagent cost
- Economical for large power plant

Legends

Process Section:
- A SO_2 absorption
- B Solid – liquid separation
- C Alkaline additive system
- D Prescrubbing

Pump Duty:
1. Spray recycle pump
2. Slurry bleed pump
3. Thickener underflow pump
4. Additive feed pump
5. Prescrubber recycle pump

MIXING OF LINER AND IMPELLER MATERIALS

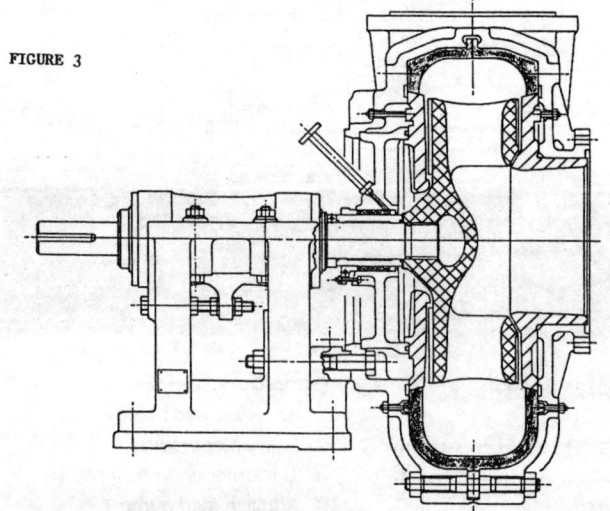

FIGURE 3

METAL PARTS may be fitted for PLANT START UP where tramp material and process upset conditions can occur.

"Trial and error" operation to determine OPTIMUM combination of materials for LOWEST RUNNING COSTS

INTERCHANGEABLE SHAFT SEALING

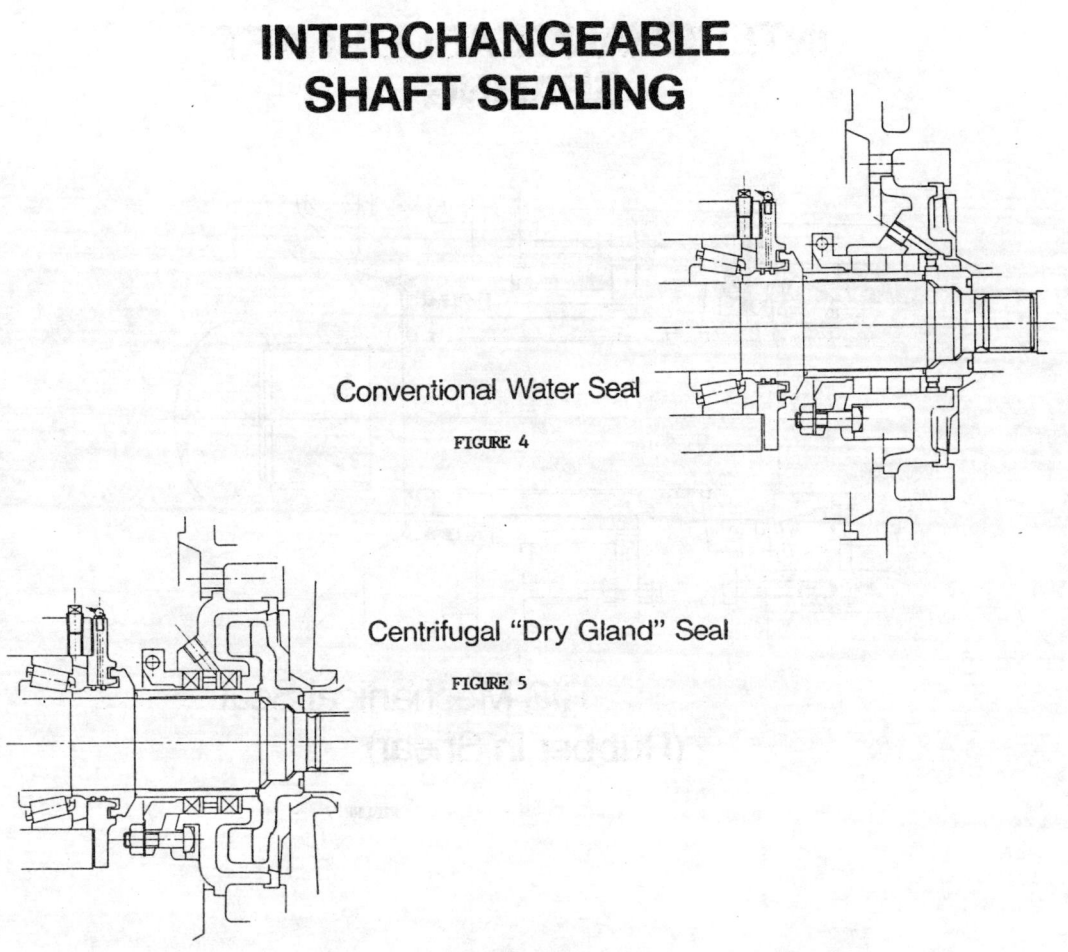

Conventional Water Seal

FIGURE 4

Centrifugal "Dry Gland" Seal

FIGURE 5

INTERCHANGEABLE SHAFT SEALING

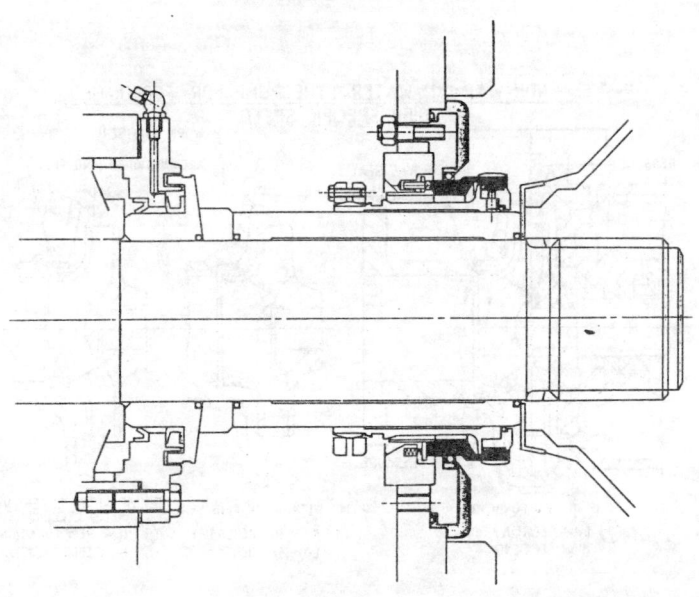

Mechanical Seal

FIGURE 6

INTERCHANGEABLE SHAFT SEALING

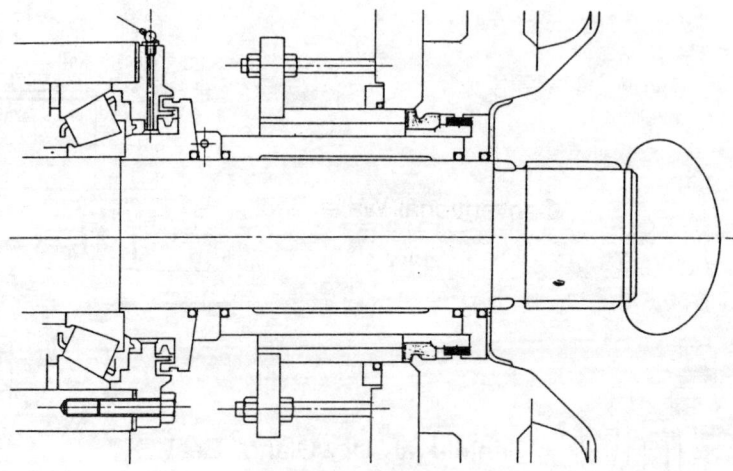

RIS Mechanical Seal
(Rubber In Shear)

FIGURE 7

FIGURE 8a

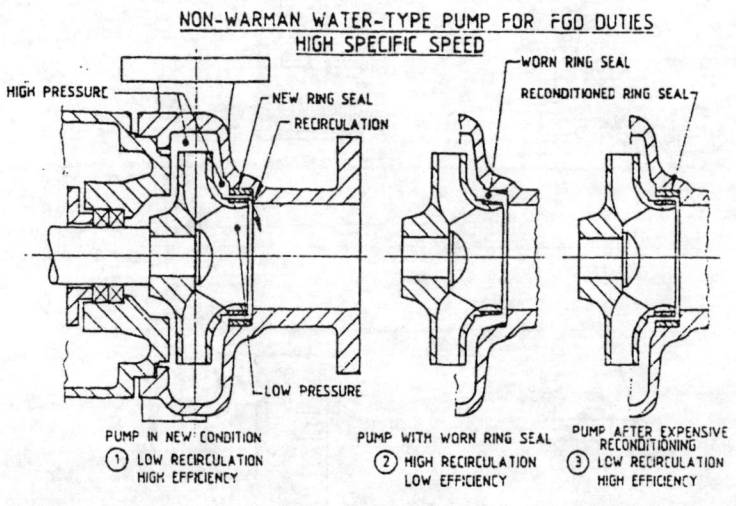

FIGURE 8b

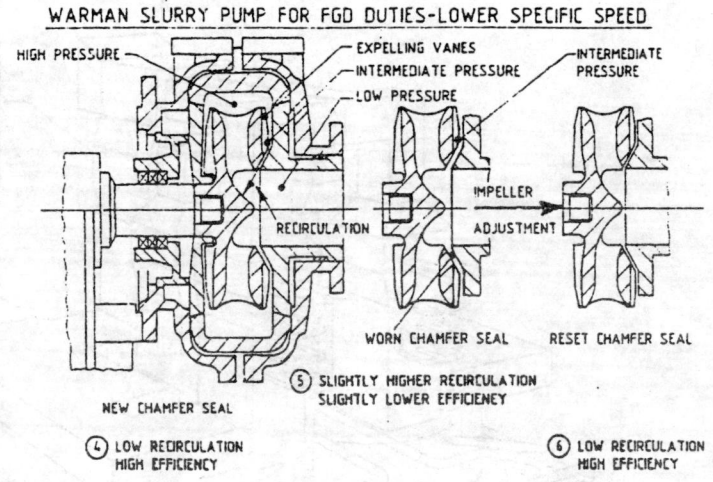

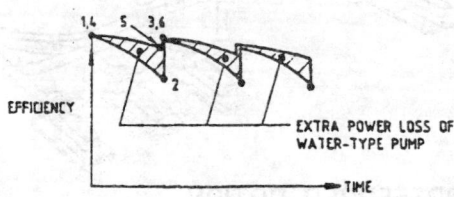

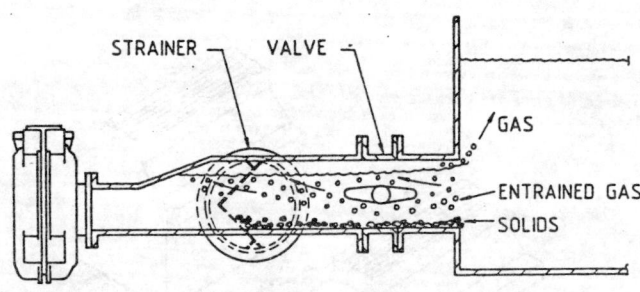

FIGURE 9

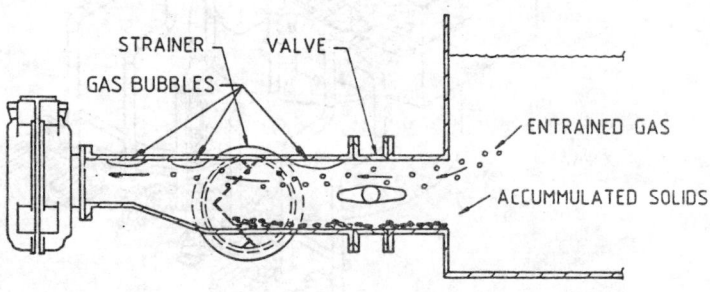

FIGURE 10

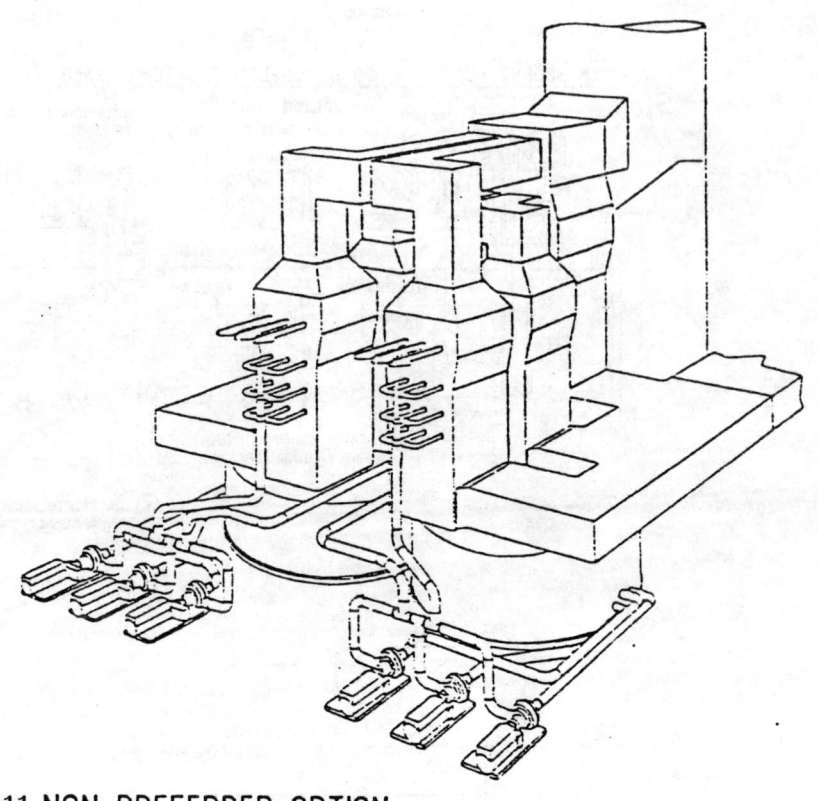

FIGURE 11 NON-PREFERRED OPTION

PUMPS IN PARALLEL FEEDING MULTIPLE BANKS OF SPRAY NOZZLES

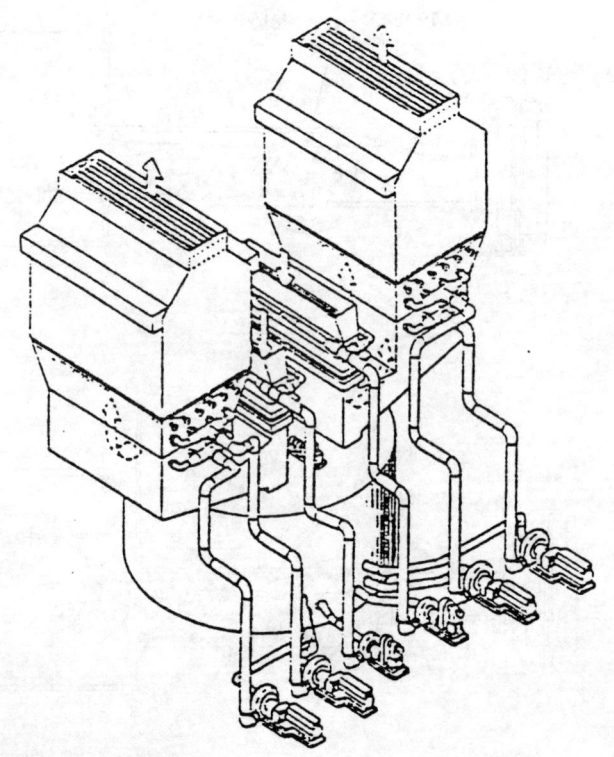

FIGURE 12 PREFERRED OPTION

EACH PUMP FEEDS ITS OWN BANK OF SPRAY NOZZLES

FIGURE 13

A03 - NO 1 NIHARD

MAGNIFICATION 500X
ETCHED IN 5 % NITAL
IRON CARBIDE
MARTENSITE
RETAINED AUSTENITE
HARDNESS - 597 HB

SPEC. - 500 MIN

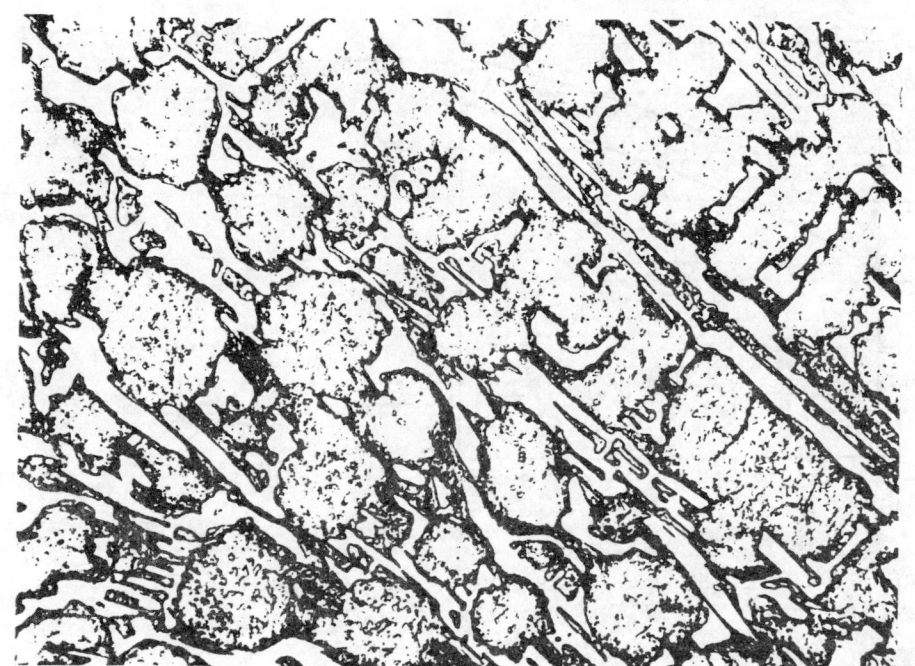

A05 - 27 % CHROME IRON

MAGNIFICATION 500X
ETCHED IN PICRAL 2 % HCL
CHROME CARBIDE
MARTENSITE
THOUGH VERY DIFFICULT
TO DISTINGUISH THERE
ARE SECONDARY CARBIDES
DISPERSED WITHIN THE
MARTENSITE HARDNESS
695 HB

SPEC. - 600 MIN

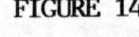

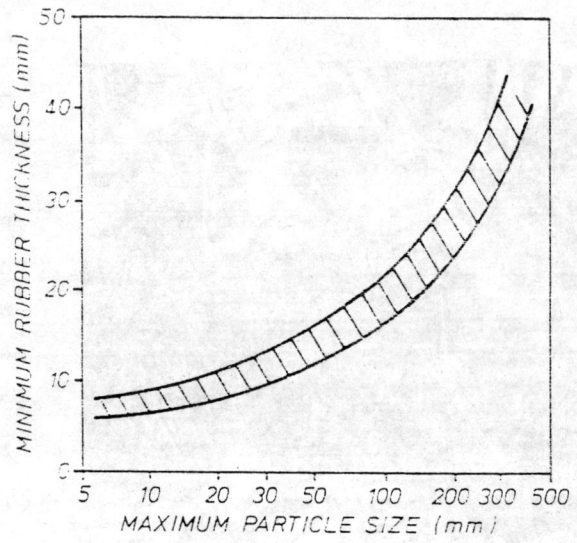

MINIMUM RUBBER THICKNESS VS MAXIMUM ABRASIVE PARTICLE SIZE; 45-55 SHORE RUBBER UNDER MILD IMPACT/SLURRY ABRASION [AFTER PINCUS]

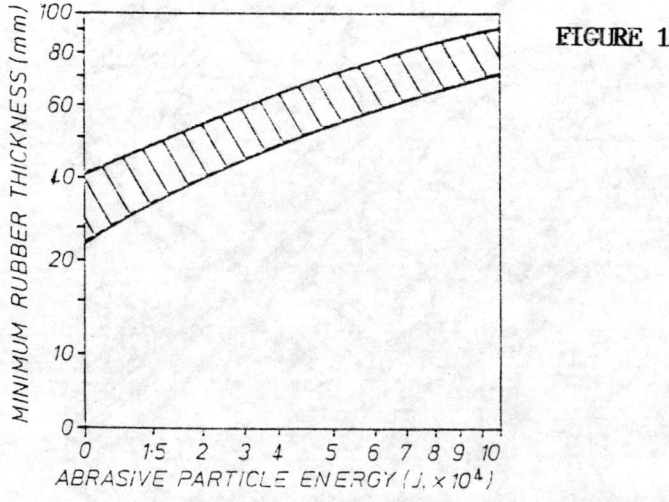

RECOMMENDED RUBBER THICKNESS VERSUS ABRASIVE PARTICLE ENERGY; 60-75 SHORE RUBBERS UNDER NORMAL (90°) IMPACT [AFTER PINCUS]

11th International Conference of the
British Pump Manufacturers' Association

New Challenges – Where Next?

18-20 April, 1989
Churchill College, Cambridge

PAPER 5

A NEW APPROACH TO THE DESIGN OF HIGH CAPACITY STORM AND FOUL PUMPING STATIONS

M. R. SINCLAIR
Technical Manager Flygt Pumps Limited

G. BJORKANDER
Senior Consultant Project Co-Ordinator Flygt AB

Summary

The worldwide up-grading of sewage and storm/surface water systems to improve environmental and social conditions and to provide flood protection, whether for domestic or industrial developments, has created a requirement for efficient high capacity pumping stations.

Traditional approaches to station design and associated plant selection have been based on the use of a relatively small number of high capacity dry well pumps discharging into pressurised pipelines. Such designs require major civil constructions to accommodate the wet well, capable of providing acceptable hydraulic conditions for the high capacity pump suctions, and large low level dry wells to house the pumps and associated valves and discharge pipework.

This paper will show that the adoption of a larger number of lower capacity submersible pumps, installed directly in a wet well, with individual siphonic discharge into a surge chamber, can result in significant reductions in capital plant, construction and energy costs, reduced hydraulic and electrical supply transients and improved flexibility of operation and maintenance procedures.

1. Introduction

The design of a pumping station cannot be considered in isolation from the up and downstream sections of the system - it is an integral part and in order to maximise system efficiency, the characteristics of the pumping station must be accurately matched to the hydraulic performance of the rest of the system. Furthermore, the civil, mechanical, hydraulic and electrical influences must be considered at every stage during the design sequence to ensure that one is not unduly compromised at the expense of another.

Unfortunately, many pump stations are still designed by first fixing the structural shape and then selecting the pump and associated equipment to fit the available space. This may have been acceptable practice in the past when the majority of stations were equipped with a standard design of pump - the traditional vertical spindle radial or mixed flow volute pump installed in a dry well. Such arguments are no longer valid today with the availability and increasing use of high capacity efficient and reliable electro-submersible pump units. It is the responsibility of the station designer to take full advantage of the flexibility of installation arrangements, which can be achieved using submersible pumps, to optimise system performance and reduce costs.

Conference organised and sponsored by the British Pump Manufacturers' Association
in conjunction with NEL and BHRA, The Fluid Engineering Centre.
Co-sponsored by the Process Industries Division of the Institution of Mechanical Engineers.

In order to compare both the engineering and financial aspects of the traditional vertical spindle wet well/dry well station with those of an optimised submersible equivalent, station designs and system performance will be estimated for two typical 'pumping duties':

- high static lift with short discharge main;
- low static lift with long discharge main;

The evaluation has been simplified by selecting the discharge system characteristics to give the same 'design total head' for each i.e., a single pump selection will suit either system.

Furthermore, on such high capacity systems, the upstream culvert, sewer or reservoir will normally have a significant influence on the pump control level selection. As it is difficult to accurately predict the 'open channel' hydraulic characteristics of the upstream system theoretically, it is recommended that it is evaluated as an integral part of a wet well/sump model test, necessary to confirm overall intake hydraulic performances, see Section 5.

Consequently, the following evaluation has been further simplified by largely eliminating the effects of the upstream system by assuming that the station suction TWL (design level) is 0.75 m above the invert of the 2.5 m diameter inlet sewer (maximum inlet velocity at 10 m^3/s = 2.04 m/s). Therefore, for a large proportion of the pumping range, free discharge from the inlet will exist.

2. System characteristics

The nominal values given in 2.1 and 2.2 and the pipework profiles shown in Figure 1a and 1b respectively have been used as the initial design criteria.

2.1 High static lift with short discharge main

Design capacity	Q_d	10.0 m^3/s
Station pipework head loss	H_{ls}	1.5 m
Static head	H_s	12.0 m
Discharge mains:		
Length	L_m	170.0 m
Diameter	D_m	2.0 m
Roughness	k_s	0.6 mm
Minor loss factor	K	1.5
Design total head	H_t	15.0 m

2.2 Low static lift with long discharge mains

Design capacity	Q_d	10.0 m^3/s
Station pipework head loss	H_{ls}	1.5 m
Static head	H_s	6.0 m
Discharge mains:		
Length	L_m	2500.0 m
Diameter	D_m	1.6 m (twin)
Roughness	k_s	0.3 mm
Minor loss factor	K	2.5
Design total head	H_t	15.0 m

2.3 Station duty

The design of a sewage (foul) pumping station differs from that of a purely storm/surface water station. For the latter, it is common practice to provide low capacity sump pumps to dispose of water and solids, which remain in the sump below the lowest main storm pump stop level at the end of a storm event.

Apart from ensuring that the sump is fully cleared of any debris, grit etc., at the end of such an event, the sump pumps can also serve to efficiently handle low inflows and minor events, resulting in reduced maximum power demand charges. In order to ensure that the sump pumps will efficiently clear the main wet well of both water and solids, they would normally be located in a low level pit fed by drainage trenches across the main wet well floor.

In contrast, sewage pumping stations, having a continous base inflow rate, do not need to be drained between pumping sequences and consequently will not be provided with fixed installation sump pumps and their associated pipework, valves and control equipment.

The comparison of pump station design has, therefore, been undertaken with the simpler sewage station, though the conclusions are equally applicable to both storm and combined (sewage and storm) pumping stations.

3. Pump types and associated equipment

3.1 Traditional wet well/dry well station

3.1.1 Pumps

For this option three (3) duty plus one (1) standby mixed flow volute pumps have been assumed, each having the characteristics shown in Figure 2 and the following duty values :

Pump capacity	3.3 m^3/s
Total head	15.0 m
Pump efficiency	86 %
Rotational speed	325 rpm (18 pole)
Motor output rating	700 kw
Total inertia	600 kg m^2

Each pump, located at the bottom of the dry well, is powered by the induction motor, mounted vertically above on an intermediate floor, via a drive shaft and associated flexible couplings. Depending on the length of the shaft, it may be necessary to provide additional intermediate bearings to limit torsional vibrations. Consequently, three (3) separate sets of bearings may be required for a single pump unit i.e., pump, motor and shaft bearings Figure 3.

Furthermore, it is not uncommon for pumps of this size to be equipped with gland packings requiring water lubrication or flushing systems, the failure of which would render the pump inoperative. Generally, increasing the number of critical components in a pump/motor assembly will not only increase the risk of failure but also the complexity of monitoring systems, and the frequency and extent of routine maintenance procedures.

3.1.2 Station pipework and valves

With pressurised manifold/discharge main systems, each pump must be equipped with discharge non-return valves and associated isolating valves, required for maintenance operations. Alternative motorised valve systems for closed valve pump start and stop may be considered. However, an auxiliary power supply for closing the valves in the event of power failure i.e., pneumatic actuators, is recommended, if flooding of the upstream system and bearing failure, due to high speed pump reverse rotation, are to be avoided. This, together with the increased complexity of control equipment, will result in a significantly higher risk factor. Consequently, the analyses have been undertaken assuming fast acting non-return valves (recoil type) and standard sluice valves are to be installed.

For dry well pumps, additional suction isolating valves are required to permit maintanance of the pump units whilst the station remains operational.

The pipework and valve dimensions considered are :

Suction pipework and valves	nominal diameter	1.4 m
	nominal velocity	2.16 m/s
Delivery pipework and valves	nominal diameter	1.2 m
	nominal velocity	2.95 m/s

3.2 Submersible pumps

3.2.1 Pumps

In comparison to the traditional concept, ten (10) duty pumps have been selected with no installed standby pump. The pumps are installed by simply lowering them down a guide bar system to locate under gravity on the permanent discharge bend secured to the floor of the sump. In the event of a pump failure, it can be quickly lifted out of the sump to be replaced by a spare pump held in reserve within the pump station. During replacement, the station capacity will only be reduced by 10%. The capital cost reduction in terms of pumps, valves, pipework and controls associated with eliminating 33% spare capacity, can be readily appreciated.

Each of the close-coupled mixed flow volute pump and submersible motor units have been selected to give the characteristics shown in Figure 2 and the following duty values:

Pump capacity	1.0	m^3/s
Total head	15.0	m
Pump efficiency	86	%
Rotation speed	590	rpm (10 pole)
Motor output rating	200	kW
Total inertia	28	$kg\,m^2$

In contrast to the vertical spindles dry well pumps, the common pump/motor shaft of submersibles requires only one set of bearings and two independant sets of interposing mechanical face seals.

3.2.2 Station pipework and valves

With submersible pumps, no suction pipework and isolating valves are required. Each individual pump delivery pipe up to the common discharge manifold is provided with a recoil type non-return valve and an isolating sluice valve for pump removal purposes.

The 700 mm diameter pipework and valves have been selected to give a nominal velocity of 2.6 m/s.

4. Station layouts

The layout and dimensions of a pumping station should be governed by the type of pump units selected, the necessary pipework configuration, the location of electric motors and the suction hydraulics.

4.1 Traditional wet well/dry well station

The layouts shown in Figure 3 (high static - short discharge main) realistically represent the standard arrangement adopted for such stations. The dimensioning of the wet well is governed by the hydraulic and pump control requirements, discussed under Sections 5.1 and 6.1. These also dictate the level of the pump volute and, together with geometric constraints of the suction and delivery pipework, determine the plan area and depth of the dry well.

For installation and maintenance reasons, the large 700 kW motors would typically be located at ground level. The height of the weather proof super structure must be sufficient to accommodate the overhead gantry crane and provide sufficient clearance to lift one motor over another during removal for major overhaul.

Figure 3 shows a common wet well for all four pumps. In the event of it being necessary to isolate the wet well from the upstream system, for maintenance/cleaning operations, the closure of the inlet penstocks will result in total loss of pumping capacity.

An alternative solution is to separate the wet well into two suction cells, each serving two pumps. This will effectively reduce the area of each cell which must be compensated for by increasing its depth, see Section 6.1.

For the arrangement shown in Figure 3 (high static - short discharge main) the total finished internal volume below ground level is approximately 10,200 m^3 with the height of the super structure above ground level being 10 m.

4.2 Submersible station

The most significant advantages of the submersible concept over the traditional wet well/dry well arrangement are the elimination of the dry well and the close-coupling of the motor to the pump, which is installed directly in the sump. This permits a wide variation of station arrangements to be considered during the design stage, one of which has been considered here.

The conventional orientation of the submersible pump in relation to the inflow is with the discharge bend adjacent to the back wall of the sump. By rotating the pump through 180° and splitting the station into two separate cells, a very compact, flexible and efficient arrangement can be achieved (Figure 4). The influence of such a design on both sump hydraulics and pump control levels is discussed in Sections 5.2 and 6.2. Maximum utilisation of plan area is achieved by positioning the discharge manifold directly above the two inlet distribution channels, which are separated from the pump suction wells by baffle walls.

With the elimination of the separate high level motors and associated drive shafts, the super structure becomes superfluous permitting, in the case of the pressurised discharge arrangement, the whole of the station to be located below ground level.

The alternative discharge configuration discussed in subsequent sections, comprising individual siphonic pump discharges into a surge chamber shown in Figure 4, can be effectively accommodated with only minor modifications.

The division of the station into two separate cells, each housing five pumps, permits the isolation of one cell for maintenance/cleaning operations whilst retaining 50% of the station design capacity.

The total finished internal volume below ground level for the submersible station (high static - short discharge main), applicable to either the pressurised or the surge chamber discharge configuration, is approximately 2,900 m^3, corresponding to 28% of the traditional wet well/dry well volume.

5. Sump hydraulics

5.1 Traditional wet well/dry well station

The design criteria for these wet wells has been established for many years. As flow enters the station, the relatively high inlet fluid energy is dissipated, to avoid 'jetting', by diffusion achieved with divergent side walls and sloping floor.

With only one pump operating for any length of time, low velocity regions can develop in parts of the wet well which, without a relatively steeply sloping floor (25°-40°), could lead to the deposition of solids. It is important to appreciate that increasing the floor gradient will result in a reduced wet well surface area and increased depth (see Section 6.1)

The center spacing of the pump suctions, bellmouth floor clearance and minimum submergence (at BWL) are dictated by the need to avoid air entraining surface vortices. If these were allowed to develop, the resulting variation in impeller loading would lead to vibration and reduced bearing life.

Relatively large submergence depths (in excess of 1.5 x bellmouth diameter) are required in these open sump configurations as various pump operating combinations can lead to bulk circulation of fluid, which, in turn, can promote the onset of air entraining surface vorticity.

Modified designs position the pump suction in a low level trench at the bottom of the floor slope to promote solids removal, but these do not significantly affect the minimum submergence requirements.

5.2 Submersible station

In contrast to the traditional sump design concept described in Section 5.1, a submersible pump station equipped with a larger number of lower capacity pumps will operate efficiently at significantly shallower depths of flow i.e., higher and more uniform velocity distribution within the sump. This is achieved through a reduction in pump operating band depth (see Section 6.2) an increased number of pump suction points and control of the flow direction and distribution by the provision of baffle walls and enlarged pump support plinths.

The generally higher degree of small scale turbulence generated with this sump configuration can improve solids transport and aid in preventing the formation of air entraining vortices.

The relatively high distribution channel velocities (of the order of 0.7 m/s) permit the use of shallow floor gradients, whilst the fall through the baffle wall ports to the channel below the pump suctions eliminates the risk of solids depositon.

The pump volute casing, directly above the suction bellmouth, acts as a vortex shroud permitting pump operation down to a very low level (typically to the top of the volute)

In order to ensure that pump suction 'pre-swirl' will not occur under certain operating conditions, a model test should be undertaken in order to determine the correct dimension of the baffle wall ports and whether suction splitter plates are required.

NB   A model test should be undertaken on all high capacity stations regardless of design, to confirm hydraulic performances and proposed operating levels.

6. Pump control levels

The pump control levels cannot be selected arbitrarily. They must be determined by evaluating the influence of the following factors :

- the capacity of each pump unit
- the maximum number of starts per hour (dictated by the design and rating of the motor and its associated electrical starting equipment)
- the sequencing of the pumps e.g., pump duty rotation
- the plan area of the wet well
- the sewer inlet conditions, either free discharge or drowned. (The former being selected for this analysis)
- the minimum pump suction submergence allowable

The control levels for a freely discharging inlet are determined from the standard formula :

$$V = \frac{T \times Q_p}{4}$$

where V = required storage volume (m^3)
T = permitted pump cycle time (mins)
Q_p = pump discharge/min

6.1 Traditional wet well/dry well station

For 3.33 m^3/s capacity pumps powered by 700 kW rated output high voltage motors, a typical maximum permissable number of starts per hour is 5, i.e., cycle time = 12 minutes.

Substituting these values in the above formula gives a required storage volume of 600 m^3. Assuming a nominal wet well plan area of 250 m^2 (at BWL) dictates that the operating band depth per pump is approximately 2.4 m.

Reference to Figure 3 will show that, as the suction water level rises the plan area of the wet well marginally increases, resulting in a minor reduction in required band depth (this has been neglected).

Providing realistic intervals of 0.5 m between successive pump start and stop levels would result in a BWL of -3.4 m, assuming that standard duty pump rotation sequencing is used in conjunction with cascade starting and stopping.

Alternatively, cyclic alternation control can be employed whereby, for any given number of pumps operating, the last pump to start is the last to stop. Allowing for one of the four pumps being out of service, cycling of the remaining three pumps permit the individual pump operating band depths to be reduced by a factor of 3, i.e., to 0.8 m, whilst maintaining a maximum of 5 starts per hour.

Combining this value with realistic start/stop level intervals (0.5 m) and minimum bellmouth submergence (2.0 m) gives the following nominal pump control levels :

Pump No	Start Level (MOD)	Stop Level (MOD)
1	-1.5	-3.0
2	-1.0	-1.8
3	-0.5	-1.3
4 (standby)	0.0	-0.8

For operational/maintenance reasons, it may be considered desirable to segregate the wet well into two separate cell. If one cells is out of service, and one of the pumps in the remaining cell fails, no cycling of pumps can occur. Consequently, in order to limit the number of starts per hour to 5, either the surface area of the cells or the individual pump operating band depths must be increased by a factor of 3, resulting in significant increases in construction costs or energy costs respectively.

Additional factors which must be carefully assessed when selecting the control levels are :

- the pump $NPSH_{re}$ characteristics, particularly with single pump operation
- the setting of the pump No. 1 start level above the impeller to ensure priming.

6.2   Submersible station

The design shown in Figure 4 effectively separates the station into two 5 m³/s pump stations having a common discharge main. Under normal operating conditions, the two halves work in unison. In the event of one half being out of service, the station discharge capacity is only reduced by 50%.

Submersible pumps having a power rating of 200 kW are typically designed for starting up to 15 times per hour i.e., cycle time = 4 minutes. With a total sump area of 162.5 m² and allowing for one pump being out of service, i.e., 90% station capacity, cycling of the remainging nine pumps results in a minimum required operating band depth of only 0.04 m per pump.

From a practical point of view, this depth is obviously insufficient to prevent spurious starting and stopping resulting from surface waves. Furthermore, in order to maintain an acceptable distribution channel velocity (nominally 0.7 m/s), the band depths and start/stop level intervals are increased to give the following nominal levels :

Pump No	Cell	Start Level (MOD)	Stop Level (MOD)
1	A	) -1.5	-2.0
2	B	)	
3	A	-1.3	-1.5
4	B	-1.1	-1.3
5	A	-0.9	-1.1
6	B	-0.7	-0.9
7	A	-0.5	-0.7
8	B	-0.3	-0.5
9	A	-0.15	-0.35
10	B	0.0	-0.2

NB The starting of pumps is alternated between each half station. The first two pumps are started and stopped together (with short time intervals between each) in order to maintain a discharge main velocity above 0.7 m/s.

The above control levels result in the pumps being started no more than 4 times per hour, i.e., only 27% of that allowable, resulting in increased bearing and motor life.

With one cell out of service and failure of one of the five pumps in the remaining cell, the maximum number of starts per hour will not exceed 14.

7. Discharge transients - high static with short discharge main

The initial pipeline profile shown in Figure 1a is clearly unsatisfactory, even when no pumps are operating. With either station design, the individual pump delivery pipes immediately upstream of the non-return valves would be subjected to a full vacuum, resulting in the formation of a vapour pocket. On subsequent pump start-up, the vapour pocket would collapse rapidly causing unacceptable pipe stress. It is therefore necessary to lower the discharge pipe level at the station end. This may not solve the transient problems further downstream, especially those associated with all pumps running and a power failure. A full analysis must be performed, not only for this fault condition, but also for the individual pump delivery pipes when one pump stops whilst the other pump continues to operate (normal stopping sequence).

7.1 Conventional pressurised system

7.1.1 Power failure

Figure 3 shows the results of the transient analysis for an initial total discharge of 10 m^3/s with subsequent power failure. Whilst these results are applicable to the submersible pump station, the pipeline pressure profiles for the traditional wet well/dry well station are practically identical.

With the initial pipe profile, column separation could occur for 50 m downstream of the pump station. One solution is to connect an air vessel to the station end of the discharge main. However, a simpler solution is to lower the pipe profile for the first 50 m as shown (incorporated in station designs Figures 3 and 4). Such a solution does not entail the additional ancillary plant and control equipment associated with the air vessel option.

Under this 'fault' condition, the discharge main flow deceleration is sufficiently slow to allow the non-return valves to close without 'slamming' and the generation of excessive positive upstream and negative downstream pressures associated with rapid flow reversal.

7.1.2 One pump stopping

The second risk condition i.e., one pump stopping with all the others continuing to deliver, can result in rapid flow reversal and valve slam due to the effective reduction in pipe length and water inertia.

A transient analysis performed on the submersible station with one pump stopping out of 10 will show that the deceleration in flow through the individual pump delivery pipe from 1 m^3/s to 0 takes approximately 0.47 seconds. Whilst the critical velocity of the non-return valve i.e., that at which the doors are just fully open under steady state conditions, is approximately 2.5 m/s (0.96 m^3/s for 700 mm diameter), the dynamic response of the valve must be known to determine the minimum time necessary to close.

For the condition being considered here, the effect of the rapid flow reversal and relatively high valve door inertia may reduce the effective available closing time to less than 0.2 seconds. Consequently, there is a risk that the valves will slam. Although the traditional dry well pump /motor sets will have a higher inertia, their lower running speed and reduced speed characteristics will give generally similar values.

The following section describes an alternative discharge configuration which eliminates these potential problems.

7.2 Surge chamber with individual siphonic pump delivery pipes

This configuration, shown in Figure 4 for the submersible concept, eliminates the requirement for high cost recoil non-return and isolating sluice valves and reduces the station pipework head losses considerably. 'Fail to open' pneumatically or electrically actuated butterfly type air release/siphon breaker valves are fitted to a 150 mm diameter branch pipe at each siphon crest to aid siphon priming at pump start and prevent backflow at pump stop.

The submersible station configuration permits the surge chamber to be located directly above the inlet distribution channels, in place of the pressurised manifold. The results of the transient analysis show that for the worst possible starting and stopping sequences, i.e., all 10 pumps together, the maximum upsurge will be +3.0 m and the minimum downsurge -2.2 m respectively, relative to the discharge water level (i.e., +15.0 m and +9.8 m above suction TWL as shown in Figures 4 and 6).

In order to prevent air entering the discharge main following power failure, the soffit of the main at the station end should be set below +10.0 m. For complete security, the invert of the delivery siphon pipes should be set at +15.0 m, i.e., surge chamber top at approximatley +17.0 m (5 m above ground level).

8. Discharge transients - low static with long twin discharge mains

For this duty, lowering of the discharge mains at the station end will not prevent the formation of vapour packets approximately half way along the main following power failure with all pumps operating. Various conventional solutions can be considered including :

- installation of double acting air valves (throttled air release) at suitable locations along the discharge mains
- increasing inertia of pump/motor sets by the addition of flywheels
- provision of air vessels at the station end of the discharge mains (one per main)

For any discharge main configuration, water column separation must be avoided. Additional requirements imposed by the use of particular pipe materials, e.g., GRP or cement mortar lined iron pipes, may limit the 'acceptable sub-pressure'. A nominal -0.5 bar gauge pressure has been assumed for comparison purposes.

8.1 Air valves

Air valves may be required to ensure complete priming of a main, e.g., expulsion of air trapped at high points or to aid in priming at initial start-up. However, they have not been considered a realistic solution to the prevention of unacceptable transient pressures in this application, owing to the large volumes of air which would need to be introduced and subsequently expelled following power failure. Regular maintenance procedures, necessary to guarantee reliable operation, may also prove difficult to implement effectively in developing countries.

8.2 Flywheels

With traditional vertical spindle dry well pumps, it is possible to increase the inertia of the system by the addition of flywheels. These serve to maintain forward momentum for a longer period after power disconnection and consequently reduce the magnitude of the subsequent transient pressures.

The results of the power failure transient analysis are shown in Figure 7, using a total pump/motor set inertia of 18,000 kg m^2 to limit the sub-pressure to satisfy this requirement, each of the three duty pumps would need to have their individual inertia increased by 5,400 kg m^2, corresponding to 11 tonne, 2 m diameter cast iron flywheels. Such a solution would prove impractical based on bearing load and motor starting limitations. Figure 9 shows the required individual pump/motor/flywheel total inertia as a function of permissible sub-pressure.

8.3 Air Vessels

The addition of air vessels to limit discharge main sub-pressure is applicable to both conventional and submersible station arrangements. In comparison to the flywheel solution, two 340 m^3 air vessels, or combinations of smaller vessels, would be required to achieve a similar result.

NB   The total weight of water in the twin mains, which must be retarded, is approximately 10,000 tonnes.

Figure 9 gives the relationship between required total air vessel volume (for twin mains) for various permissible sub-pressures.

The results of these transient analyses would indicate that a practical remedy cannot be achieved using only one of the 'conventional solutions'.

8.4 Surge chamber with individual siphonic pump delivery pipes

As with the short discharge system, the surge chamber concept also proves to be a compact and practical solution for the submersible station. For this duty, however, operational flexibility can be improved by segregating the upper surge chamber into two halves, each serving one discharge main.

The results of the transient analysis for the worst possible starting and stopping sequences, i.e., 10 pumps together, 5 per main (Figure 8), show that the maximum upsurge will be +10.9 m and the minimum downsurge -5.8 m respectively, relative to the discharge water level (i.e., +16.9 m and +0.2 m above suction TWL).

In order to prevent air entering the main following power failure, the soffit level of the discharge mains should be set at or below 0.0 m. This requirement can be accommodated simply by the construction of low level pits at the discharge ends of the surge chambers.

In order to prevent air/gas bubble formation within the siphons under normal operation, the siphon crest inverts should not be higher than +14.0 m, i.e., 8 m above the discharge water level. The height of the upsurge, following power failure, and consequently that of the surge tower structure can be limited by the provision of overflows into the lower distribution channels at the inlet end of the station.

With such an arrangement, the surge chamber will rise approximately 9.5 m above ground level (+6.0 m).

9. Electrical supply transients

The rating of the power supply to a pumping station is governed not only by the maximum duty load but also by the transient characteristics associated with pump/motor starting. To ensure that the pumping station will not induce voltage reductions in the supply not in excess of the stipulated limit, the most arduous load condition must be evaluated i.e., the starting of the last duty pump.

9.1 Traditional dry well pumps

In general, 18 pole induction motors exhibit low maximum/rated torque ratios and low power factors. The following values, typical of such machines, used in a computer simulated starting analysis, indicated that it was unlikely to be able to start the pump/motor/system with less than 70% of the nominal voltage applied to the motor i.e., 70% auto-transformer start.

Values for each of the two running pumps:

Shaft power	P_s	548	kW
Pump efficiency	E_p	86	%
Motor efficiency	E_m	94	%
Input power	P_{in}	583	kW
Corrected power factor		0.9	
Running kVA		648	(duty kVA for 3 pumps = 2020)

Values for the third pump starting:

DOL start current	750 A	(3.9 x rated current) at 3.3 kV
Starting power factor	0.2	
70% start current	367 A	(I_{dol} x 0.7^2)
Starting kVA	2098	

Vectorial addition yields a resulting maximum transient kVA value of 3066, which is 58% in excess of the maximum duty condition. The corresponding power factor value falls to a minimum of 0.52.

9.1 Submersible pumps

The most arduous load condition for this arrangement is for 9 pumps running and the 10th pump starting. The following comparative values assume that simple low cost Direct-on-Line starters are employed.

Values for each of the 9 running pumps:

Shaft power	P_s	171	kW
Pump efficiency	E_p	86	%
Motor efficiency	E_m	94	%
Input power	P_{in}	182	kW
Corrected power factor		0.9	
Running kVA		202	(duty kVA for 10 pumps = 2020)

Values for the 10th staring pump:

DOL start current	905 A	(3.8 x rated current) at 660 v
Starting power factor	0.2	
Starting KVA	1034	

Vectorial addition yields a maximum transient kVA value of 2582 i.e., 28% in excess of the maximum duty condition and a corresponding minimum power factor of 0.72.

10. Energy

The energy required for the various station and discharge configurations considered has been determined using the flow duration diagram (typical of many foul or combined systems) shown in Figure 10.

As would be expected, for the high static, short discharge main system there is little difference between the traditional dry well and the submersible pressurised systems - approximately 7.5 million kWh per year. Also, the use of individual siphonic discharges into a surge chamber, reducing the duty total head by approximately 1 m, correspondingly reduces the yearly energy consumed by 7%.

More significant, however, is the following comparison between the traditional dry well pumps utilising a pressurised discharge and the submersible alternative utilising the siphonic/surge chamber configuration, for the low static, long delivery main system :

Total yearly volume pumped = 156,494,000 m^3
Yearly energy consumed :
    Dry well pumps - pressurised = 7,005,000 kWh
    Submersibles - siphonic/surge chamber = 5,913,000 kWh

Assuming identical efficiencies for the two types of pump, the 15.6% reduction in energy consumed by the submersible option can be accounted for simply, firstly, by the accurate matching of the station discharge to the inflow, achieved by the use of 10 pumps and, secondly, by the reduction in station pipework losses from the elimination of the recoil non-return valves and delivery manifold.

Simplifying the electricity supply scale of tariffs, by assuming a nominal charge of 3.5 p/unit, will result in an energy cost saving of approximately £38,000 per year.

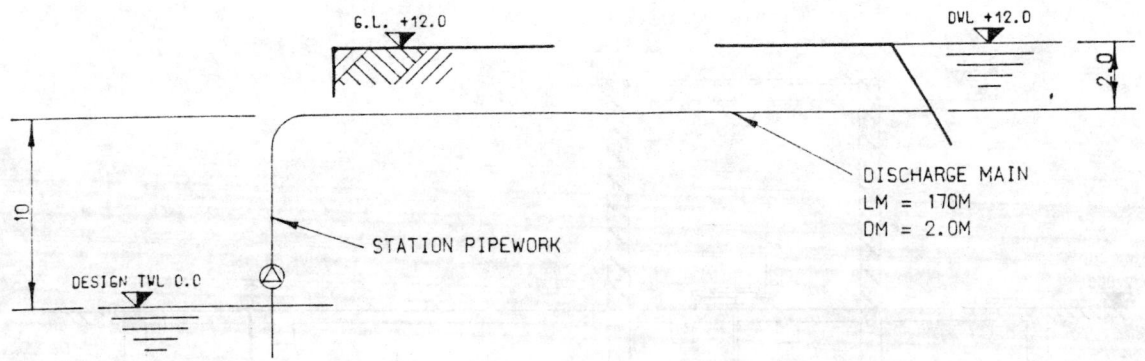

Figure 1a : Schematic of high static, short discharge main system

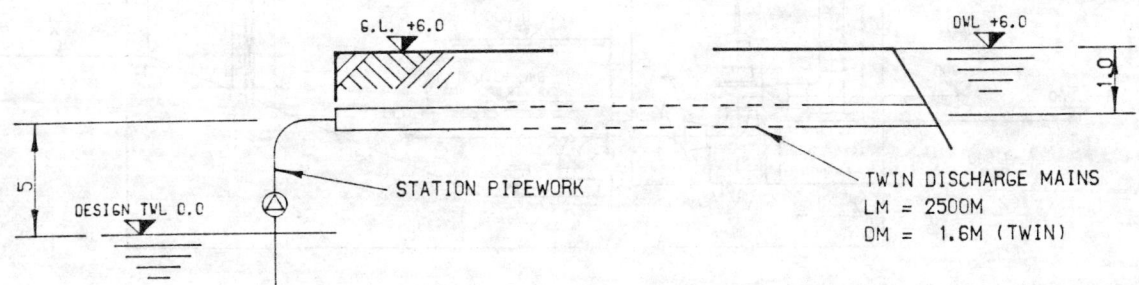

Figure 1b : Schematic of low static, long twin discharge main system

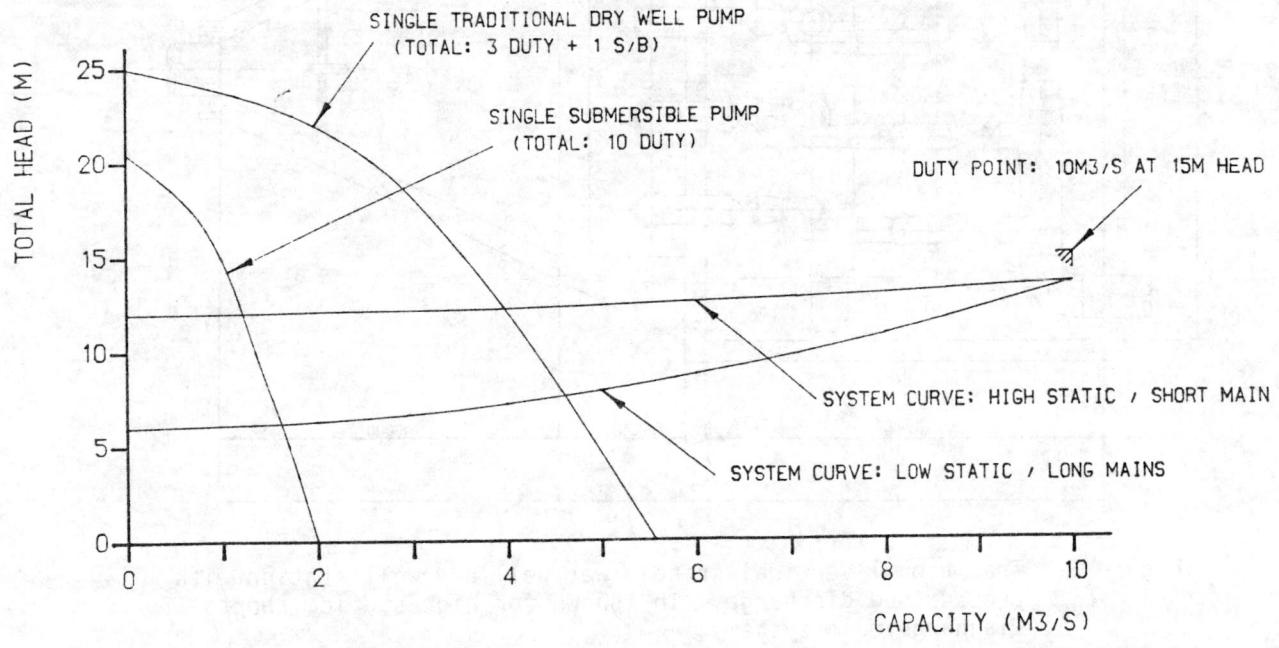

Figure 2 : Pump performance and system curves

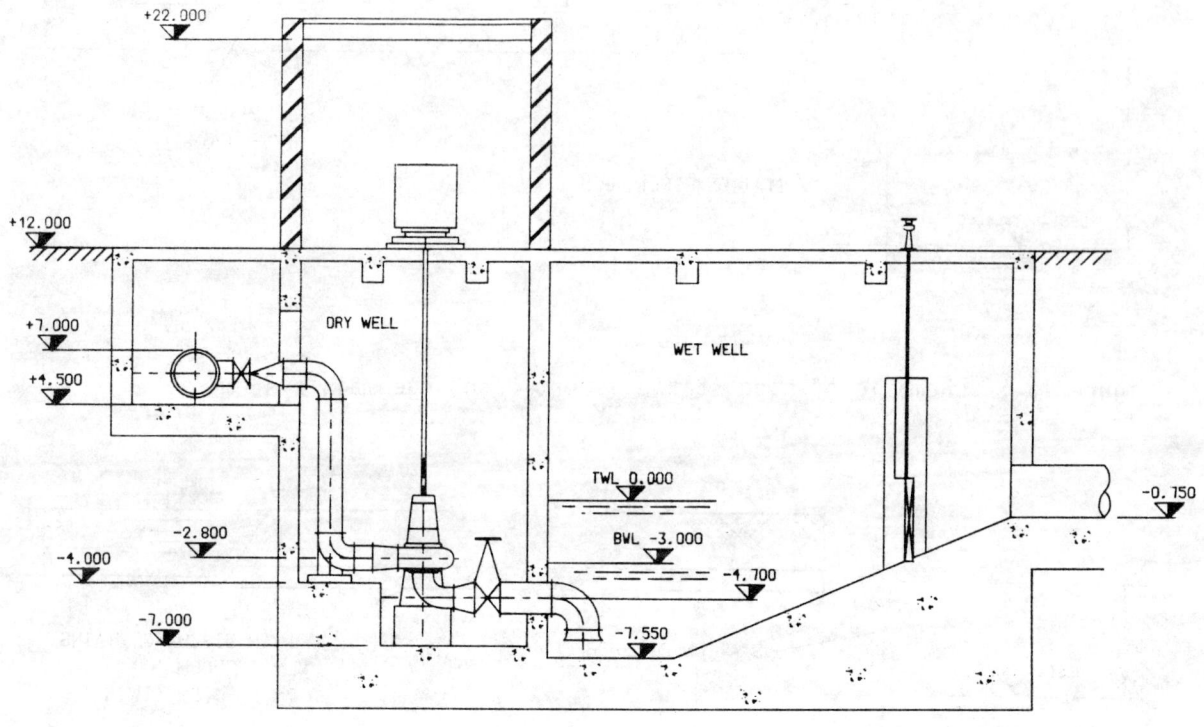
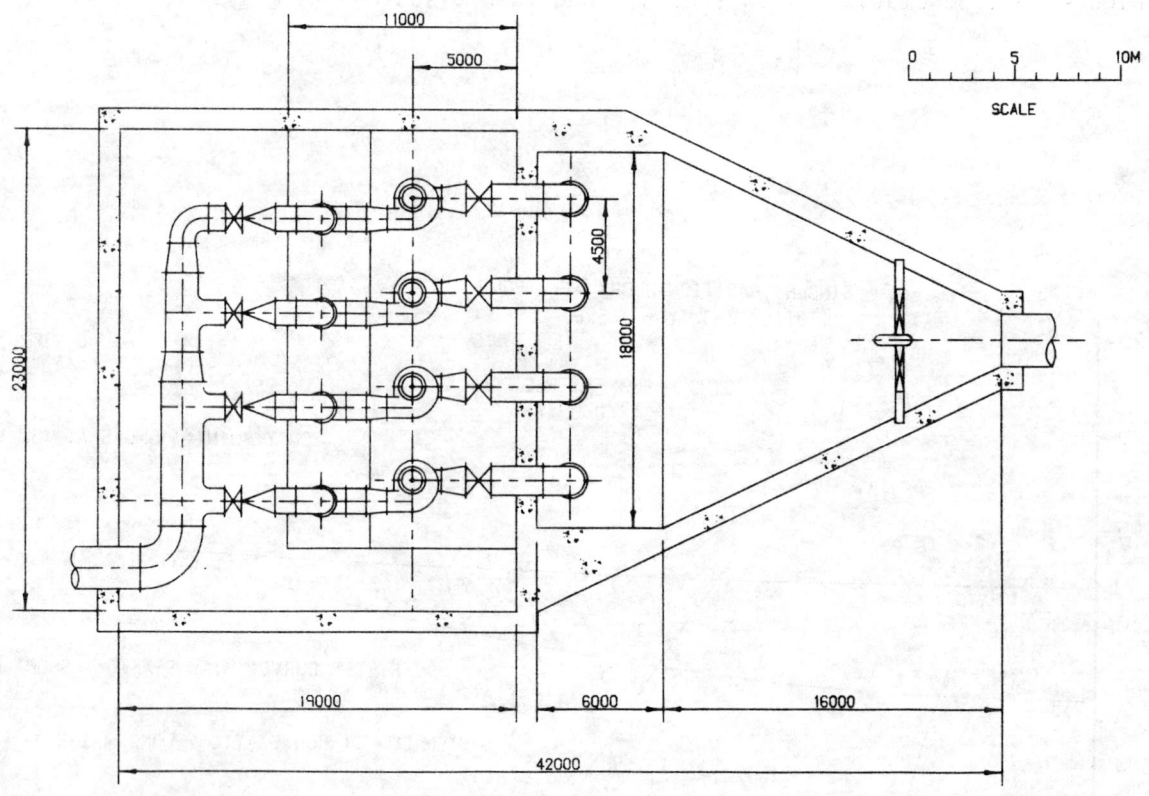

Figure 3 : Traditional vertical spindle wet well/dry well station with pressurised discharge main (shown for high static, short discharge main system)

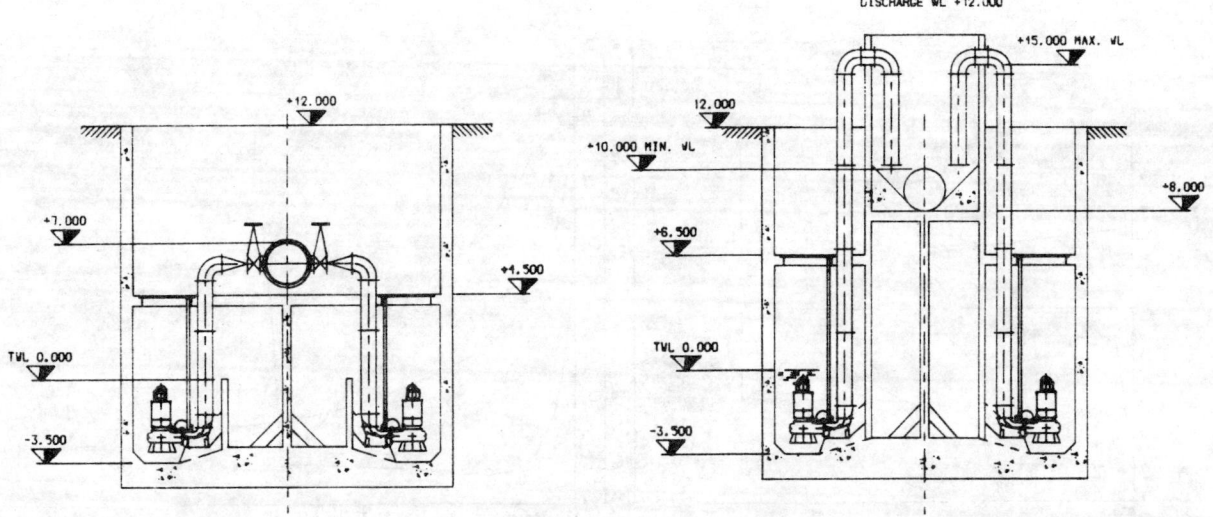

Pressurised Discharge Arrangment   Surge Chamber/Siphon Arrangement

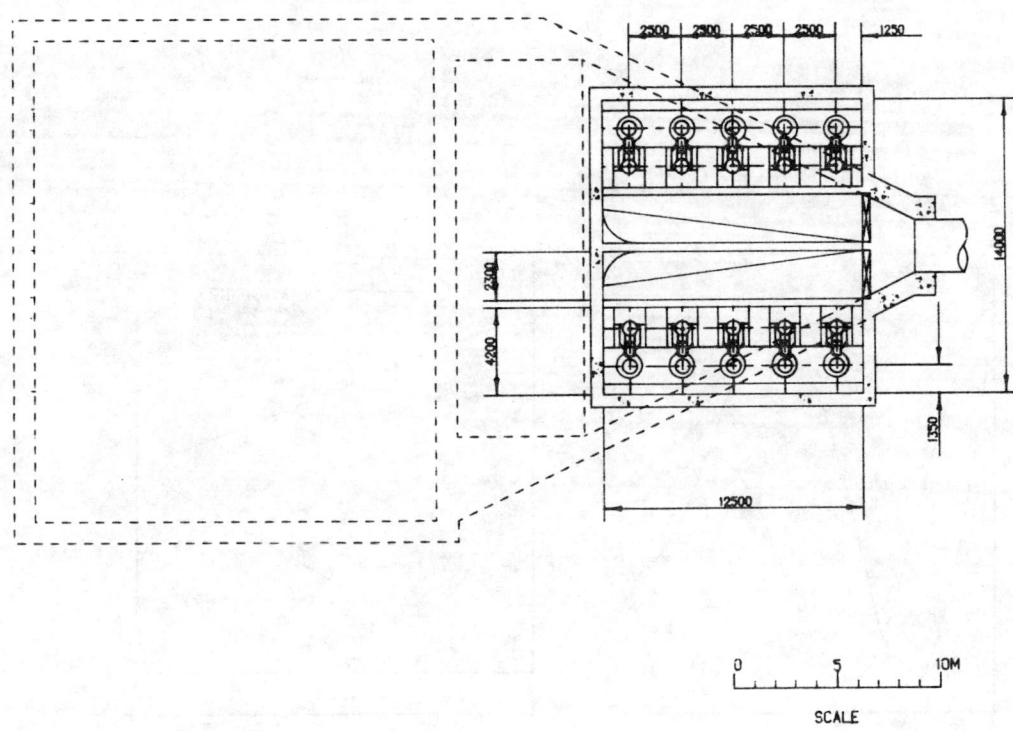

Figure 4 : Submersible pump station with pressurised and alternative siphon/surge chamber discharge configurations (shown for high static, short discharge main system)

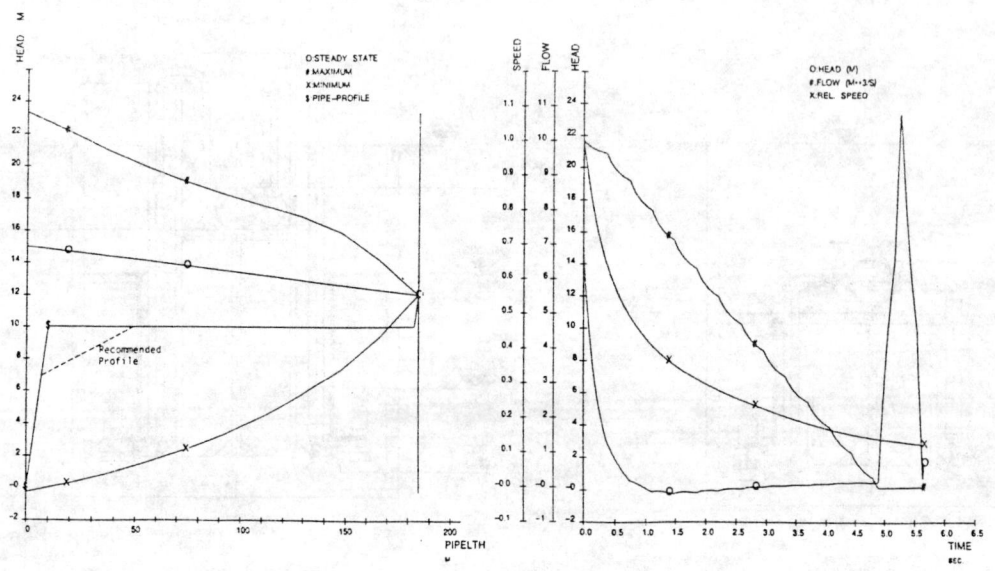

Figure 5 : Transient analysis results for high static, short pressurised system following power failure with 10 m³/s discharge (submersible option)

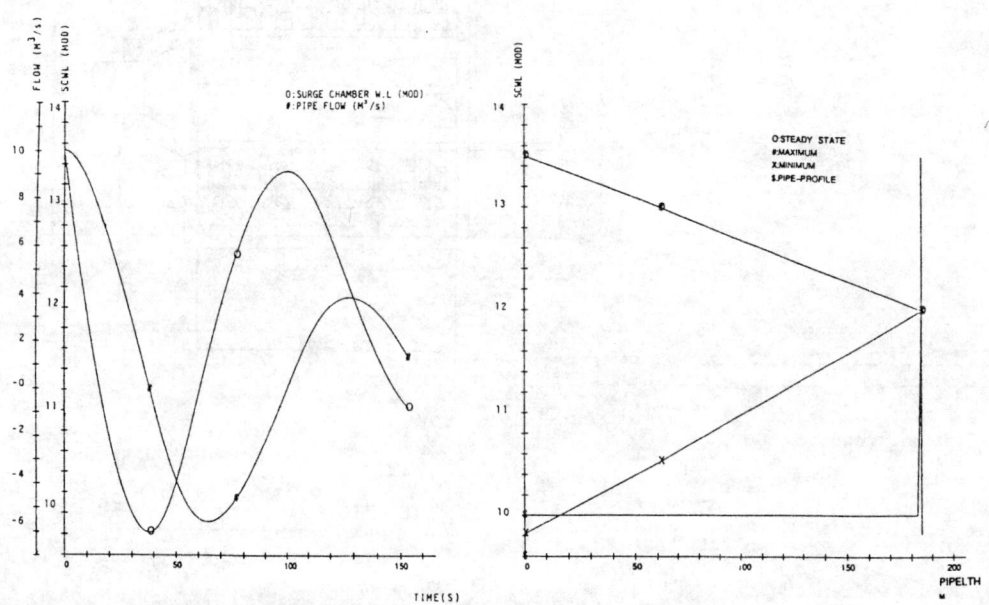

Figure 6 : Transient analysis results for alternative siphon/surge chamber system following power failure with 10 m³/s discharge

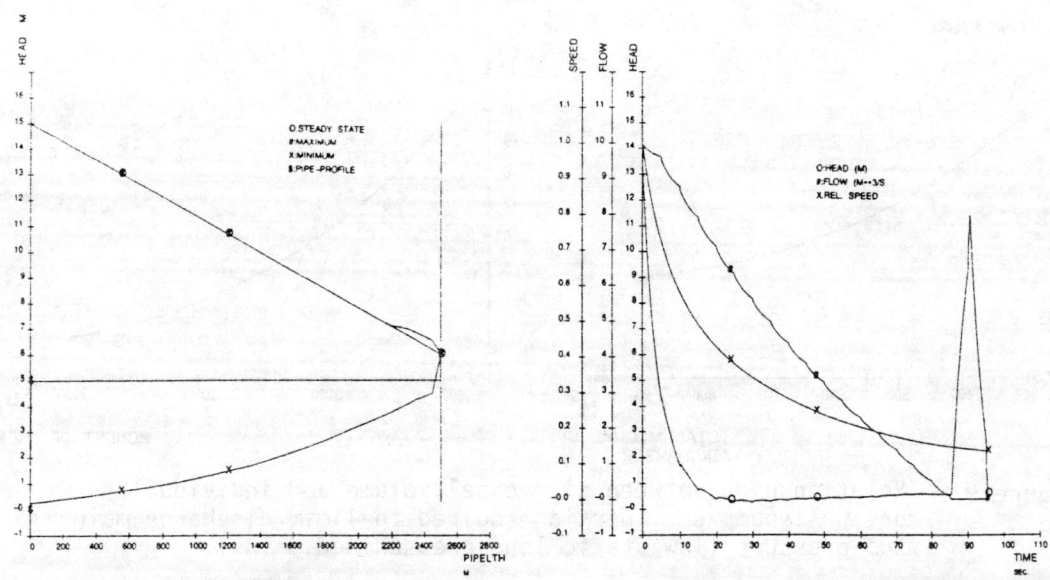

Figure 7 : Transient analysis results for low static, long pressurised systems following power failure with 10 m³/s discharge - traditional pumps fitted with flywheels

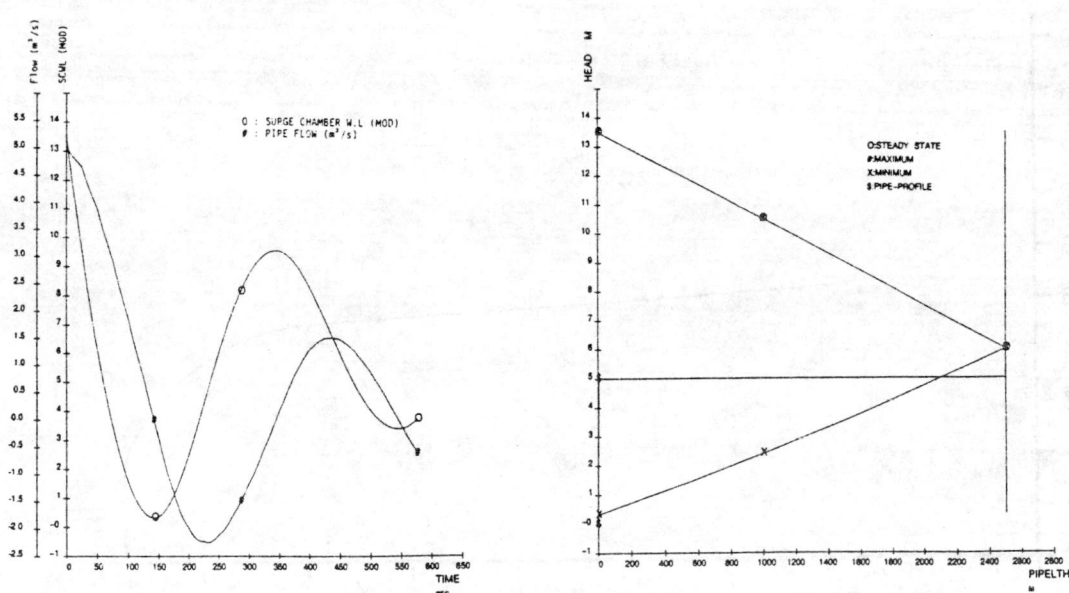

Figure 8 : Transient analysis results for alternative siphon/surge chamber system following power failure with 5 m³/s discharge per main

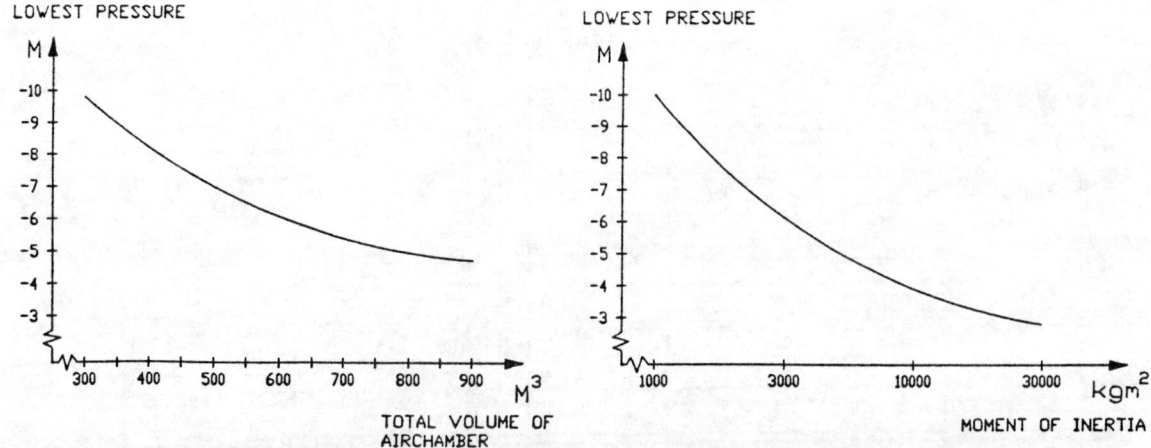

Figure 9 : Relationships between air vessel volume and individual dry well pump set inertia required to limit discharge main sub-pressure (low static long pressurised mains)

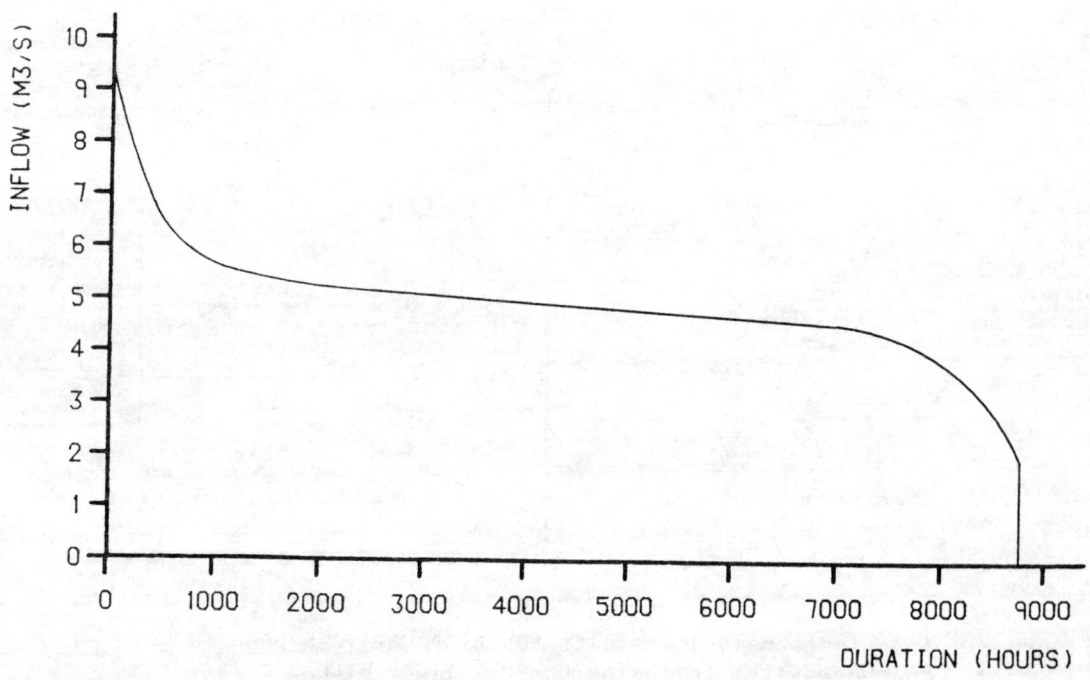

Figure 10 : Duration diagram used in energy analysis

11th International Conference of the
British Pump Manufacturers' Association

New Challenges – Where Next?

18-20 April, 1989
Churchill College, Cambridge

PAPER 6

ENERGY COST MANAGEMENT IN WATER SUPPLY

N Cullen
WRc Engineering Centre, Swindon SN1 8YR, UK

Summary

Past research on leakage control and pumping system optimisation is already making big contributions to energy cost management. Current studies show that direct improvement of of pump and motor efficiency will offer large returns on investment provided that investigation costs are contained. Cheap, accurate measurement of efficiency is important and the thermodynamic method is therefore being developed for use in a Water Industry context. Low-cost methods of improving pump and motor efficiency are being studied. Investigation strategies are being developed and the existing implementation programme adapted to incorporate pumpset efficiency opportunities.

1. Introduction

Water is the lifeblood of every nation. In most cases, particularly in developed countries, pumping is literally the heart of the operation.

In the UK Water Industry the infrastructure is well established, and the key operational topics are now efficiency, maintenance and renewal.

This paper concentrates on the work being undertaken by Water Research Centre aimed at enabling Water Undertakings to improve pumping efficiency and sustain it at a high level. The cost of the work is being shared equally by the Energy Technology Support Unit (ETSU) as an indication of the importance attached by central government to the studies.

1.1 Power consumption and costs

Some round numbers give a feel for the scale of water pumping in England and Wales. The electricity used by the Regional Water Authorities in 1986/1987 cost about £130 million and the total water supplied averaged just over 330 litres per head per day, made up of 130 litres/head for domestic consumption and 200 litres/head for industrial and other uses[1].

This power consumption can be accounted for by pumping through a total head of about 200 metres at an average energy cost of 4 pence per kilowatt hour. This total head of 200 metres includes both the static and friction head components of pumping for water supply and distribution and for double or multiple handling of clean and dirty water at treatment works, booster stations, etc. Some areas of the country are able to use gravity transport extensively. Sewage pumping by District Councils is not included in these figures but is thought to cost just over £10m pa.

Conference organised and sponsored by the British Pump Manufacturers' Association
in conjunction with NEL and BHRA, The Fluid Engineering Centre.
Co-sponsored by the Process Industries Division of the Institution of Mechanical Engineers.

A rough estimate of the installed value of pumping equipment to do this work is £66 million, assuming that each pumpset is in use 50% of the time and has an average purchase cost of about £90 per kilowatt. Amortised over 10 years at an interest rate of 10% pa, £66 million represents £11 million pa.

These figures form a foundation for two major points which colour the research programme:

(1) The energy consumed is worth over 10 times the plant which uses it;
(2) Much of the energy goes to waste.

1.2 Wasted energy

Waste takes various forms. The remedies have been systematically tackled by WRc and can be classified as in Table 1.

Table 1: Classes of energy waste

Class	Remedy	Programme
Wasted Product	Pump Less Water	Leakage Control
Wasteful Processes	Operate System Efficiently	System Optimisation
Wasted Power	Use Efficient Plant	Pumpset Efficiency

1.3 Research programmes - past and present

The Leakage Control studies, undertaken in conjunction with the industry's Sewers and Water Mains Committee, in the late 1970s[2], led to substantial implementation programmes which are still in progress.

System optimisation, including pump scheduling, permits more efficient use of the nation's power generation and distribution facilities. A water undertaking can save money by taking advantage of cheap rate tariff periods, limiting the power levels reached, utilising tank storage capacity and operating the most efficient available pump at any given point in the day. The WRc computer program 'OPTIONS' was designed to provide optimal pump scheduling for large interconnected supply systems. Its use at Essex Water Company demonstrated potential savings of 10% of the electricity bill[3]. The program was not designed specifically for Essex and can be applied to most supply systems, small and simple or large and complex.

Pumpset efficiency is the topic of current research activity. This is concentrating on the efficiency of pumps and motors and on the direct measurement of this efficiency. The techniques involved are generally well known and much of the research work revolves around the steps necessary for technology transfer into a Water Industry context.

The savings to be made by improving pumpset efficiency are expected to be significant. For instance, in the Essex study, information on pump efficiency and performance obtained during the scheduling investigation let to recommendations worth an additional 6% or more of the electricity bill. Equally important, such savings often represent a high return on investment.

1.4 Structure of paper

The remainder of this paper outlines 6 major topics involved in the pumpset efficiency research:

- Investment appraisal;
- Measuring hydraulic efficiency using the thermodynamic method;
- Improving pump efficiency by surface finish and coatings;
- Choosing motors and pumps;
- Investigation strategies;
- Implementation of research results.

2. Investment appraisal

In view of the high energy/plant cost ratio, small improvements in efficiency will produce big returns on investment. In a pumping system meeting a given demand, causes of inefficiency can be classified as:

(i) physical i.e. because of some characteristic of the pump, motor or pipework;
(ii) operational e.g. pumps being run at inefficient points of their duty curves.

The same type of investment appraisal can be applied to both classes of inefficiency, because all that matters is the amount of improvement, the cost of making the improvement, and the life of the improvement. An allowance for investigation costs must also be made. A simple spreadsheet computer program is used by WRc to permit rapid evaluation of a large range of options.

Two examples based on a 100 kilowatt pumpset valued at £10,500 running 50% of the time, at 3 pence per kilowatt hour will illustrate key points. Three pence per kilowatt hour represents the mean between normal and off peak electricity tariffs. In investment appraisal of this kind the tariff used should represent the marginal cost of the electricity saved and not necessarily an "actual" tariff.

2.1 Low cost improvements

Figure 1 shows the benefit of improving the efficiency by 5%, if this can be done at low cost, say 25% of the value of the pumpset. Return on investment starts at 24% per annum if the pump is already 70% efficient, rising to 50% per annum if the pumpset is only 50% efficient. The corresponding payback periods range from 3.6 to 1.9 years. If the investigation costs can be made less than £2,000 then the benefits are even greater, as the bottom curve shows.

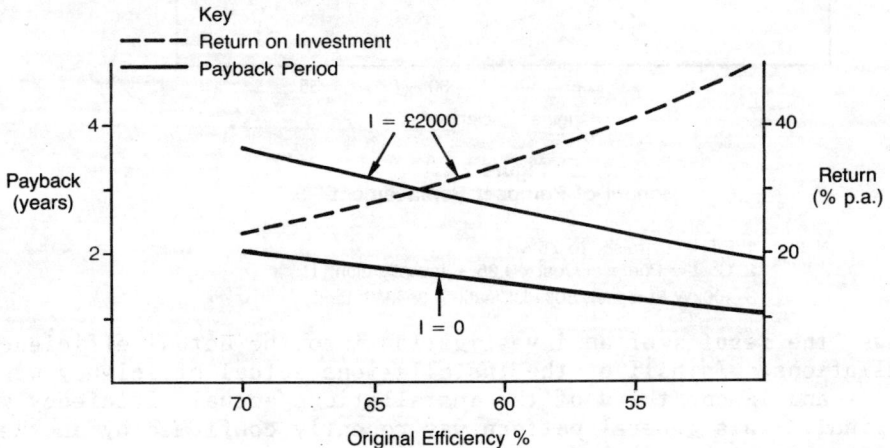

Figure 1
Benefit of Low-Cost Efficiency Improvement

Notes: 1. Efficiency raised **by 5%**
2. Cost = Pumpset Cost x 0.25 + Investigation (I)
3. 100kW Pumpset; 50% Utilisation; 3p/kWh tariff

Such savings might be achievable by coating the pump impeller and/or bowl, replacing or refurbishing the impeller, rescheduling the operation of a pump or group of pumps or modifying pumps so that the actual demand is met at or near their best efficiency point (BEP). The high return on investment, even for pumps already close to maximum efficiency, indicates that this type of option should always be considered.

2.2 Pumpset replacement

Figure 2 addresses cases where a more expensive solution is required to improve efficiency. It allows for full replacement of the pumpset, plus associated civil works, or alternatively for instance an expensive investigation, review and control scheme. It assumes that the efficiency of the improved system has been raised to 75% and that the improvements last only ten years, which is a pessimistic assumption.

A return on investment of 12% per annum is obtained if the current efficiency of the system is 65%, and returns rise progressively to about 60% per annum if the starting efficiency and investigation costs are lower. The payback period is in the range 1.5 - 6 years.

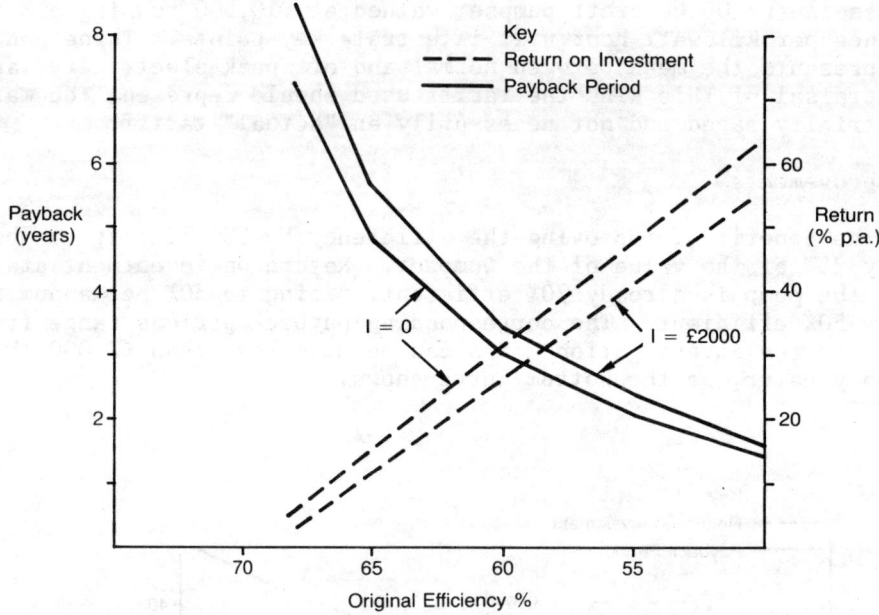

Figure 2
Benefit of Pumpset Replacement

Notes: 1. Efficiency raised **to 75%**
2. Cost = Pumpset Cost x 1.25 + Investigation (I)
3. 100kW Pumpset; 50% Utilisation; 3p/kWh tariff

Figure 3 shows the results of an investigation into the actual efficiency of 12 pumping installations. In half of the installations actual efficiency was 5 - 9% below original, and in one third of the installations actual efficiency was 10 - 19% below original. This general pattern was recently confirmed by another study covering over 300 pumps in several Water Industry organisations.

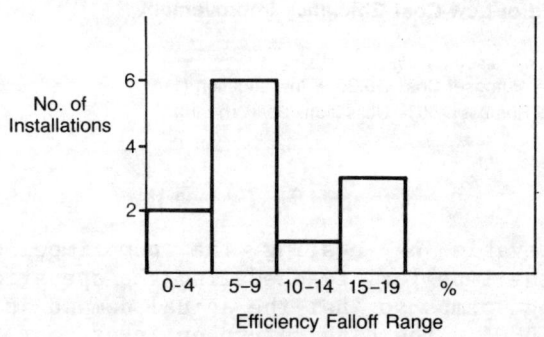

Figure 3
Sample of Measured Efficiencies

Suppose for illustration that the efficiency of the installations could be improved by use of either low-cost methods or pumpset replacement. Table 2 gives some possible figures from a preliminary economic assessment for pumpset improvement.

Many points can be drawn from this table; here we indicate the high return on investment obtainable, dependent on the amount of investigation costs, the pump utilisation and the electricity tariff. Extrapolated to a national scale, illustration suggests that an investment of up to £44 million on pumpset investigation and improvement could bring returns of 25% pa resulting from the associated savings in electricity cost.

The WRc research is intended to make the technology and methods for the realisation of these savings available as far as is possible. Specific cases obviously will require a more detailed economic and technical appraisal.

Table 2: Illustrative economic assessments for pumpset improvement

	Case 1	Case 2	Case 3
No of installations	6	4	4
Efficiency falloff	5% - 9%	15% - 19%	15% - 19%
Tariff	3p	3p	4p
Capital value (say)	£120,000	£80,000	£80,000
No of pumpsets	12	8	8
Improvement cost	£30,000	£100,000	£100,000
Investigation cost (max)	£17,800	£11,600	£11,600
Efficiency improvement	5%	15%	15%
Electricity saving*	£13,400 pa	£31,200	£41,700
Return on investment+	35% pa	25% pa	35.5% pa

* based on 50% utilisation
+ based on a 10-year life for the improvements.

3. Measuring hydraulic efficiency by the thermodynamic method

The thermodynamic method is an elegant means of measuring directly the efficiency of a pump liquid end, because it avoids the need to measure flow, which is notoriously difficult to do accurately, particularly within the confines of a pumping installation, with convoluted and confined pipework, multiple pump operation and a range of duties.

The method measures the developed head and temperature rise across the pump. The gain in useful energy per unit time is proportional to the developed head x flow and the losses are proportional to temperature rise x flow and hence in simple terms:

$$\text{Pump efficiency} = \frac{\text{output energy}}{\text{input energy}} = \frac{\text{developed head}}{\text{developed head} + \text{constant} \times \text{temperature rise}}$$

The thermodynamic method has been in successful use for many years in high-head installations. The current work is to ascertain its applicability, refine its use and produce a Code of Practice for use in the Water Industry, where most pumps operate at a head of less than 100 metres.

The method is important for two main reasons:

(1) Lower cost;
(2) Increased accuracy.

The lower cost is important because when low cost improvements are being contemplated, the investigation cost is a significant part of the equation, as Figure 1 has shown. Investigation will, in the majority of cases, include a determination of actual efficiency.

Increased accuracy is crucial. Current efficiency determinations depend on the measurement of flow which typically, given all the difficulties attendant on site, can only be measured within ±10%, unless extensive in-situ calibration work is carried out. Such a large uncertainty blankets the range of current efficiencies covered by Figures 2 and 3, and makes the estimation of financial returns pure guesswork. It is a prime reason why more work on efficiency improvement has not been undertaken to date.

Table 3 shows the uncertainty of pump hydraulic efficiency measurement due to uncertainty in the measured parameters and is valid for pumpsets operating at up to 80% hydraulic efficiency. The uncertainty of differential temperature measurement is assumed to be 4 milliKelvins and the relative uncertainty of each pressure transducer is assumed to be 0.1%.

Table 3: Expected accuracy of thermodynamic efficiency assessment

Pump delivery head	(metres water)	30	50	100+
Uncertainty in efficiency measurement	(±%)	5	3	1.5

The uncertainty of measured parameters is dependent upon installation effects which may be significant for absolute efficiency measurement, and a knowledge of these installation effects is needed to make a confident assessment. However, it is currently considered that the uncertainty of measurement tabled above will cover most Water Industry applications.

The thermodynamic method is a central part of the research package. The output from this part of the study will include:

- A theoretical examination of the method and a sensitivity analysis with regard to existing pump efficiency, pump input power, operating head, and the accuracy of temperature and pressure measurement instruments;

- A test rig based examination of the method to assess sensor requirements, limits of applicability and pipework effects;

- Sitework in conjunction with the test rig work to establish practical guidelines on input power measurement and instrument siting for the categories of pump common to the Water Industry.

Some potential applications of the method are listed in Table 4 with comments.

Table 4: Potential applications of thermodynamic method

APPLICATION	REMARKS
(1) Direct measurement of efficiency of individual pumps	Low cost, better accuracy than other methods
(2) Determination of flowrate by combining efficiency and power measurements	Low cost, difficult to obtain by other methods Requires assumptions about motor efficiency
(3) Real time measurement	Useful for analysis of varying duty and multiple pump operating conditions, and for ICA schemes Dependent on long-term stability of instruments
(4) Pump performance monitoring	To identify falloff in performance and assist in optimising costs

4. Surface finish and coatings

Case work by one Water Authority has shown a 3% improvement in pump efficiency using coatings. An initial scoping study undertaken for WRc was less optimistic, suggesting an average 2%, less on smaller pumps and more on larger pumps. Efficiency improvements from coating may extend across the full range of a pump performance curve. Caution is required in some cases where alteration of the pump characteristics by coating may result in an efficiency reduction at some duty points. Increased "know-how" on the effects of coating is needed.

The WRc programme includes:

- A review of currently available materials and methods of coating pump liquid ends;
- Study of the effects of different coating thicknesses and surface roughnesses;
- Investigation of the benefits of coating pumps "from new";
- Development of Water Industry guidelines.

5. Choosing motors and pumps

As already indicated, a few percent energy saving represents big money and will often more than justify expenditure to obtain the correct, rather than the cheapest, plant. Matching plant to duty is an inexpensive exercise which is highly cost effective. Initial over-sizing of pumps to allow for uncertainty or future expansion is likely to be more expensive than alternative solutions.

The efficiency of a pump and motor depends upon the duty which they are being asked to perform[4]. Many installations in water supply are amenable to operation using one or a set of pumps each running at only one set duty; some must unavoidably use one pumpset to cover a range of duties.

The steps in choosing pumps and motors are thus as follows:

(1) Identify the required duties;
(2) If possible, choose a mix of pumps to meet these at fixed duty points;
(3) Ensure that the fixed duty pumpsets have a very high efficiency;
(4) Where use of variable duty pumpsets is unavoidable use the best overall method of achieving this.

There have been significant developments in motor drive technology over the past 20 years which address steps (3) and (4) in particular. The Water Industry is generally more aware than many industries of the energy cost savings offered by this new technology. However, a survey conducted for WRc[5] indicates that there are opportunities for greater penetration. There are various obstacles to this: inappropriateness, unreliability or uncertainty about reliability, and inadequate or out of date knowledge and expertise.

The project is making a start at removing these obstructions. Three specific classes of technology are being considered: high efficiency motors (HEMs), fixed speed motor controllers (FSMCs), and variable speed drives (VSDs).

High efficiency motors incorporate improved designs and materials in the magnetic core to reduce losses. Motors of the size used in water supply typically have an efficiency of 1 - 2% better than standard designs not only at full load but also at part loads[6].

Fixed speed motor controllers are microprocessor-based variable voltage devices which reduce the power absorbed by a motor at low loads, and improve power factor. In practical terms, this would reduce very substantially the energy used by a pumpset which has to operate with a very variable head. There is limited scope in water supply for this application although a few opportunities exist e.g. booster pumps used to even out supply pressure variations, or river extraction which may exhibit low head/high variability.

Variable speed drives offer the user the ability to match the pump characteristic with the demands of the system. In water supply they would be most useful when friction is the main component of head, e.g. on booster pumps used for pressure control. The alternative of using a set of fixed speed pumps is often better[7].

A commonly type of VSD is the static inverter, which involves rectifying the AC supply to DC then inverting it to an AC supply for which the voltage and frequency can be varied to suit operating requirements. A new generation of electronically controlled variable speed drives has already obtained significant penetration of Water Authorities and Companies, but there appears to be a need for monitoring and reporting back from specimen installations on a "before and after" basis in order to disseminate operating experience.

At two trial sites, overall savings up to 30% were obtained. These are special cases where the use of a VSD enables more effective pressure control in part of a network, reducing the amount and cost of water pumped from other points. The electricity saved on the pumpset in question is only a proportion of the total saving. The benefits come from reduced head and pumping less water as well as better efficiency.

6. Investigation strategies

The preceding example of 30% savings illustrates the interplay between pump efficiency improvement and system optimisation. Both activities will save energy costs, but tackling both together can result in even greater savings. However, care must be taken not to optimise the system within such narrow constraints that it cannot meet emergencies or exceptional circumstances, and plant of adequate reliability must be used. Data collected for one type of study will be largely applicable to the others.

Figure 4 illustrates an approach to investigating the energy costs of a water supply network involving several pumping installations. It provides for the optimisation to complete several cycles until further improvements are minimal, followed by reliability checks. The decision steps are built primarily around financial benefits.

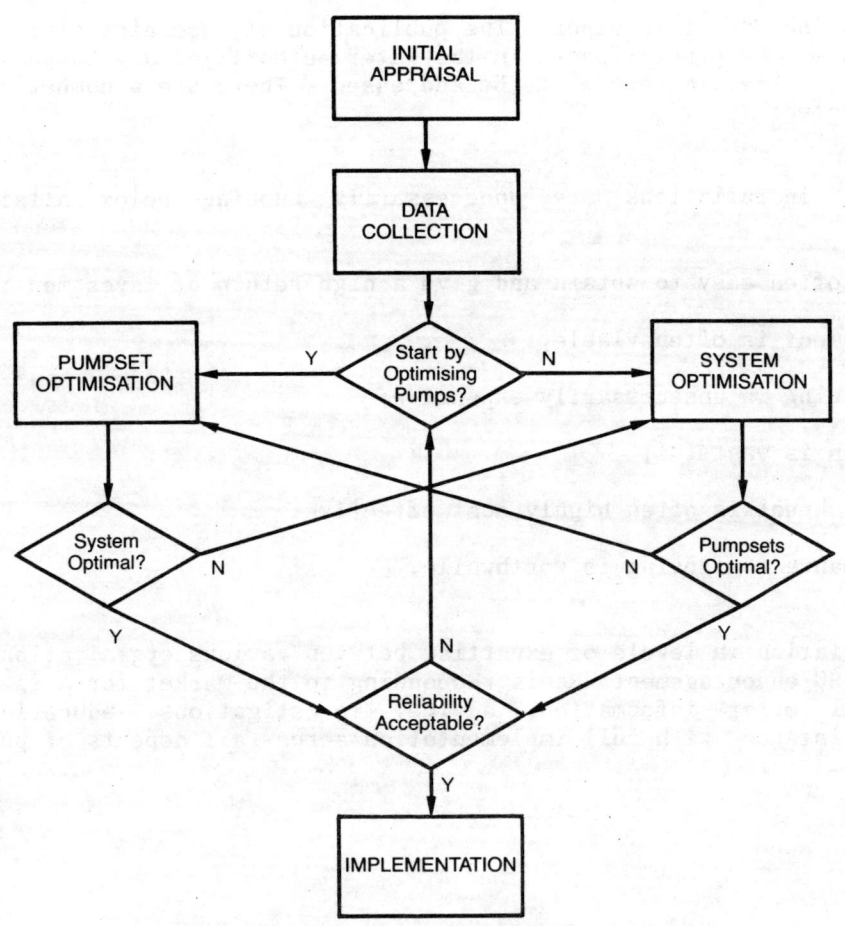

Figure 4
Energy Costs in Water Supply
An Approach to Investigation

As indicated in Section 1.2, WRc have considerable expertise in Systems Optimisation acquired during a number of of investigation contracts and supported by a range of software including the program 'OPTIONS' which is founded on several years of research and development. The aim is now to develop a parallel investigation strategy for Pumpset Optimisation which will complement the technical developments outlined in sections 3 - 5 above.

7.   Implementation of research results

The problem of translating research into useful changes in operational practice is a long- standing concern of the Water Industry, as in many other industries. Factors affecting the success of implementation include:

Uncertainty   - "What risks am I taking?"
Ignorance   - "I never realised there was so much to be saved!"
Inertia   - "This is the way we've always done it."
Lack of expertise - "We haven't got the trained manpower"

The WRc Energy Cost Management programme is aimed at enabling the Industry achieve real financial savings rather than intangible benefits. An integrated package of Research and Development outputs is being prepared to assist the Industry to obtain these savings.

The Thermodynamic Method is a key to reducing uncertainty about actual efficiencies. Awareness of the possibilities is being increased by a programme of publications, which includes this paper. The publication of case histories - both success and failure - by practitioners in the Water Authorities and Companies is to be encouraged. Inertia needs to be addressed. There are a number of key messages to be repeated:

- Many pump installations are unnecessarily running below attainable efficiency;

- Savings are often easy to obtain and give a high return of investment;

- Pump replacement is often viable;

- Pump over-sizing is unnecessarily expensive;

- Pump mismatch is wasteful;

- Pump refurbishment is often highly cost effective;

- Pump performance monitoring is worthwhile.

There is a wide variation in levels of expertise between various organisations and therefore with ETSU encouragement WRc is responding to the market for a flexible service which will offer information, advice, investigations, education and training, and assistance with full implementation across all aspects of pumping for water supply.

8. Conclusions

* The Water Industry, supported by WRc, is already making a big effort to manage its energy costs.

* There are some examples of serious inefficiency of existing installations. The high cost of electricity will often justify substantial investigation costs and wholesale plant replacement in appropriate cases.

* There are further worthwhile savings to be made by small improvements in efficiency but some of these require use of efficiency measurement and improvement techniques which are still being developed.

* The thermodynamic method of efficiency measurement looks promising for use in Water Supply.

* Greater awareness of the opportunities is required, supported by investigation strategies targeted on cost savings, and by improved information about technical performance and plant selection.

9. References

1. Waterfacts. Water Authorities Association. November 1987, pp 10 and 31.

2. Leakage Control Policy and Practice. National Water Council 1980.

3. Saving Money by Pump Scheduling at Essex Water Company. Water Services magazine, July 1988, pp 292 - 293.

4. Davidson, J (Ed). Process Pump Selection - A Systems Approach. Institution of Mechanical Engineers, 1986. Page 70, Fig 23.2.

5. Energy Efficient Motor Drives in the Water Industry - A Preliminary survey. Kingsmen Consultants, March 1988. Contract report to WRc.

6. Bateman, B C. The Search for Improved efficiency in Motor Drives. In Industrial Electric Drives Today. Seminar documentation. Kamtech Publishing Ltd, Surrey SM5 3AX, 1987.

7. Hobson J A. Sewage Pumping - An Energy-Saving Manual. WRc TR 250, August 1987. Page 34.

References

1. Waterfacts, Water Authorities Association, November 1987, pp 10 and 11.

2. Leakage Control: Policy and Practice, National Water Council 1980.

3. Saving Money by Pump Scheduling at Essex Water Company, Water Services magazine, July 1988, pp 172-173.

4. Davidson, J. (ed.). Process Pump Selection - A Systems Approach, Institution of Mechanical Engineers, 1986, Page 10, Fig 2.2

5. Energy Efficiency Know-how in the Water Industry - A Preliminary Survey, Kilgannon Consultancy, March 1988, Contract Report No 82.

6. Tatarek, R C. "The Search for Improved Efficiency in Motor Drives, In Industrial Electric Drives Today". Seminar Documentation, Kempton Publishing Ltd, Surrey SM3 8AT, 1987.

7. Hobson, J.A. Speed Changing: An Energy-Saving Manual, ETSU ER 150, August 1987, Page 24.

98

11th International Conference of the
British Pump Manufacturers' Association

New Challenges – Where Next?

18-20 April, 1989
Churchill College, Cambridge

PAPER 7

RETROFITTING SILICON CARBON BEARINGS TO A PROCESS PUMP.

O.v.Bertele
Consultant

SUMMARY.

This paper first describes the difficulties experienced with a pump on a CO_2 Removal system. These were mainly shafts breaking, excessive wear on the wear-rings and loss of pump efficiency as well as seal failures. To overcome these problems the originally fitted, oil fed, bearings, including the thrust bearings, were replaced with Silicon Carbide bearings operating in the pumped fluid. The modification resulted in a much shorter shaft with a critical speed well above the running speed, and a pump with only one instead of two seals. A pump has been converted to SiC bearings and has been tested on water, but has, so far, not been installed on the plant. The reasons which are behind the design and compromises which had to be made in the modification, which could be avoided in a completely new design, are described in this paper.

1. Introduction.

Continuous high maintenance demands, both in time and material cost, of pumps handling Potassium Carbonate initiated a review on how to best overcome the maintenance problems in future. Various possibilities were investigated, including the use of robust multistage pumps which are used successfully on other installations, as well as modifying the installed pumps in various ways. One possibility considered was to use the existing pump casings and impellors with internal ceramic bearings. The modification of one pump is the subject of this paper. Besides the saving in new equipment, by prolonging the life of the existing pumps, another advantage envisaged was that, because of the much shorter overall length of a pump with internal bearings, the use of individual electric drives becomes feasible. Finally the development would help to gain experience of process pumps with internal ceramic bearings.

2. Duty of pump

A duty frequently encountered in the chemical and gas industry is the removal of Carbon Dioxide from a gas. The gas is absorbed at high pressure in a suitable liquid, the saturated liquid is then let down and the gas is stripped of the liquid. The liquid, which is now free of CO2 is then repressurised with a pump and used again in the absorption tower. On the system discussed both the absorption as well as the stripping is done in two stages (Fig. 1), and 'Lean' as well as 'Semi-lean' pumps are installed. The pumps are coupled together and driven by a 1000 kW steam turbine. Three pump trains were installed on one plant- two running one spare. This paper concentrates on the 'Lean' pump, though most of the problems experienced apply equally to 'Lean' and 'Semi-lean' pumps. The duty of the 'Lean' pump discussed in this paper is:

Rate: 0.1 m³/sec (350 m3/Hr) of Potassium Carbonate solution, a fluid with a density of 1300 kg/m3. The available NPSH is only 10 m and the required head is 250 m.

To achieve the head an operating speed of 68 rps (4000 rpm) was required; to keep the suction specific speed within reasonable limits a double entry pump was chosen.

3. Original design

For the 'Lean' duty a barrel pump, originally designed for use with packed glands, was selected. The pump was equipped with rolling element bearings and mechanical seals.(Fig 2). The 1st critical speed of the shaft assembly in air was about 48 rps (2900 rpm). However the actual critical speed when operating is raised due to the support provided -on the original concept-by the packed glands and by the wearrings(A packed gland is shown on the drive end.). Once the pump is fitted with mechanical seals the support given in the area of the glands is lost and only the support given by the wearrings remains. This support however depends on the condition of these rings - in practice they wear rapidly, fail to provide any support and thereafter these pumps tend to run exceedingly rough, with vibration velocities at the bearing housing of 20mm/sec not uncommon.

4. Problems encountered

The seals-Crane 109's- were designed to operate with a water flush. Condensate was pressurised with a 'Seal flush' pump and the glands of up to six pumps (and two hydraulic turbines) were supplied from one source. This demanded good control of seal water rates to individual seals, particularly as the seals of 'Lean', 'Semi-lean' pumps and the hydraulic turbines operated at three different pressures. Any interruption in seal water supply does lead to rapid seal failure, a common occurrence before a reliable and well controlled seal water supply was available. A recent investigation shows that the mean seal life (MTBF) of the 'Lean' pumps is now in the order of 3 years, with an L10 live of about 6 month. (1) However, during the original commisioning stage of these pumps seal failures plagued the installation, and each seal change required both coupling and rolling element bearing to be removed and replaced. Bearing failures were frequently experienced, mainly due to the loss of internal clearance. This, with 50 mm bearings and a driver rated at 1MW, nearly always lead to gross overheating of the bearing and shaft, followed by comprehensive failures!

Replacing the original rolling element bearings with white metal journal and thrust bearing and changing from self contained bearings to an oil circulation system cured all bearing problems. However the wearrings- both casing and impellor- kept on deteriorating due to a combination of corrosion and mainly erosion caused by the ceramic particles in the pumped liquid. Trials with wearrings of different materials-stainless, Colmaloy, Stellite, both solid and on stainless and others, were unsuccessful in finding a combination with good wear resistance. This was suprising as carbon steel was used in many parts of the system, where the flow velocities were low, without excessive wear problems. An increase in radial clearance from a nominal 0.25 mm to 1 mm can occur in a period as short as 6 month, and such an increase in clearance leads to a lowering of the critical speed and increased vibrations and deflections of the shaft which in turn lead to shaft fractures. Such fractures always start from the key way under the impellor and occur after about 2 years of service.

Impellors damaged by cavitation were frequently observed, particularly on the "Lean" pumps. This is attributed to the complex drive arrangement of these pumps. As stated above there are always two pumps operating in parallel; the speed of the two sets is controlled such that the 'Semi-lean' flows are equal. This results, when an efficient, slow running, 'Semi-lean' pump is coupled with a worn out 'Lean' one on one stream and on the other stream the opposite combination occurs, in one 'Lean' pump operating at a very small flow with all the problems of rough running which are associated with part load running (2)

5. Modification.

The continuous and large maintenance required by these pumps led to an investigation on how best to replace the existing pumps. One option investigated was to fit the existing pump housing with internal bearings in Silicon Carbide. These bearings could be fitted into the space of the seal throttle bushes (Fig 3) in the existing pump casing with only minor modifications.
The following requirements were specified:
  Reliable operation- 20.000 Hours continuous. Twice that time without major overhaul.
  Capable to operate without seal/bearing flush though it was planned to supply a clean water flush during normal operation.
  Use of the existing casing and impellor.

6. Advantages of internal bearings.

The following advantages are likely to occur when internal bearings are used: The span between bearings is halved and the much short shaft will result in a high critical speed and a corresponding reduction in shaft deflection , shaft stresses and errosion wear in the wear-rings. Further only one mechanical seal will be needed. (With the low suction pressure, about 2 bar, at which these pumps operate the unbalance force due to the shaft terminating within the liquid are small and can be supported by the thrust bearing.) Because of the heavy constuction of the pump casing pipe forces will not affect the internal alignment of the pump. Finally considerably less space will be required by pumps with internal bearings. This will permit separate drivers (Electric motors) for every 'Lean' and 'Semi lean' pump to be used. This is not feasible at present due to the limited space available. Separate drivers will simplify the control of sharing the flow between two pumps operating in parallel.

7. Use of SiC bearings

The use of product lubricated Silicon Carbide (SiC) bearings is well established in canned pumps, both electric and magnetic coupled ones.
Hydrodynamic bearings operate by forming a liquid wedge shaped film between the two surfaces, so that these do not come into contact. The film thickness depends on the rubbing speed, the bearing load- pressure/ unit area-, and the viscosity of the fluid. Normally this film has to be thicker than any solid particles in the lubricant as otherwise rapid wear occurs if such particles bridge the gap. Due to the great hardness of SiC any abrasive particles which are in the pumped fluid do not affect the bearing, it is the particles themselves which experience the wear as they are ground down, though it has been known that abrasive parts softer than SiC can cause SiC to be grooved. Bearings with SiC running against SiC are capable to operate at high specific loads (3) with lubricants of low viscosity, as long as the lubricant does not evaporate, as in spite of its superb characteristics at elevated temperatures, SiC must not be allowed to operate dry against SiC as this leads to instant failure of the bearing. This is due to the high coefficient of friction of such a rubbing pair.

8. Design.

The above outlined advantages of ceramic bearings made it highly attractive to evaluate their use. One of the critical decisions which had to be taken was where to place the thrust bearing. To minimise the shaft length between bearings it would have been advantageous to position the thrust bearing outside of the two journal bearings. Placing it at the free end would have meant a large relative movement - about 1mm-at the seal when the pump is warmed up- its operating temperature is about 120 °C. Placing the thrust bearing on the drive end was also considered. This however would have meant that the overhang at the coupling would have been large, leading to a modeshape of the first critical speed where the deflection is not between the bearings but in the overhang. Furthermore the thrust bearing would be at the maximum distance from the pumped liquid and this would have made it difficult to operate without a fresh water bearing & seal flush, one of the original requirements. The final design, with the thrust bearing inboard of the drive end journal bearing resulted in a shaft with the 1st critical at about 110Hz (7000 rpm). The shaft diameter at the eyes of the impellor was not altered so that the hydraulic performance of the pump was not changed. It should be noted that in a new deign the shaft at the impellor could be of a smaller diameter than that of a conventional pump and therefore permit a larger area ratio with a higher head coefficient at part load (4). The critical speed of the rotating element is influenced mainly by the drive end overhang and not the distance between the two journal bearings. Therefore the two parameters with the greatest influence in keeping the first critical frequency above the running speed are the length of the shaft overhang beyond the journal bearing and the mass concentrated at its end. To keep the shaft extention short it was decided to use a special mechanical seal with both faces in Silicon Carbide. This combination makes it impossible to run the pump dry without damaging or destroying the seal faces and for this reason is not favoured. However, as the same applies to the bearings in this design it was considered acceptable on this pump, as it is essential to prevent the pump from starting or running when dry. With the seal being mounted so close to the thrust bearing, which provides axial location, a simple 'O' ring with reduced travel was selected as the secondary seal. To minimise the mass at the overhang, a light gear coupling with a long and light spacer was selected. The overall length of the pump was reduced from 1.40 m to 0.80 m. In a new design of a pump with process fluid lubricated bearings the overall length could be reduced even more by eliminating the bearing brackets from the pump casing. It should be noted that the free end bearing can be made very small; that on the drive end has to be of a diameter as large as possible so as not to lower the critical speed.

The bearings were designed to be able to operate without a clean water purge. This was relatively simple on the non drive end where a pumping ring mounted outboard of the bearing ensures circulation through a channel in the rubbing faces of the journal with a return past the stationary outer ring. A more complicated arrangement was required at the drive end where the circulation not only has to cool the journal but also the thrust bearing and the mechanical seal. This was done by incorporating a pumping ring in one face of the tilting pad thrust bearing which circulates process liquid through the seal, journal and both thrust bearings.

9. Experience

To date the modified pump has only been operating on water. The tests are encouraging, in particular the vibration level measured on the pump housing is very low, with maximum velocities of 2.2 mm/sec. At part load (16% of flow at BEP) the rotor tends to shuttle axially without significantly increasing the vibration level. One surprising observation was how quiet the pump operates, proving that oil lubricated journal bearings contribute significantly to the noise generated by pumps.

It is hoped that the pump will soon be installed on a plant so that practical experience, in particular operation without flush water, can be gained.

10. Conclusion

Process pumps with product lubricated Silicon Carbide bearings have considerable advantages over conventional pumps. The pump described in this paper runs much smoother in its modified form and hopefully will require less maintenance. However the usual advantage of 'stiff' shaft pumps, namely to be able to operate dry will not be achieved till ceramic bearing materials capable of dry running are available.

11. Acknowledgments

The author wishes to express his thanks to C&P Fertiliser Works (Formerly ICI Agricultural Division) for permission to publish this paper and to INGERSOLL-RAND Pumps who carried out the modification and provided the test data.

12. References.

1. v.Bertele, O — 'Why do seals fail unpredictably?' 10th Fluid Sealing Conference BHRA Insbruck 1984
2. Stanmore, L.K. — 'Field problems relating to high energy centrifugal pumps operating at part load'. Part load pumping conference. I.Mech E Edinburgh 1988
3. Kamelmacher, E. — 'Design and performance of silicon carbide product lubricated bearings.' Power Industries Division I.Mech E 1983
4. Thorne, E.W. — 'Design by the area ratio method' BMPA 6th Conference Canterbury 1979

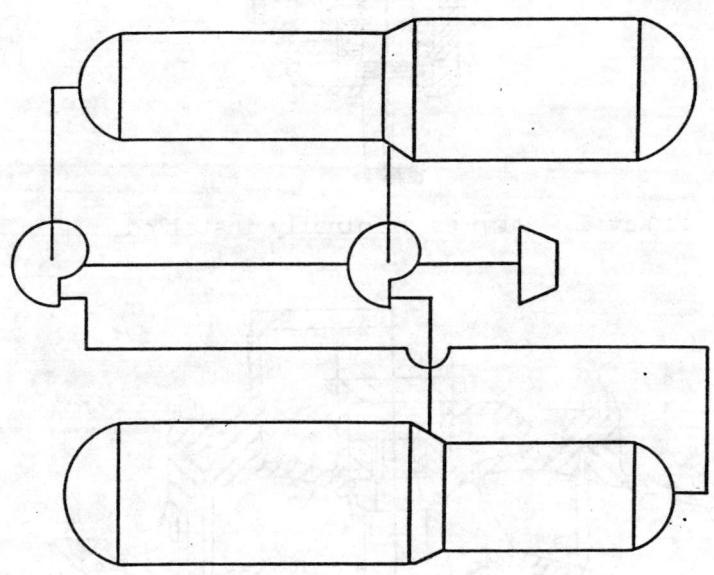

Figure 1. Flow sheet of absorption plant

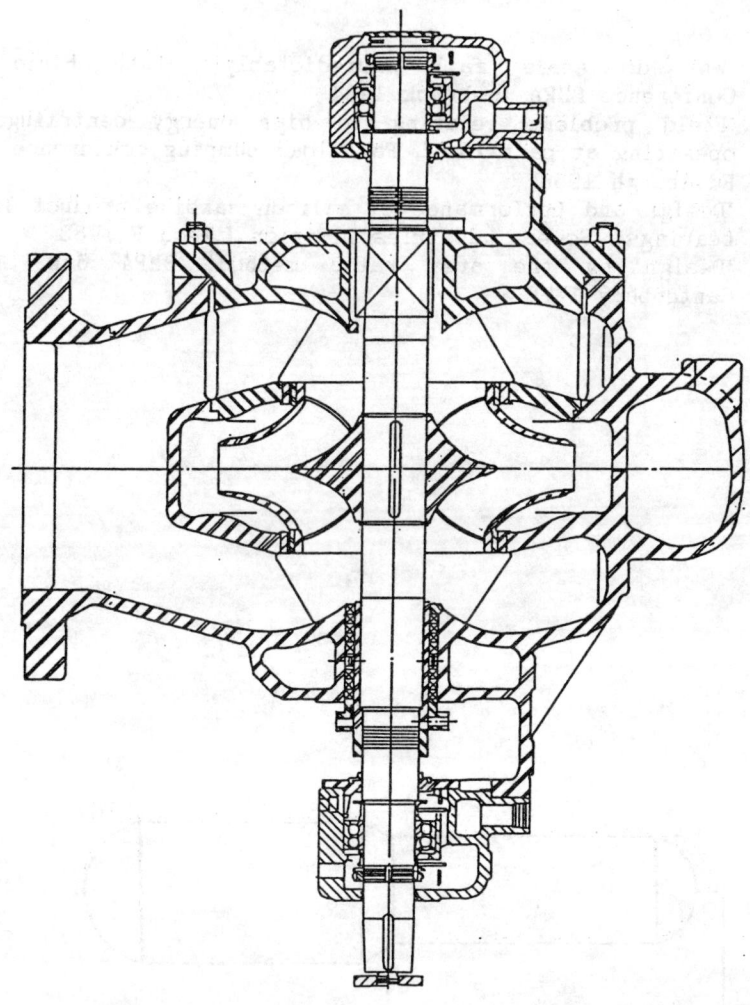

Figure 2. Pump as originally installed

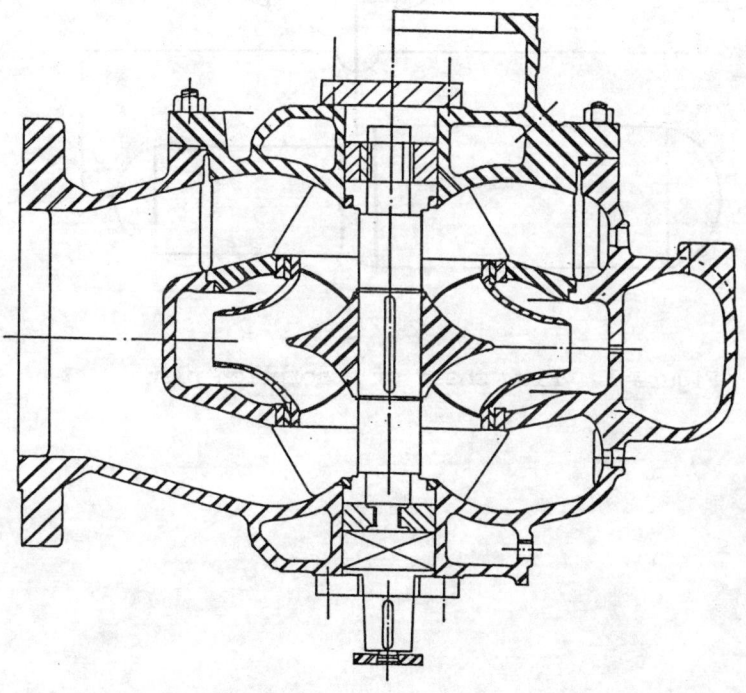

Figure 3. Modified pump

11th International Conference of the
British Pump Manufacturers' Association

New Challenges – Where Next?

18-20 April, 1989
Churchill College, Cambridge

PAPER 8

SEALS FOR HYDROFLUORIC ACID APPLICATIONS

R. Wallis
John Crane UK Limited, Slough, England

Summary

Of its various applications in industry hydrofluoric acid is used as a catalyst for the alkylation process in the refinery and petrochemical industries. Because of the nature of the chemical it is essential that the seal installation should be effective in preventing the acid from escaping.

This paper discusses the results of trials, on hydrofluoric acid alkylation plants, with various counterface materials and elastomers.

A seal arrangement with a reverse pressure capability and using the latest materials is proposed as a more effective and safer alternative to the installations traditionally used.

1. Hydrofluoric Acid

Aqueous hydrofluoric acid is principally used in industry for metal cleaning and the etching of glass. In some process industries, for example the manufacture of phosphoric acid and hydrochloric acid, it may occur in trace quantities as an impurity.

Anhydrous hydrogen fluoride is the backbone of the vast fluorocarbon chemical industry. It is also used in the refinery and petrochemical industry as a catalyst for the alkylation process.

Both aqueous hydrofluoric acid and anhydrous hydrogen fluoride are highly toxic and corrosive, contact with even small amounts can lead to loss of limbs and under certain circumstances to death. It is essential that seal installations should be effective in preventing the acid from escaping.

Conference organised and sponsored by the British Pump Manufacturers' Association
in conjunction with NEL and BHRA, The Fluid Engineering Centre.
Co-sponsored by the Process Industries Division of the Institution of Mechanical Engineers.

2. Alkylation Process

In the refinery industry alkylation is the combination of isobutane with low molecular weight olefines (propylene, butylene), which are too volatile for petrol, into a mixture of highly branched paraffins. The product has a high octane rating and is used as a blending stock for petrol. Alkylation is carried out in the presence of an acid catalyst which is either sulphuric acid or anhydrous hydrofluoric acid.

Alkylation is also used to make a number of petrochemical intermediates, for example detergents, by the alkylation of benzene with olefines of selected molecular weights. Hydrofluoric acid being used as the catalyst.

3. Historical Seal Installations

For 'normal' chemical processes handling hydrofluoric acid of various concentrations, and when mixed with other acids, a standard double seal arrangement has been successfully used. The seal counterface materials have either been a filled PTFE or a standard carbon/graphite against a cobalt/chromium alloy deposit on Monel, high purity alumina or carbon/graphite. Metal parts are normally of Monel.

For the hydrofluoric acid alkylation process in the refinery industry there are two principal licensors of the process, Phillips Petroleum and U.O.P. Both make recommendations concerning the seal installation and the materials of construction for those process streams containing hydrofluoric acid.

One licensor specifies single seal installations the other is not specific about single or double installations. In one case specific grades of carbon have to be used against a silicon free cobalt/chromium alloy. The other just asks for carbon against either high purity alumina, cobalt bonded tungsten carbide or sintered silicon carbide.

A much simplified process chart is shown in Figure 1. More detailed flow charts for both processes can be found in Reference (1).

It is difficult to obtain just what the acid concentrations are in the process lines. The process streams normally being described as hydrocarbon plus a trace of hydrofluoric acid. It is thought that trace in some instances could mean upto 5%. In the line from the settler to the reactor and to the acid regenerator it is of a much higher concentration and in the stream from the acid regenerator it is of the order of 90%.

With the single seal applications most commonly used on these plants the seal is theoretically isolated from the hydrofluoric acid by the use of an external flush of alkylate or isobutane at a pressure higher that that of the pump service. The flush flowing continuously over the seal toward the pump inlet and thus preventing access of the hydrofluoric acid to the seal. Installation of a throttle device in the vicinity of the stuffing box throat prevents too much dilution of the pumped liquid, Figure 2.

On the linear alkyl benzene detergent applications the seal arrangement is normally chosen together with the processor with double back to back seals usually being used. The sealant between the two seals is at a higher pressure than the pumped liquid giving a better seal environment and minimising dilution of the pumped liquid, Figure 3. However, with such an arrangement the hydrofluoric acid containing pumped liquid will be in contact with the inner surfaces of the seal counterfaces which in consequence, need to be of suitably corrosion resistance materials.

4. Hydrofluoric Acid Alkylation Problems

The single seal arrangement with clean flush should operate in an acid free environment. If the flush is inadequate then acid would come into contact with the seal and hence the reason why the licensors specify corrosion resistant materials.

When isobutane is used for the flush there is a greater risk of dry running (and hence high wear), due to vaporisation occurring across the seal faces, than when alkylate is used.

It is thus essential that an ample flush is maintained at all times to keep the seal cool and prevent contact with acid. The use of multipoint injection and the incorporation of calibrated instrumentation will all aid to flush integrity and better seal life.

About six years ago very short seal lives were experienced on some single seal installations handling hydrocarbon containing hydrofluoric acid. On other seals handling similar products more acceptable seal lives were being experienced.

Analysis of the failures showed that the problem was due to chemical attack of the carbon counterface. The attack of the carbon was characterised by roughening of the surface, swelling and loss of shape due to delamination, Figures 4 and 5. Sectioning of the carbons for microscopic examination of the structure showed the chemical attack to form numerous cracks, Figures 6 and 7.

The microscopic examination also confirmed that the carbon grade (Grade A) was the correct grade as specified by the licensor.

A carbon grade (Grade B) which had been used on similar plants in the U.S.A. was tried on the problem pumps and found to give a much more acceptable life.

Corrosive attack of carbon in hydrofluoric acid environments has been attributed by McNulty (2) to its ash content, that is, the inorganic impurities such as silicates, which can occur in the raw materials used in the manufacture of carbon/graphites. In his investigations samples of various carbons were immersed in 50% aqueous hydrofluoric acid in the laboratory for 60 days. Delamination was only reported with one grade of carbon which had a particularly high ash content. No obvious delamination was observed for carbon grade 'A' although extraction of silicon was thought to have occurred. He comments that there was no evidence that permeation of the hydrofluoric acid into the carbon or the extraction of silicon would in themselves lead to early carbon failure. His conclusion however was that ash content should be restricted to less than 4% for such services.

Even with McNulty's understanding his observations did not find support in actual operating plants where gross delamination of the very low ash content carbon grade 'A' continued to occur with consequently extremely short seal lives.

Although it was felt by operators that by replacing the carbon 'A' with the different carbon grade 'B' that acceptable seal lives were being obtained, there was a general feeling that greater improvement might be obtained by considering other counterface materials.

An additional complication, associated with a static seal, was caused when failures were experienced when non-black filled fluorocarbon elastomer 'O' rings were installed.

The uncertainties raised with the selection of all materials on this type of service prompted the initiation of a series of trials in co-operation with the plant operators. The objective being to establish the optimum combinations in hydrofluoric acid.

5. Counterface Materials

5.1 Plant Immersion Tests

Early immersion tests by one of the operators, comparing a hard carbon, grade 'C', and a soft electrographite, grade 'D', suggested on visual appearance that of the two the electrographite would be chemically preferable.

Actual operational use of carbon 'B' was showing it to be chemically acceptable.

Slight corrosion had occasionally been experienced with the low silicon containing cobalt/chromium cast alloy but it was never considered to be a significant problem. Availability, however, has been a problem and it was considered advantageous to consider more modern, readily available counterface materials to assess whether these could further enhance seal life. Four alternatives were proposed for a further immersion test on plant.

i) The most obvious choice of material for the application would be sintered alpha silicon carbide since it would be considered virtually inert to all chemicals.

ii) Reaction bonded silicon carbide contains a distribution of a free silicon phase. Although it is well known that silica is attacked by hydrofluoric acid, the available literature showed a mixed opinion as to whether the silicon phase was susceptible to attack or not in hydrofluoric acid. A grade containing about 10% free silicon was included for test.

iii) Under normal circumstances the cobalt or nickel bonded tungsten carbide materials would never be considered as suitable for use in mineral acids because of attack of the metal binder phase; this eventually leading to the component falling apart. However, it was known that such materials were apparently working in similar applications and thus samples of each were also included for trial.

The samples were placed in a pot which was piped into the acid re-run line. The process stream was given as 90% anhydrous hydrofluoric acid at 25°C. After removal from the process stream the various samples were visually examined and sectioned for microstructural examination to give an assessment of the amount of corrosion.

In all cases, other than a dulling of the surfaces, there was no obvious visual appearance of corrosion, but under microscopic examination the following observations were made:-

i) There was some slight attack of the silicon phase in the reaction bonded silicon carbide on the surface but no deep penetration, Figure 8.

ii) Both the nickel and cobalt bonded grades showed attack but not as much as had been predicted. The nickel grade, Figure 9, showed a greater level of attack than the cobalt grade, Figure 10. On both grades the corrosion was in patches rather than a totally uniform layer.

iii) As expected the sintered alpha silicon carbide showed no signs of attack.

5.2 Plant Seal Trials

As previously mentioned it was felt by a plant operator that the electrographite 'D' was chemically preferable. It was anticipated, however that this softer electrographite was likely to have a shorter life than the hard carbon 'C' and so it was decided that seal trials should be carried out with both styles of material using the currently preferred carbon 'B' as the basis for comparison.

To complement the carbons it was decided to run the seal trials with sintered alpha silicon carbide and the cobalt bonded tungsten carbide.

The combinations tested were thus:-

   Carbon B vs. low silicon cobalt/chromium cast alloy
   Carbon C vs. low silicon cobalt/chromium cast alloy
   Carbon D vs. low silicon cobalt/chromium cast alloy
   Carbon C vs. cobalt bonded tungsten carbide
   Carbon C vs. sintered alpha silicon carbide

Interpretation of the seal trials proved not to be straight forward making it very difficult to be definitive that one combination was better than another. The trials were carried out with the single seal plus flush arrangement. Three plants were involved but each had slight differences in operating practice which was thought to influence the integrity of the flush being used and hence the life of the seal. In addition pumps would breakdown because of other reasons and the seals replaced before a meaningful life had been achieved.

Running against the cobalt/chromium alloy the electrographite 'D' on one plant gave very poor performance but on another the combination gave a good wear life. The consensus of opinion was that it was competitive with the baseline material, carbon 'B', on the less demanding applications. However, despite the optimistic immersion tests some samples of the electrographite faces returned for microscopical examination showed some internal cracking, Figure 11, questioning the long term chemical resistance of the material.

Trials with the hard carbon 'C' against the cobalt/chromium alloy resulted in very early failure of the seat due to thermal cracking, proving such a triboligical combination as being inadequate for this service. The hard carbon 'C' against the sintered alpha silicon carbide and cobalt bonded tungsten carbide showed to be more promising tribological combinations. It has, however, been found that the hard carbon 'C' has swelled and jammed in the metal retainers. Delamination and gross loss of shape has not occurred as it did with the originally specified carbon 'A'. Microscopical examination has shown the presence of internal cracks, Figure 12, and hence also thrown doubt on the long term integrity of this material.

The use of the replacement carbon, 'B', running against the cobalt/chromium alloy counterface has increased seal life from a matter of a few days to in excess of twelve months.

Examination of numerous samples of carbon 'B' from the various plants after operating for periods from a couple of months to twelve months against the cobalt/chromium alloy has not shown any external signs of chemical attack, any tendency to swell, or any internal cracking. In some instances there has been evidence of some removal of the impregnant but this has not proved to have affected the wear performance of the component.

5.3   Laboratory Immersion Tests

As additional support to the observations of chemical attack of the carbons from the operating plants, samples of various carbons were also immersed in commercial concentrated hydrofluoric acid in the laboratory for 60 days. The carbons tested were the four grades already discussed,

   'A' - the originally approved carbon
   'B' - the replacement carbon used in the U.S.A.
   'C' - the hard carbon
   'D' - the soft electrographite

together with a carbon/graphite (grade 'E') more commonly used in mechanical seals.

After the immersion period the carbons were cut in cross section and examined with the microscope.

It was also considered that if internal attack of the carbons was due to extraction of silicon then it might be possible, by means of energy dispersive X-ray analysis, to map impurity element concentrations which would correlate with the observed form of internal cracking.

After removal from immersion in the hydrofluoric acid none of the carbons appeared visually to have suffered any chemical attack. Even carbon 'A' which had consistently lost shape in the various plants.

Examination under the microscope showed no internal cracking of the type observed with the plant failures. Some slight attack was observed on the edge of carbon 'A', Figure 18, where it appears to be associated with the resin impregnant. Examination of carbon 'E', known to have a high ash content, showed some removal of internal particles, Figure 14.

These observations, to some extent tend to agree with those of McNulty, that no gross attack occurred on the carbons in laboratory tests. They also lend support to the field experience in aqueous hydrofluoric acid in chemical process applications where carbon has not apparently been a problem.

The X-ray distribution images on the various carbons did not show any elemental segregation which would correlate with the internal cracking. Figure 15, shows the X-ray images for carbon 'A'. It can be seen that there are some traces of calcium, iron and sulphur within a graphite grain, but not in a distribution to explain the crack morphology.

Figure 16, shows the X-ray images from carbon 'E'. As it can be seen there are concentrations of aluminium, silicon, sulphur, potassium, calcium, and iron, in a particulate defined distribution. This type of particle would be considered to correlate with the particles observed to have been removed after the immersion test, Figure 14.

From these observations it can only be concluded that the anhydrous hydrofluoric acid in a hydrocarbon environment is more aggressive to the carbons than is the aqueous hydrofluoric acid.

Whilst it is totally accepted that hydrofluoric acid will attack silicate phases it is very difficult from this series of tests to believe that this can be the sole explanation for the mechanism of internal cracking observed on the carbons from the operating plants.

Examination of the micrographs of the carbons from the plant failures, Figures 7, 11 and 12, indicate that the cracking tends to follow the graphite/resin impregnant boundaries. Perhaps the hydrocarbon containing hydrofluoric acid is more able to penetrate along these boundaries and attack any slight impurity enrichment, although not observed, which would normally be expected in such regions. Any reactions to form salts would contribute to swelling of the carbons. The ability for penetration would be dependant upon the structure and manufacture of the carbon grade.

5.4   Laboratory Wear Tests

In a double seal system installed for a linear alkyl benzene application there was a requirement for the pressure of the sealant between the two seals to maintain a safety margin above the maximum potential system conditions. This has increased the performance requirements of the seals for the application with two particular duty conditions being required, 22 bar at a face velocity of 8.9m/s and 43 bar at a velocity of 14.3 m/s with clean alkylate as a sealant.

In operation on the plant, at the lower pressure condition, using the normal industry standard counterface materials of carbon 'B' against cobalt/chromium alloy, overheating and high wear of the carbon was obtained. It was concluded that the counterface material combination had inadequate thermal and lubricity properties and it was proposed to the customer that the cobalt/chromium alloy should be upgraded to sintered alpha silicon carbide.

To demonstrate the suitability of the material choice, wear tests at the required duty conditions, but using water as the test medium, were carried out at the vendors seal test facility. Carbon 'B' and the hard carbon 'C' were both evaluated. The results demonstrated a marked improvement in seal performance by using the sintered alpha silicon carbide although there was no significant difference in wear performance between the two carbon face materials, Table 1.

Doubts on the chemical resistance of the hard carbon 'C' favoured the choice for carbon 'B' against the sintered alpha silicon carbide to be installed on site and which subsequently is operating successfully.

Further wear tests at the vendors seal test facility were subsequently carried out at the higher pressure duty conditions, again using water. Sintered alpha silicon carbide was in this instance used against all three of the competing carbon grades, the standard 'B', the hard carbon 'C' and the electrographite 'D'. The electrographite, as expected, failed very shortly after start up, giving seal overheating and consequently an unacceptably high wear rate. Carbon 'B' showed a better wear life than carbon 'C', Table 1.

6. Elastomers

Although PTFE secondary seals are commonplace in hydrofluoric acid applications, historically, a standard fluorocarbon elastomer had also been supplied in dynamic as well as static applications and had apparently been acceptable.

For ease of identification, there has been a preference to use coloured elastomers, which means using a mineral filler rather than carbon black during the compounding of the elastomer. Some of these coloured compounds, notably green materials, were installed on a hydrofluoric acid alkylation plant. Very short lives were obtained due to the 'O' rings skiving and blistering. This led to an investigation for a more suitable compound.

It was known that litharge cured fluorocarbon elastomers had been used successfully on hydrofluoric acid applications but it is a compound which is not easily commercially available. A range of compounds were obtained to cover variations in polymer type, filler and cure system. An immersion test on the plant was carried out under the same conditions as previously mentioned for the counterface materials.

Table 2 gives the results of the analysis of the various 'O' rings together with the results of the immersion test. It should be noted that the polymer types identified are of the proprietary fluorocarbon polymer Viton from Du Pont. Chemical resistance of the polymer is related to its fluorine content, the higher the fluorine the more resistant. The GF polymer has better fluid resistance than the B polymer (not available for test), which in turn has more resistance than the A polymer. The perfluoroelastomer is a very highly fluorinated elastomer having exceptional chemical resistance.

Surprising the perfluoroelastomer did not give as good a performance as perhaps would have been expected. Of the fluorocarbon compounds some which were black suffered attack as badly as the green compounds, Figures 17 and 18. This would be attributed to the relevant amounts of carbon black to mineral filler. Of the compounding variables examined it would be concluded that the curing system used rather than the polymer type had the greatest influence on the chemical resistance, the litharge being the best, Figure 19. This would be in agreement with data published by Du Pont (3) which showed that a litharge cure had the greater influence than the type of polymer used in compounds of Viton A and Viton B.

Litharge cured fluorocarbon is now being used in plants on a regular basis without any problems.

FEP covered 'O' rings are conceptually an attractive option where chemical resistance is required. This style of 'O' ring has been used on hydrofluoric acid alkylation plants but with a particularly disturbing occurrence. An operator was examining one of these components after use and when it was flexed some acid was seen to eject from an undetectable crack in the encapsulation. It is unknown if a crack was present on assembly or if the acid vapour had been able to penetrate the encapsulation. Such components would not be recommended for hydrofluoric acid duties.

7. Installation Design and Development

Material developments must be co-ordinated with a functional design integrity and double back to back seals should be the preferred installation. This type of installation lacks support by some users because of its reliability being dependent on an ancillary circulation system. The weakness is attributed to the inboard seal, which normally has an external pressure differential in favour of the ancillary system, being incapable of withstanding a reverse pressure in the event of failure of the ancillary system.

Innovative designs are readily available which can and are proven to eliminate this problem and such an arrangement is shown in Figure 20. The inboard seat is axially trapped and the carbon face is retained by a clever reversal of the hydraulic thrust with the secondary seal being restrained from moving axially by means of a step in the sleeve. An API plan 32 arrangement at the rear of the inboard seal can be included if required to give added protection to the seal components.

The advantage of such a reverse pressure capability is that if a failure should occur in the sealant system, or an unpredictable pressure surge occurs in the process, the seal will not open and create a major contamination of the sealant system.

The incorporation of any seal into a cartridge unit is to be strongly recommended for a greatly improved ease of installation, especially in environments such as hydrofluoric acid where cumbersome protective clothing has to be worn.

8. Conclusions

i) Although single seals with an uncontaminated flush continue to be recommended on the basis that the presence of a flush protects the seal, and seal life in plants can be much improved by increasing its integrity and quality, experience has shown that this is not an ideal installation. For all duties where hydrofluoric acid is present a double seal should be mandatory to eliminate any potential hazard. In addition it would be recommended that in such an installation a reverse pressure capability be incorporated and careful attention paid to the quality and design of the ancillary sealant system.

ii) An aqueous hydrofluoric acid environment would appear not to be as destructive to carbons as the anhydrous hydrofluoric acid environment experienced on alkylation plants.

iii) Of the carbon grades examined the carbon 'B' has consistently shown itself to be the least affected chemically and to have a satisfactory wear performance over a wide range of duty conditions and would overall be the preferred face material.

iv) Seat materials such as hardmetal and silicon carbide are well established in other industries for producing improved reliability and seal life than that achieved with the cobalt/chromium alloy. In hydrofluoric acid applications the sintered alpha silicon carbide would be preferred for safe long-term operation.

v) A litharge curing system in the fluorocarbon has been found to predominate over the choice of fluorocarbon polymer. Such compounds are working satisfactorily as secondary seals.

vi) In the course of the examination of the various materials especially the carbons, it was observed even after neutralisation that weeping of acid from within occurred for some considerable period after removal from the plant. It is strongly recommended that carbons and elastomers should never be reconditioned but should always be scrapped. They should never at any time be handled without protective clothing.

9) References

1. Hydrocarbon Processing, September 1984, pp 123-124.

2. McNulty, K.F., "Seal Failures in Hydrofluoric Acid Alkylation", World Pumps, November 1984, pp 392-394.

3. Moran, A.L., "Viton B", E.I. Du Pont De Nemours and Co. (Inc.), Report Number 59-4, October 1959.

Table 1 Results of Laboratory Seal Tests.

Carbon Grade	Test Pressure Bar	Face Velocity m/sec	Mean Carbon Wear After 150h mm
B	22	8.98	0.003
C	22	8.98	0.001
B	43	14.4	0.010
C	43	14.4	0.020
D	43	14.4	0.075 *

* Test stopped at 30 h because of excessive seal overheating.

Table 2 Elastomer Evaluation.

Sample	Colour	ANALYSIS OF ELASTOMER COMPOUND				RESULTS OF IMMERSION TEST.	
		Polymer	Curative	Carbon Black	Non-Black Inorganic Filler	Volume Swell %	Comments
1	Green	A	Calcium Hydroxide & Magnesium Hydroxide	-	41	-	Swollen, badly cracked.
2	Black	A	Calcium Hydroxide	16	12	-	Swollen, badly cracked.
3	Brown	A	Calcium Hydroxide	3	39	-	Swollen, cracked
4	Black	A	Calcium Hydroxide	13	13	-	Swollen, little cracking.
5	Green	A	Calcium Hydroxide	-	29	-	Swollen, badly cracked.
6	Black	GF	Magnesium Hydroxide	28	13	-	Some large blisters.
7	Black	A	Lead Oxide	17	11	9.3	No obvious effect.
8	Black	Perfluoro-elastomer		-	-	-	Some large blisters.

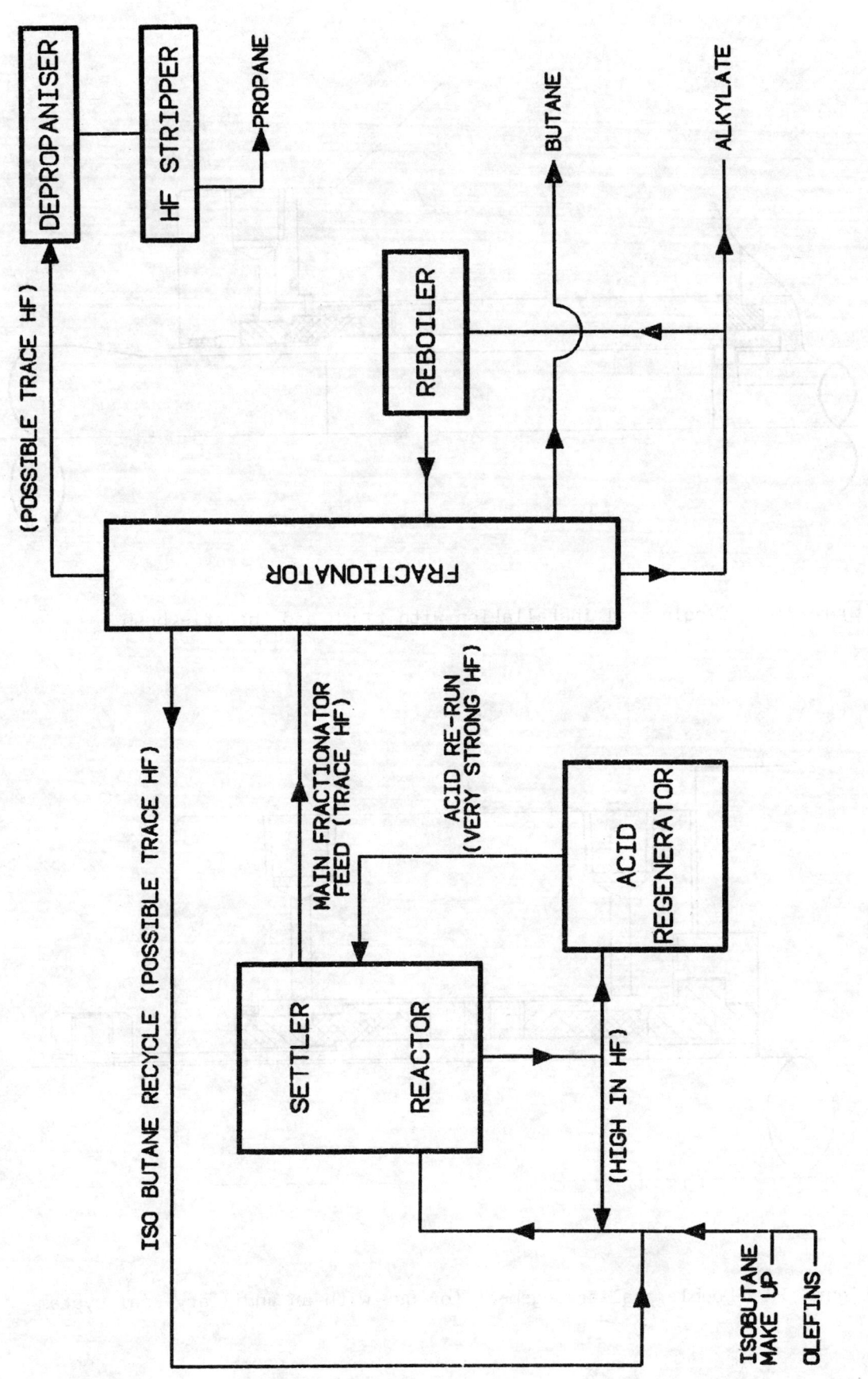

Figure 1: Simplified process chart for hydrofluoric acid alkylation.

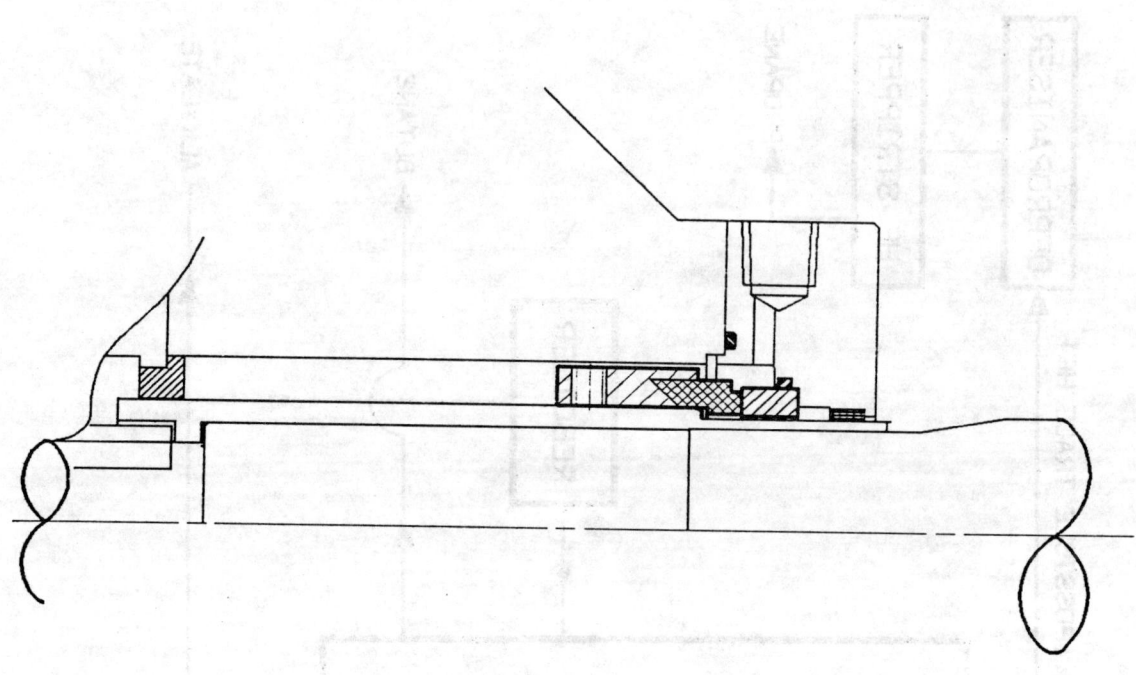

Figure 2:   Single seal installation with flush and throttle bush

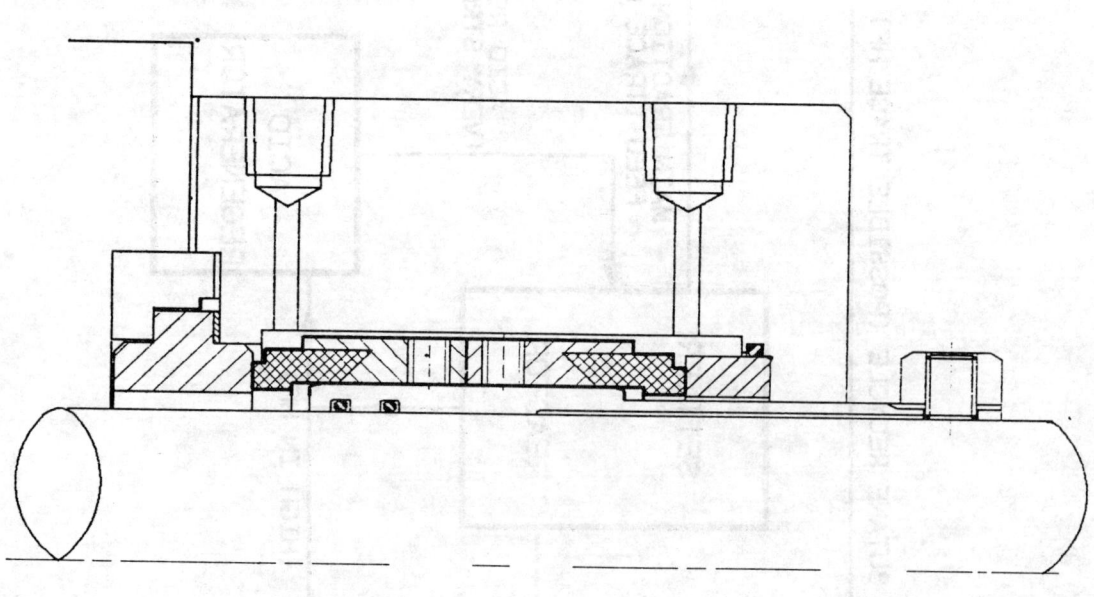

Figure 3:   Double seal arrangement for use with an ancillary seal system

Figure 4: Carbon face from hydrofluoric acid alkylation plant showing loss of shape and flaking due to chemical attack.

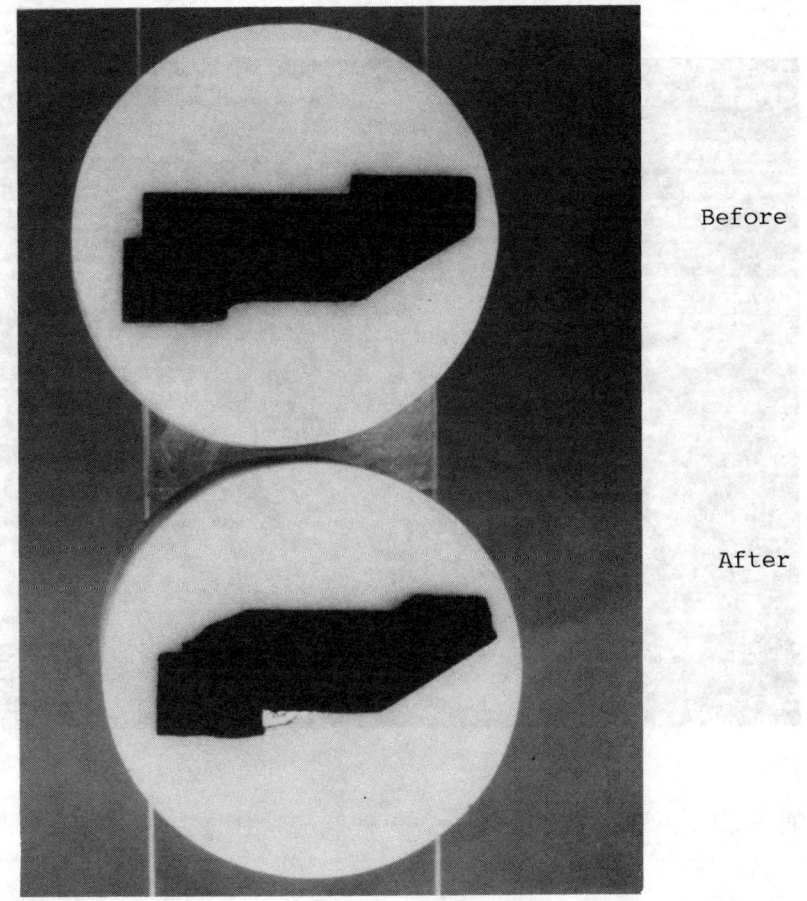

Before

After

Figure 5: Cross section of carbon face before and after use showing general loss of shape.

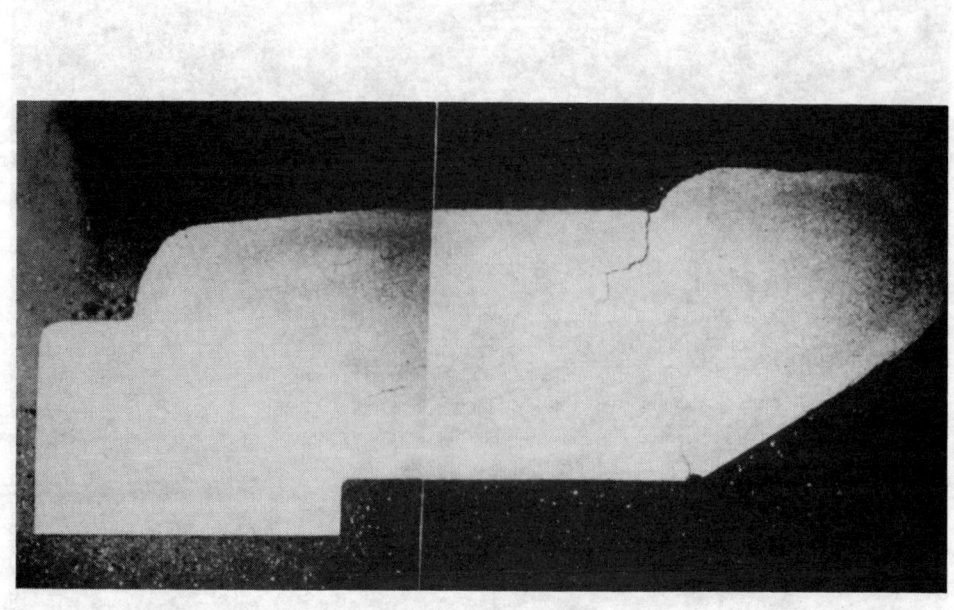

Figure 6:   Macrograph of carbon showing internal cracking.

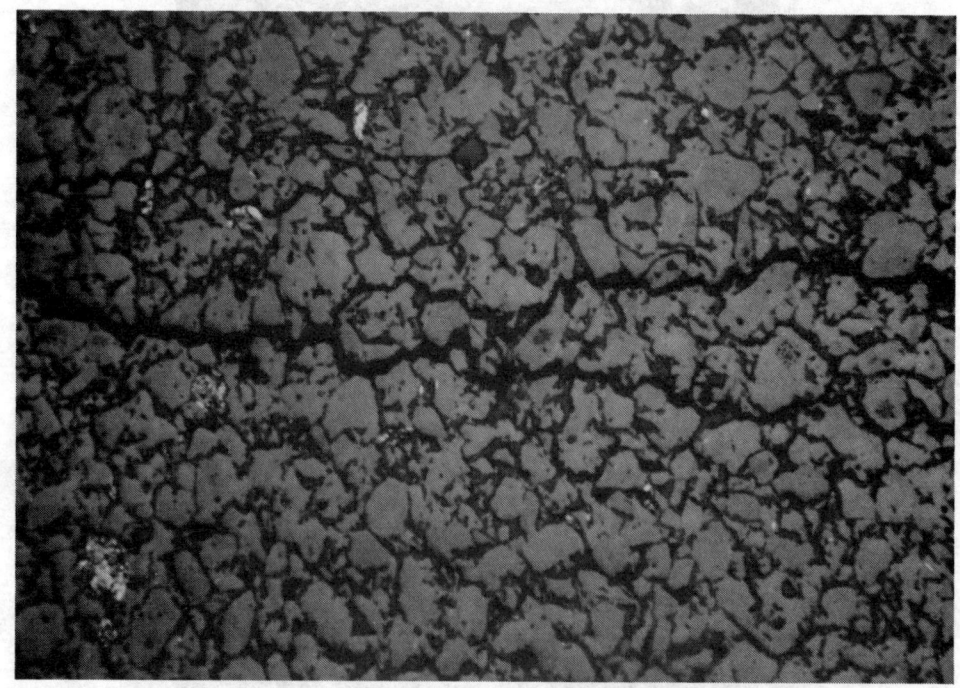

Figure 7:   Micrograph of carbon showing an area of the internal cracking.

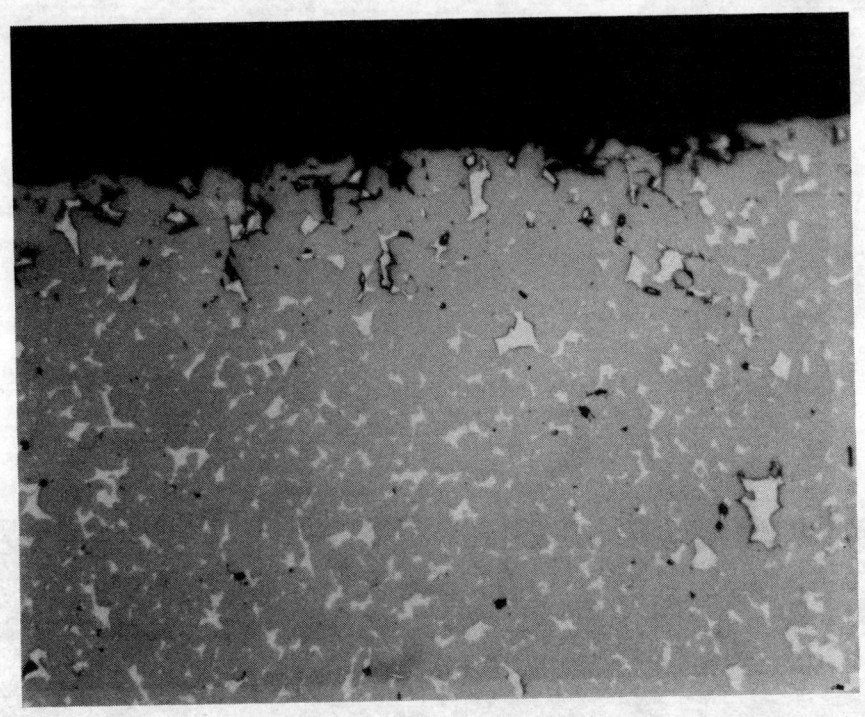

Figure 8: Reaction bonded silicon carbide showing slight attack of the silicon phase (white) at the sample surface.

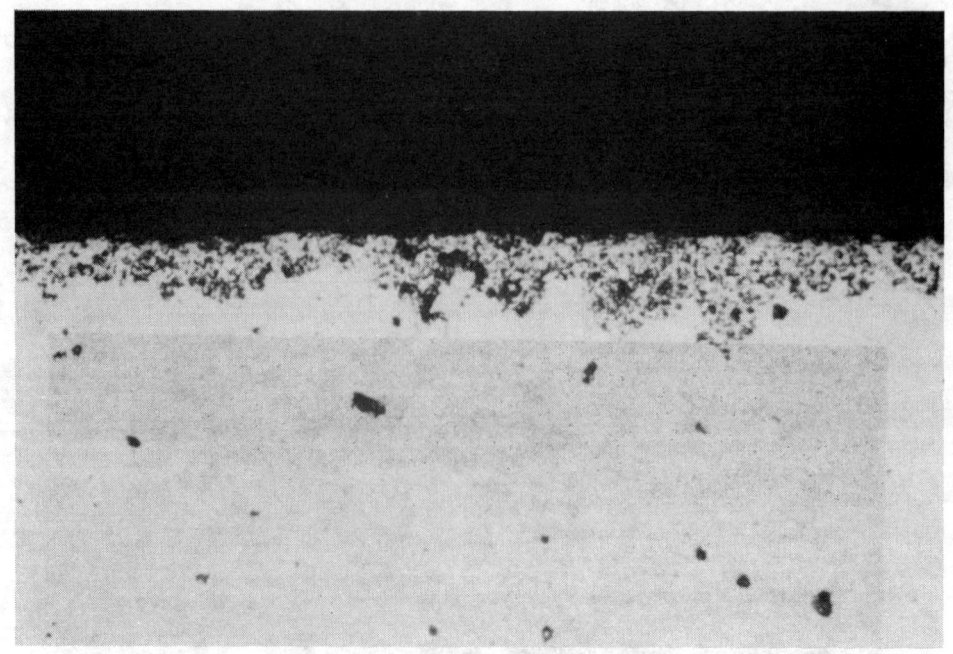

Figure 9: Nickel bonded tungsten carbide showing chemical attack at the sample surface.

Figure 10: Cobalt bonded tungsten carbide showing slight chemical attack at the sample surface.

Figure 11: Internal cracking due to chemical attack in electrographite 'D'. Mag. x 200.

Figure 12: Internal cracking due to chemical attack in hard carbon 'C'.

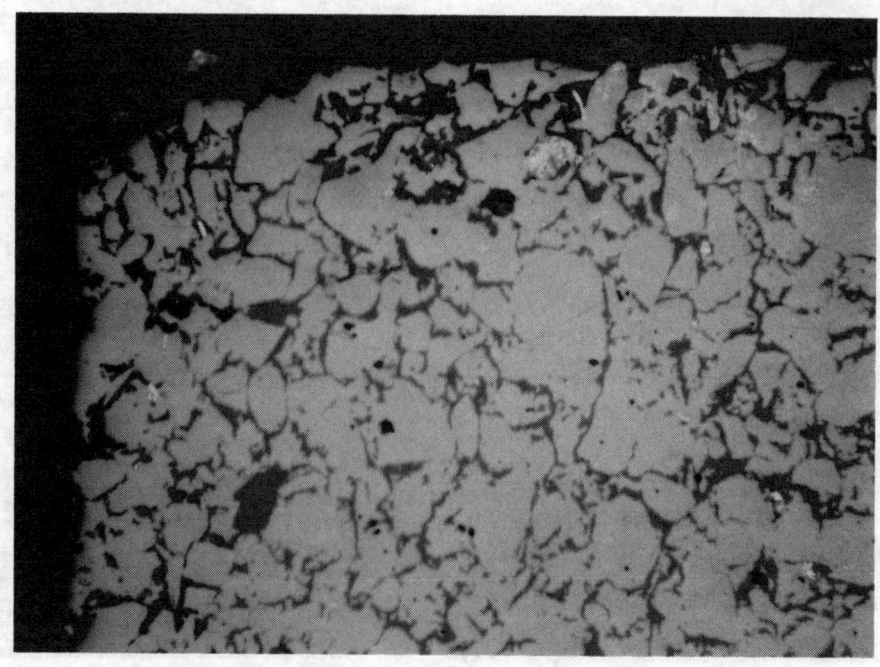

Figure 13: Slight surface attack in the resin impregnant for carbon 'A' after the laboratory immersion test.

Figure 14: Some particulate removal (black areas) from carbon 'E' after the laboratory immersion test.

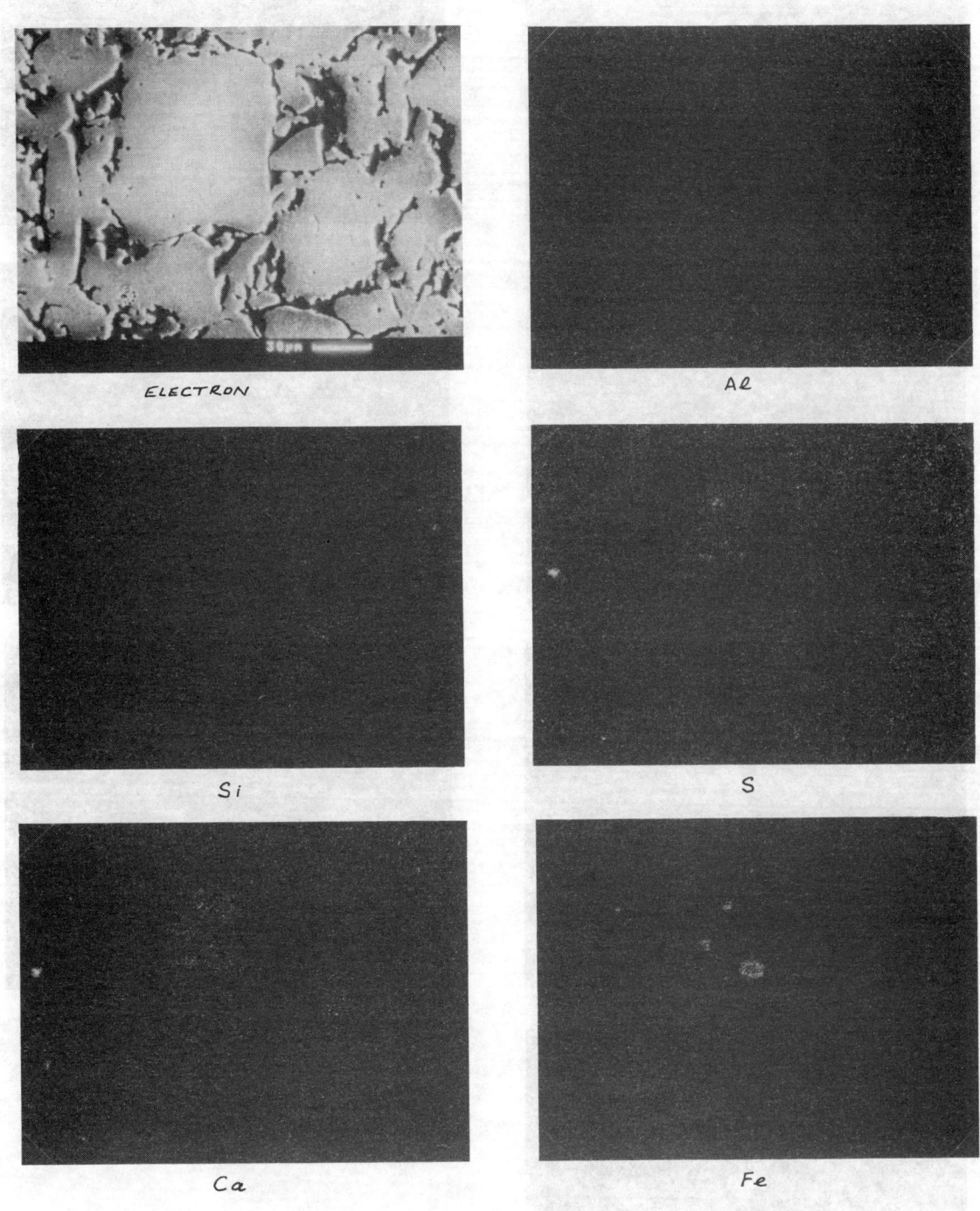

Figure 15: Electron image and X-Ray images showing element concentrations for Carbon 'A'.

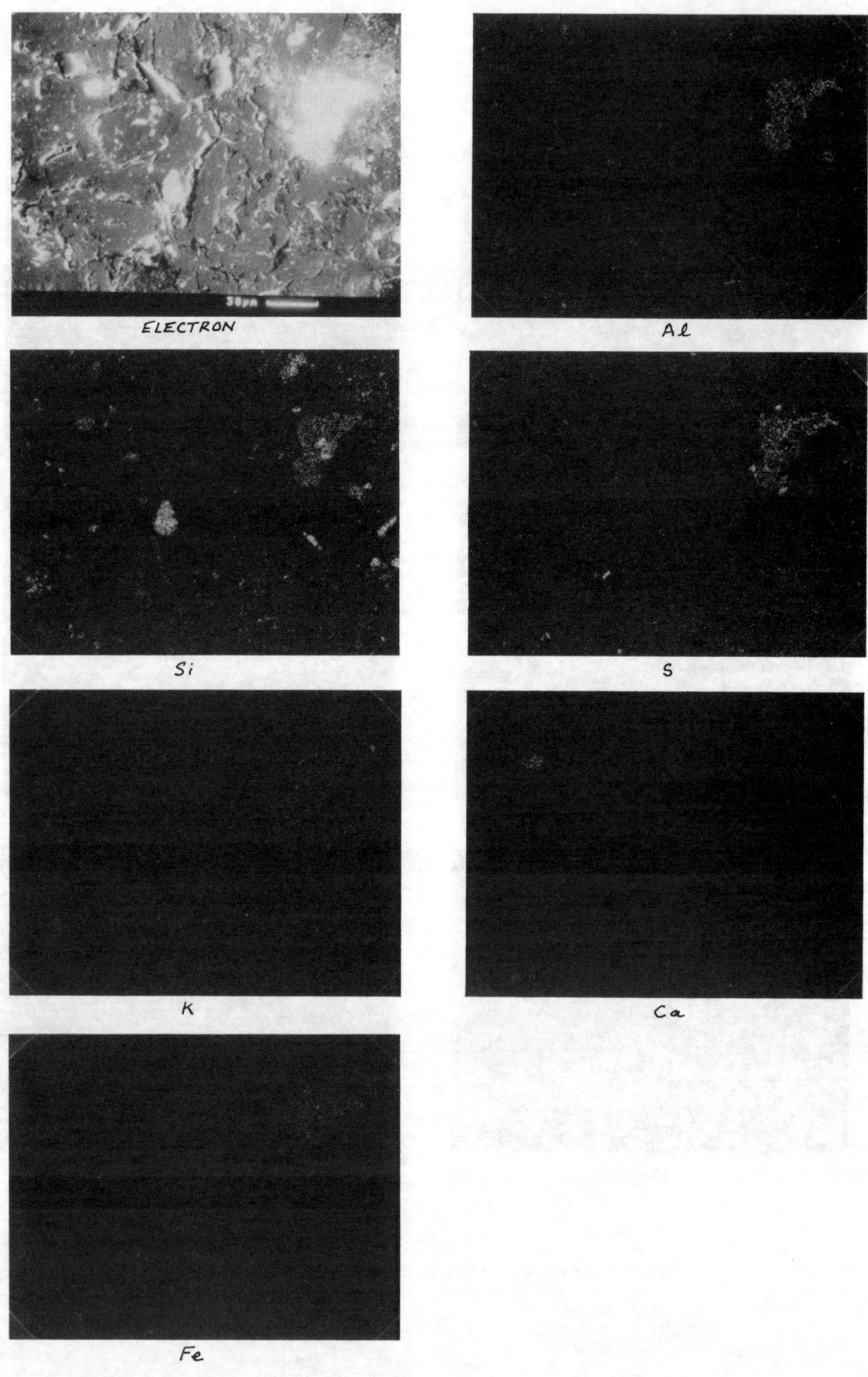

Figure 16: Electron image and X-Ray images showing element concentrations for carbon 'E'.

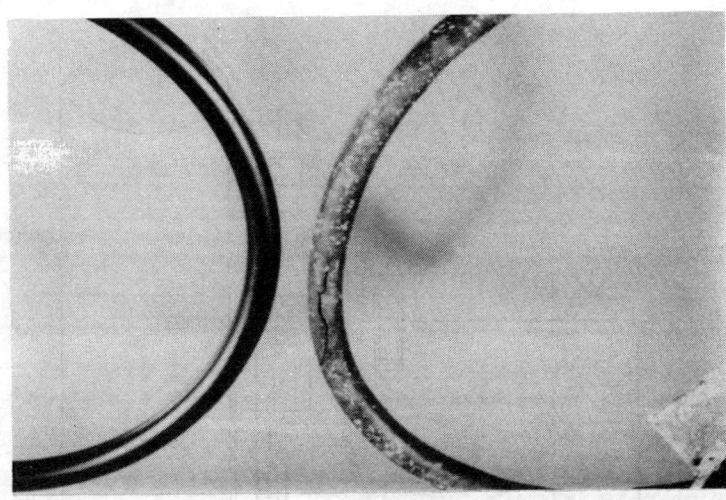

Figure 17: Black fluorocarbon elastomer before and after immersion on plant.

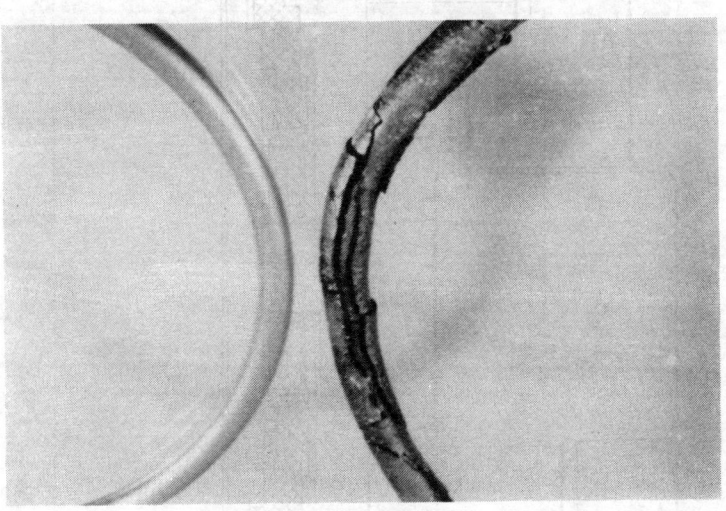

Figure 18: Green fluorocarbon elastomer before and after immersion on plant.

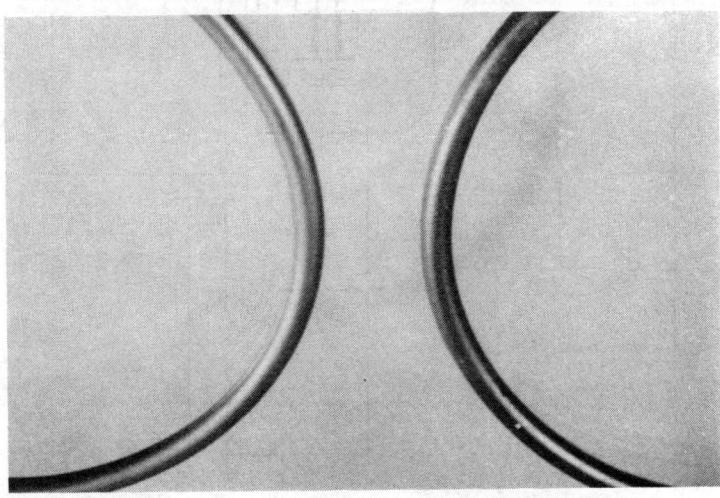

Figure 19: Black litharge cured fluorocarbon elastomer before and after immersion on plant.

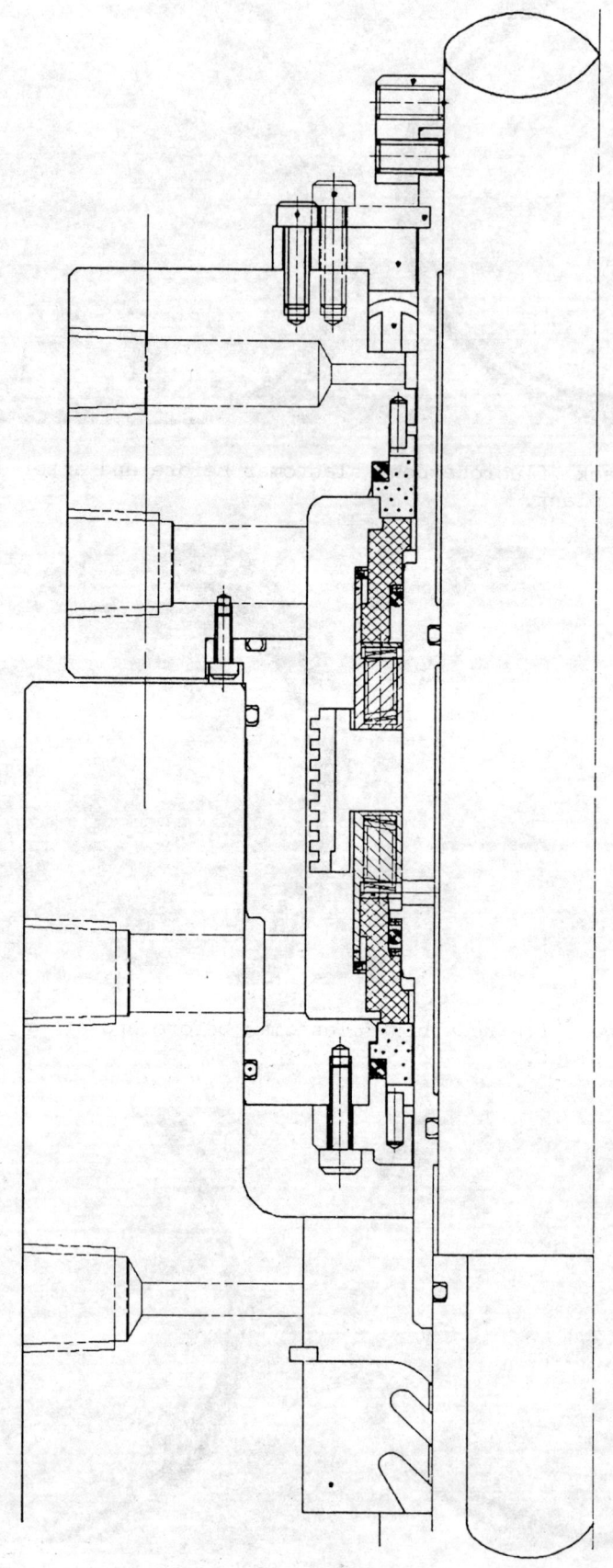

Figure 20: Double seal arrangement with reverse pressure capability

THE GAS-LIQUID PERFORMANCE OF A CENTRIFUGAL PUMP: PRIMING USING THE SHROUD REFLUX METHOD

R.K. TURTON
Senior Lecturer
Department of Mechanical Engineering
Loughborough University of Technology
Loughborough, Leicestershire, England

SUMMARY

The shroud reflux method of priming a centrifugal pump is argued to give an economical design, as the casing is the only special component. A simple design based on the reflux method is described and its priming capability is examined. Casing modifications and internal clearance reduction, and their beneficial effects on priming are described. Concluding discussion indicates a need for a better discharge zone to promote gas separation, and a need to study impeller action thoroughly is mentioned as being part of a continuing programme of research.

1. **INTRODUCTION**

The poor gas-handling capacity of a conventional centrifugal volute pump is well understood, and has two implications for the pump user. The first is the difficulty of inducing the pump to pass liquid (prime) in the tanker unloading mode or empty suction line situation, the second is the breakdown of pumping action when the gas liquid ratio gets above 15 to 20%. Both effects are related to the inability of the impeller to expel air. This contribution examines systems available that enable pumps to prime when running at normal speed.

External aids, like the jet ejector, or the liquid ring pump will not be discussed, as these act as positive evacuators for the impeller of the pump and their design for use is well understood as the article by Sterling[1] and while effective, are expensive. Aids that are integral to the pump are of more interest, particularly because despite the number of designs used little useful design data seems to be available, though there are many articles on trouble-shooting.

Of most interest, on cost and simplicity grounds, are designs that use conventional impeller, shaft, bearing and seal systems with a modified casing. Recirculating systems of various types, (Figures 1 and 2, give two examples) return retained liquid to the impeller suction eye, where gas is entrained from the suction chamber and pumped by the impeller to the discharge separator. The priming action in Figure 1 is clear. In the second system flow re-circulates from impeller discharge through the valve to pump suction until the differential pressure over the valve allows it to close. This occurs when a large proportion of discharge pressure is achieved. These designs have a poor volumetric efficiency, and the second relies on a valve (which may "stick") to reduce recirculation when the pump is primed.

Of rather more interest are those designs where a passage system directs the liquid against the impeller which energises the resulting gas/liquid emulsion to draw gas through the wear ring clearance and through the impeller during priming. Three known designs are shown in Figure 2, the Hannibal design in particular provides a venturi produced purging jet, and efficiencies of 75% are claimed [Rachman[2][3]]. These all require large casings and impellers with few blades and no shrouds to be effective. The two upper designs in Figure 2 operate during priming by returning liquid to the impeller, where mixing extracts air/gas from the impeller and the resulting flow is passed up to the separation zone. Liquid is retained as in the designs of Figures 1 and 2 with the suction pipe above impeller centre-line. The Hannibal system is claimed to have positive extraction assistance during priming, and to have better efficiency than the other designs.

Rachman and Welte[4] studied a fourth design, Figure 3, which is a simple adaptation of a normal design, the shroud reflux system. Rachman compared the designs, and worked on the reflux design also. He comments that the separator area and the size and placing of the reflux holes are crucial to effective design, but remarked that a theoretical approach to design is not possible, and that further study of the flow phenomena should be undertaken. Such a study has been undertaken and some of the observations made will be discussed.

2. THE EXPERIMENTAL APPARATUS

The pump impeller used is of conventional centrifugal shrouded backward curved design, with a simple divided casing as illustrated in Figure 4, loading dimensions being in Table 1. All the dividing walls and internals were manufactured from perspex, and the impeller had a perspex shroud to allow flow in the blade passages to be studied. A pressure transducer was inserted in the suction line, two diameters upstream from the suction flange.

Figure 5 depicts the experimental system with which the results discussed below were obtained, the pump being placed over an open sump, whose water level was capable of variation, and the pump elevation could also be varied. Flow patterns were recorded during priming, using video and short exposure photography.

3. EXPERIMENTAL PROGRAMME

The very simple design of casing in Figure 4, with the reflux hole position close to the cut-water as shown, was first tested, with the pump raised on a platform. The priming times are shown in Figure 6; the upper point of the curve at 6.5 m suction lift represents the priming limit, as above this lift the pump would not prime. Figure 7 shows a typical suction pressure trace. The horizontal axis is atmospheric, positive pressure being below, and negative (below atmosphere) above, just as traced from the UV record. In region I the impeller provides a fairly steady suction in region II some water begins to be drawn along the line, and regions III and IV show how the water flows erratically as prime begins to occur. The rapid variations in region III are difficult to explain, but appear to relate to water cascading from the suction line into the suction chamber. This suggests that turbulence is present in the suction as well as in the separator volume. Flow was observed in the casing and Figure 8 consists of sketches of bubble flow patterns using three reference planes. The clearance space flow pattern, (Section DD) was typically seen as the pump began to prime and was repeatable. The main flow from the diffuser into the separator volume (Section BB) appeared to follow the casing wall giving rise to strong vortexing round the discharge connection as sketched in Section CC. Examining the video indicated quite strong frothing two-phase flow leaving the diffuser and it was noted that the water phase followed the inner wall of the diffuser). It was also argued that the very strong vortexing around the discharge did not allow good separation, and that better diffusion and more separation volume was needed. As a first step the diffuser was lengthened by inserting a straight section up to the separator plate height, giving a reduced divergence angle of 7 degrees. This did not give prime.

The diffuser was modified to give a double curvature, and the resulting shape and slightly improved priming times and suction lift are shown in Figure 9. The flow pattern in the separator region was not so confused, but the expected improvement did not occur, so consideration was given to the side clearance between separator wall and impeller and its influence on flow back through the reflux hole. Since it was argued by Rachman[1] and Baibakov[5] that side clearance for open impellers was important, the side clearance was reduced from 13 mm to 5 mm, and the pump tested. As Figure 10 shows, there was very little change in suction lift or priming time. The effect of changing the reflux hole position was briefly considered, and with the hole close to the cutwater plugged a hole the same size and at the same radius was drilled spaced 180° away from the first hole. The priming time reduction for a suction head of 3.35 m was 3 seconds, but a much greater flow to suction through the wear ring clearance was observed. Opening up the hole by 25% in diameter resulted in no prime, so it is believed the hole size needs careful examination, but no further tests on this configuration were done, as the other parameter, increased height of the discharge above centre-line to allow better separation was considered important. This new casing is being constructed, and will be tested when available.

CONCLUSIONS AND FURTHER WORK

A very simple reflux type self-priming casing design has been studied using a standard, backward curved, shrouded impeller, and it has been shown that the diffusion and separator zones interact between one another, good diffusion that slows liquid velocity down to allow gas separation at the top of the separator improves priming time and the suction lift that can be handled. It is considered that a larger separator volume is desirable, and a new casing design is to be studied embodying the findings reported. Side clearance does not appear to affect priming action significantly, but the reflux hole size chosen appears to be of importance, and this and the hole position will be worth further study.

This contribution does not cover the action of the impeller passages, but since this is related to gas-liquid behaviour the literature available has been studied and a study commenced using the same impeller size and design initially. Work is at an early stage, and it is hoped will be reported later.

The overall performance and efficiency of the pump in both its self-priming and normal forms will be studied and reported.

ACKNOWLEDGEMENTS

The contribution of Mr. Hammouche Anseur supported by an Algerian Government grant, and the provision of the pump and casings by HMD Seal-Less Pumps of Eastbourne is gratefully acknowledged by the writer.

REFERENCES

1. L. Sterling, Various Priming Methods for Marine Pumps, Pumps, Pumpes, Pompen, 1975: 109: pp 936-944.

2. D. Rachman, Physical Characteristics of Self-Priming Phenomena in Centrifugal Pumps, BHRA, SP 911, 1967.

3. D. Rachman, A Study of the Priming Process in a Centrifugal Pump. MSc Thesis, Glasgow University, 1966.

4. A. Welte, Untersuchungen an Selbstansaugenden Kreiselpumpen, Doctoral Thesis, TH Hanover 1959.

5. D.V. Baibakov, A new Re-Circulating-Type Centrifugal Pump. Chem.Pet. Eng., V16: 9-10: Sept-Oct 1980, pp 543-544.

TABLE 1

Design data of the standard impeller:

Outside Diameter	180 mm
Suction Diameter	50 mm
Passage Width at Outlet	7 mm
4 blades, outlet angle	23.5°

Duty at 2600 rpm. 60 imperial gallons per minute at 95 feet head.

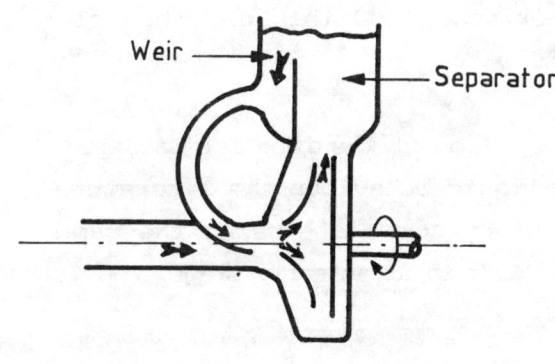 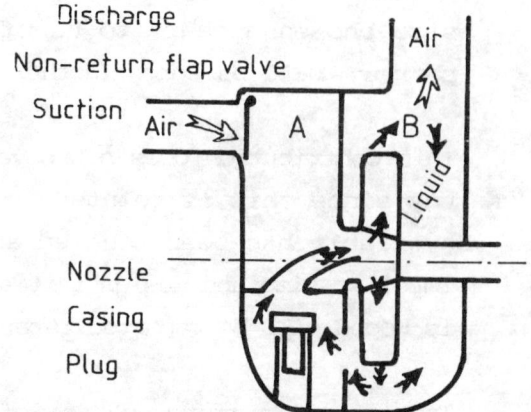

FIGURE 1

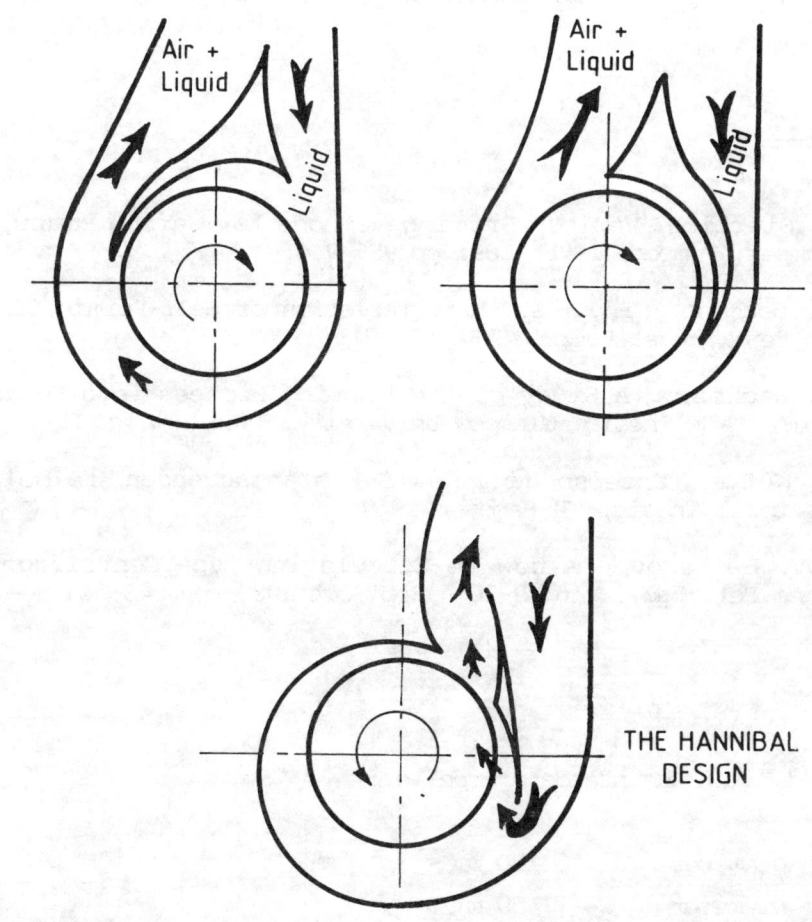

FIGURE 2

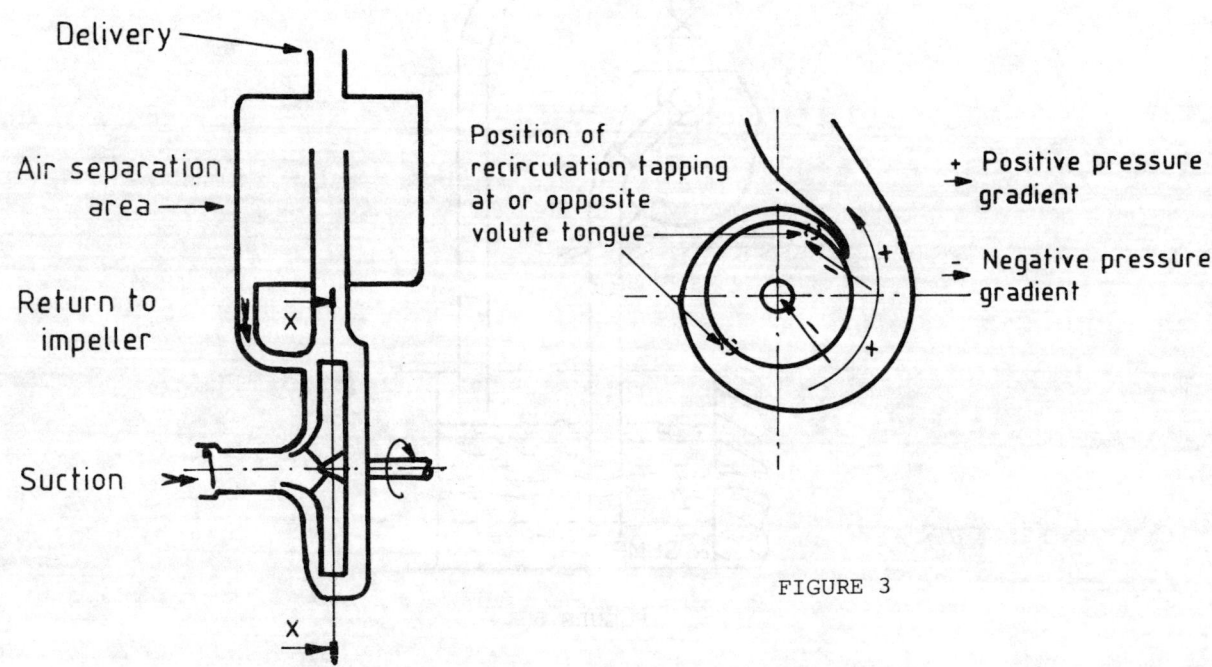

FIGURE 3

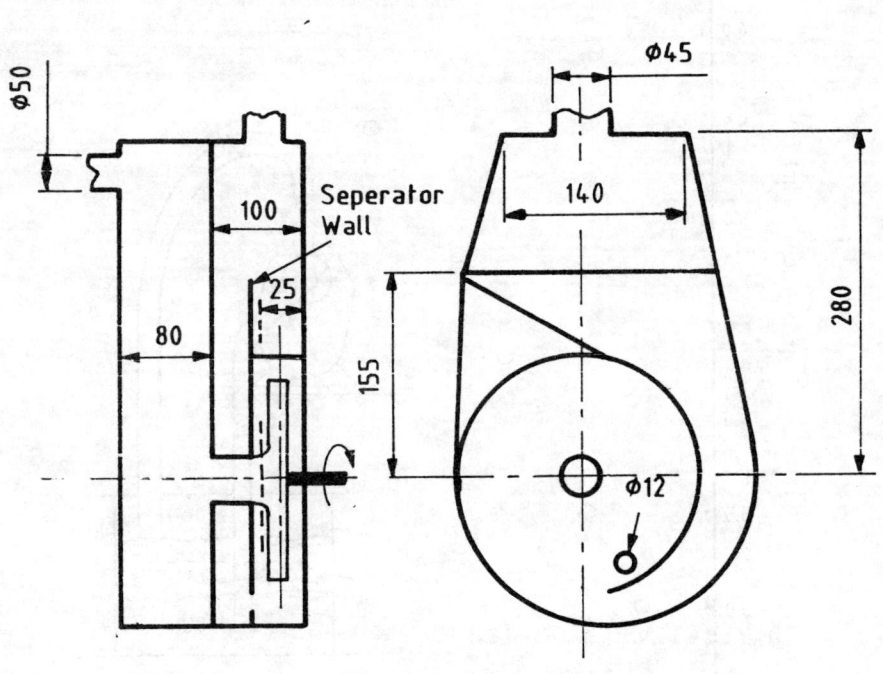

FIGURE 4

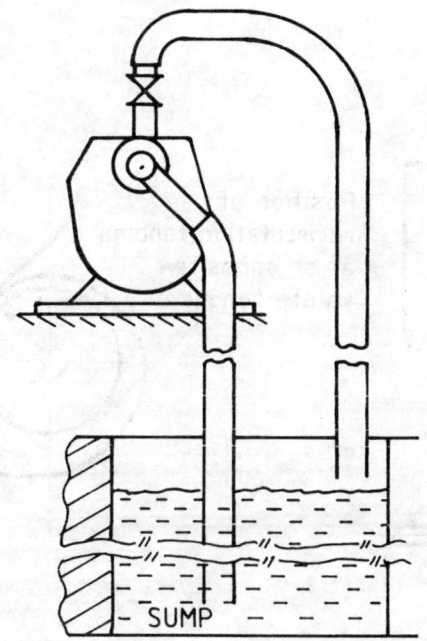

FIGURE 5

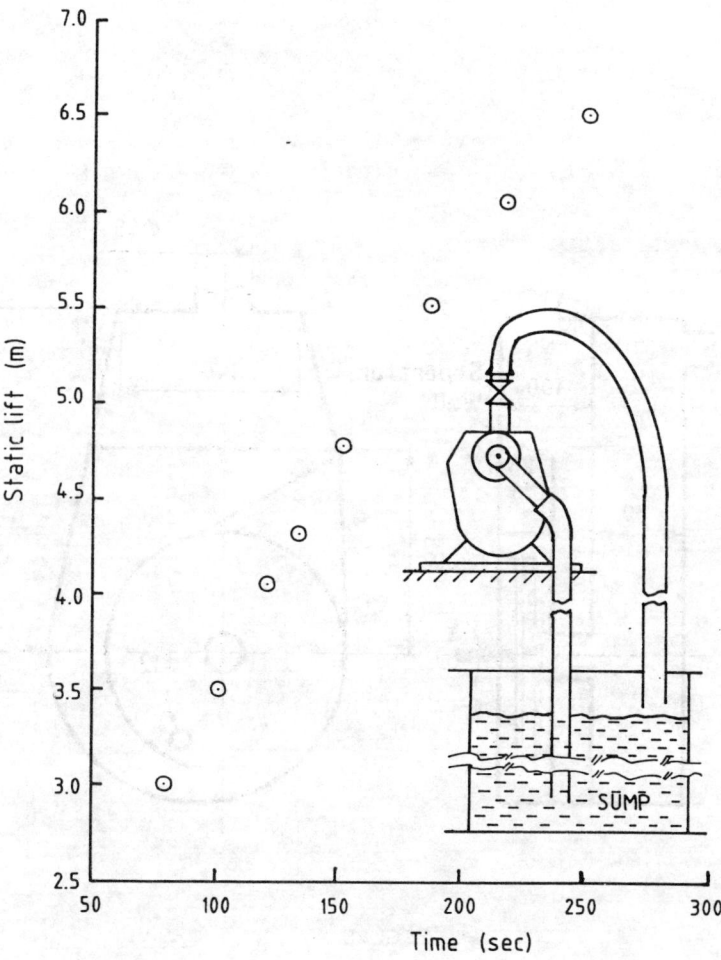

FIGURE 6

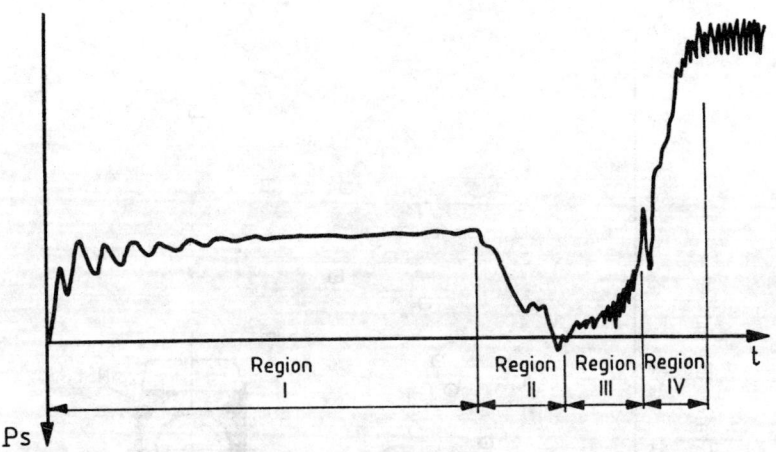

FIGURE 7

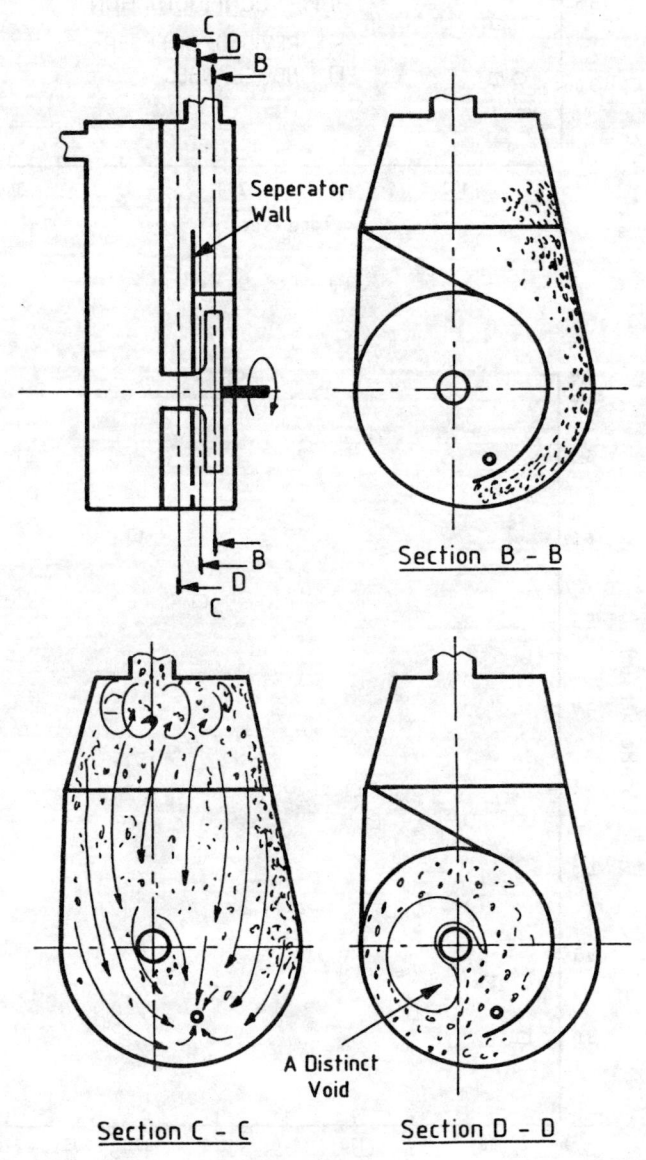

FIGURE 8

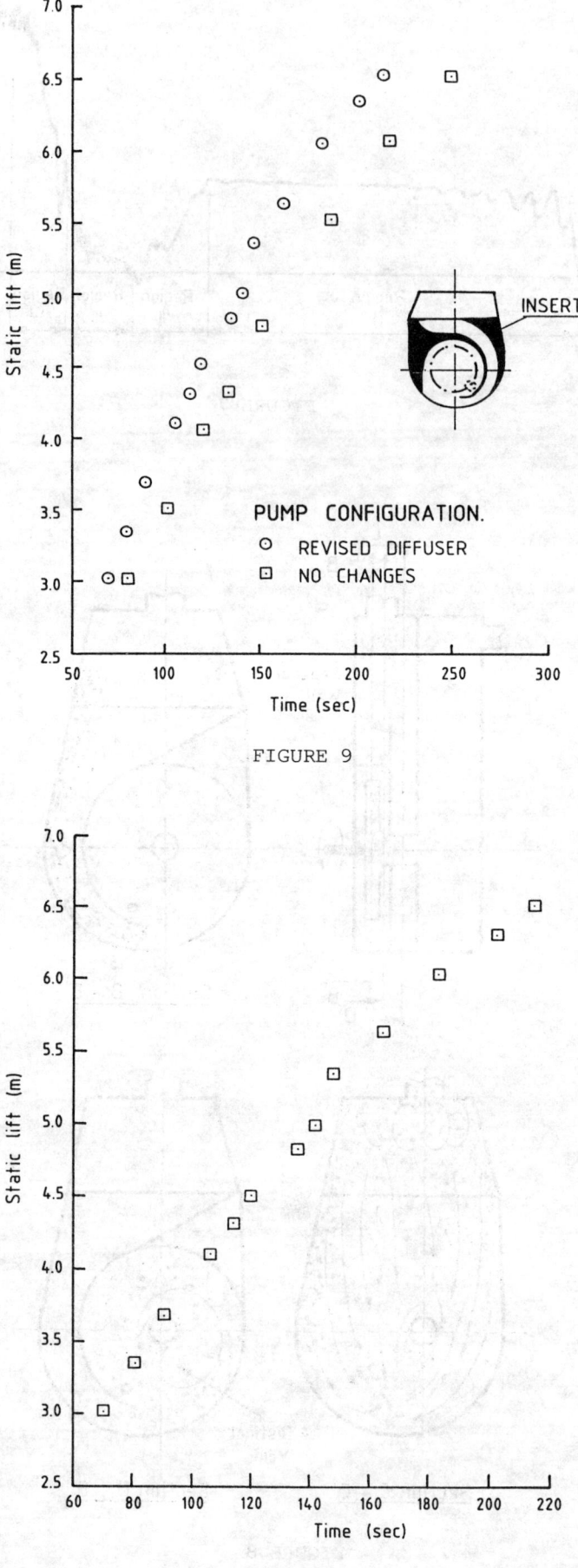

FIGURE 9

FIGURE 10

11th International Conference of the
British Pump Manufacturers' Association

New Challenges – Where Next?

18-20 April, 1989
Churchill College, Cambridge

PAPER 10

MEETING THE CHALLENGE OF THE MODERN CHEMICAL PROCESS PUMP

Pierre Fabeck   - Product Manager - Europe
Kevin H. Sierwald - General Manager - UK

Durco Process Equipment Ltd., Milton Keynes, England

SUMMARY

This paper is designed to illustrate the development of a new process pump in terms of market and technical needs. It is divided into two sections.....market requirements presented by myself (Kevin H. Sierwald) and technical requirements presented by Pierre Fabeck.

THE MARKET

Until the mid 70's the design and marketing of new products had been PRODUCT DRIVEN by manufacturers. Instructions to sales - go out and sell this product; we have developed it; it is good and customers need it. In Europe there were many standards, in American there was, of course, API 610 but also a voluntary standard AVS. Customers bought varying standards and consequently had little or no interchangeability. DIN and/or ISO were introduced in the mid 70's and the Americans defensively made the AVS a national standard - ANSI.

In the late 70's the markets changed from being PRODUCT DRIVEN to MARKET DRIVEN - i.e. the customer decided what he wanted and refused to buy non-standard products. It was at this stage (early 80's) that our company decided to truly enter the market as a European market driven company by developing a European ISO pump.

Having determined that there is a market for DIN/ISO pumps.... how do we get into it?

Our strategy is shown in table 1.

Our sales and engineering personnel visited users in many countries with standarised questionnaires to assess market needs, prepared designs, repeated the exercise and developed prototypes. Following testing internally, units were made and sited at

Conference organised and sponsored by the British Pump Manufacturers' Association
in conjunction with NEL and BHRA, The Fluid Engineering Centre.
Co-sponsored by the Process Industries Division of the Institution of Mechanical Engineers.

potential users for running tests under arduous site conditions. Development continued during these tests until the range was proved to be successful and ready for market.

Main market needs are shown on table 2 - "market needs".

In our case we had discovered that a "me to" product was not what the market needed but certain design features in the ANSI pump produced by our American parent were requested by users in several major European countries. These needs were met as Pierre will show later in the presentation.

At this stage, the glossy sales literature was designed closely followed by the production of detailed technical manuals plus installation and operation manuals.

The product is ready to launch at selected venues in Europe and various suitable magazines.

Now to the technical requirements......

TABLE 1   MARKET STRATEGY

a)   BUY INTO MARKET BY BUYING ONE OF THE MANY AVAILABLE COMPANIES IN EUROPE.

b)   INVESTIGATE MARKET AND ITS REQUIREMENTS.

c)   VISIT POTENTIAL CUSTOMERS IN EUROPE.

d)   DISCUSS TECHNICAL AND COMMERCIAL REQUIREMENTS.

e)   PREPARE BASIC DESIGNS AND REPEAT c) and d).

f)   DEVELOP PROTOTYPES AND TEST TO DESTRUCTION,

g)   DEVELOP UNITS FOR FILED/ENDURANCE TESTING.

h)   CORRECT ANY DEFICIENCIES AND BRING TO MARKET.

TABLE 2   MARKET NEEDS

a)   MEETS A RECOGNISED EUROPEAN STANDARD

b)   IS OF PERCEIVED HIGH QUALITY

c)   IS COMPETITIVE

d)   IS AVAILABLE

e)   IS INTERCHANGEABLE

       1)   dimensionally with other manufacturers

       2)   has interchangeable parts to reduce stock holding

f)   HAS F.A.B's

THE CHALLENGES FACED BY MODERN CHEMICAL PROCESS PUMPS

Several surveys conducted with process industries show that 90% of the reported pump failures are related to either mechanical seal or bearing problems.

These two types of failures are inter-correlated, as a failure on a bearing can induce a seal failure and vice versa.

It is generally admitted that the bearing problems are very seldom related to inherent failure of the rolling elements, but rather to inadequate operation or selection of the bearings (high loadings, thrust reversal, excessive temperature, reduced internal clearances, oil contamination).

Seal manufacturers have made these last years tremendous improvements in the design of seals able to cope with the tough requirements of modern process duties.

It is generally recognised that a considerable amount of seals and bearing failures are due to inadequate pump design. The main causes of these problems are related to the following weaknesses:

- too short shaft diameters
- too large overhang

- high and uncontrolable seal chamber pressure
- high and uncontrolable axial loadings on the shaft
- axial thrust reversal
- poor seal environment
- insufficient concentricity
- detrimental effect of wear
- oil contamination

Many of these weaknesses are due to the fact that the majority of process pumps were designed 20 or 30 years ago, in the days of shaft sealing by packed gland. Their original design did not meet the specific requirements of modern mechanical seals.

Furthermore, the standards that have governed the chemical process pump market until recently were purely dimensional: ISO 2858, BS 5257, NF E 44.121, DIN 24.256. The pump manufacturer had freedom of design, provided outline dimensions were met. In many cases this led to a poor compromise on quality and reliability.

That is why many users in the chemical industry have started to specify API 610 "petroleum" design for the critical duties where increased reliability was needed.

Fortunately there is today another standard for process pumps that brings additional reliability: ISO 5199 "Technical Specifications for Centrifugal Pumps - Class II". Officially released in 1986, after having existed several years under the "draft" status (DIS), this standard specifies most of the important design parameters such as maximum shaft deflection, casing rigidity, bearing life, vibration level and corrosion allowance, as well as a lot of mechanical features that ensure a trouble-free operation of the pump.

It must be pointed out that this standard is not yet mandatory, and that its requirements come in addition to the "dimensional" requirements of ISO 2858 (BS 24256, DIN 24256, NF E 44.121), so there is no conflict between the "old" standard and the "new" one.

Very few chemical process pumps meet the recommendations of ISO 5199.

The CHEMSTAR pump, developed by DURCO is one of the first process pumps that incorporates in its original design the toughest requirements of ISO 5199.

The purpose of this article is to show that specifying ISO 5199 is the ideal solution when increased reliability is needed, and that it is more suited to the needs of the chemical industry that the API 610 concept designed for petroleum applications. In order to elaborate on the benefits of ISO 5199; it is first necessary to describe the design of the CHEMSTAR pump.

CHEMSTAR DESIGN

Fig. I shows a sectional drawing of the CHEMSTAR pump.
Fig. II and III show respectively a cut-away view and a bare shaft pump.

IMPELLER

The main feature of the pump is the setting of the impeller against the rear cover plate. The back of the impeller is open so that this impeller has outstanding solids-handling capabilities. Fig. IV shows the distinctive features of the impeller, while Fig. V shows a picture of the impeller.

Balance holes located in the impeller eye allow an internal leakage to take place on the back of the impeller. The controlled clearance between impeller and rear cover controls the amount of leakage. An external shaft adjustment system, consisting of an adjustable bearing carrier (Fig. I, 1) enables the clearance between impeller and rear cover to be restored quickly and accurately without having to open the pump.

A "blind" adjustment is made possible by the use of calibrated notches located on the carrier. This external adjustment system is required by ISO 5199 (para 4.8.3) for those impellers which need an axial adjustment. It must be noted that nothing is mentioned in API 610 with respect to external adjustment.

The balance holes combined with the controlled rear cover setting reduce and control both the seal chamber pressure and the axial thrust. So, seal chamber pressure is outstandingly low and repeatable, so curves showing seal chamber head versus capacity can be provided (Fig. VI). This results in extended seals and bearings life.

Most chemical pumps are subject to seal chamber and axial thrust increases with internal wear. In the case of double seals, an uncontrollable increase in seal chamber pressure can cause the latter to become higher than the buffer fluid, which provokes an inversion of the leakage flow between the seal faces, allowing the pumped liquid to contaminate the buffer fluid.

The need for reducing seal chamber pressure is recognised by API 610 (para 2.1.3), which bans the use of pumping vanes or back ribs (para 2.6.1) to establish axial balance of the rotor.

The Chemstar investment cast impeller has an extremely smooth surface finish which results in better and more repeatable performances than a regular sand-cast impeller.

The standard impeller material for most duties is duplex stainless steel ASTM A 744 Gr CD4M Cu, which is 80 HBN harder than regular 316 stainless steel and thus less likely to wear due to erosion or cavitation. It is also more resistant to chemical attack in most applications. Due to the difference in hardness most of the wear is concentrated on the rear cover whose wearing area is about 4 to 5 times the area of a regular wear ring. This rear cover is therefore the only major part susceptible to wear and up to 3 mm of wear can be compensated by re-adjusting the impeller as the design includes a 3 mm corrosion allowance.

Another feature of the CHEMSTAR impeller is that it is screwed against a shoulder on the shaft which eliminates the need for a locknut. This allows the inlet of the impeller to be profiled for optimum hydraulic flow, resulting in lower NPSH values and longer pump life when operating under low suction pressures (Fig. IV, 7). The PTFE gasket confined between impeller and shaft sleeve protects both the shaft threads and the shaft/sleeve interface (Fig. I, 6). This confined "O-ring" type gasket is much less subject to potential failure than a classical locknut gasket such as shown in Fig. VII, 6. This addresses the general need of chemical pumps to avoid threads in wetted parts.

The absence of back wear ring or pumping vane allows the shaft overhang to be reduced which lowers the shaft deflection and also the loadings on the inboard bearing due to radial forces exerted on the impeller (Fig. IV,9).

Last, the inside of the casing is machined in front of the impeller (Fig. IV, 6), so that a reduced impeller/casing gap is created on the whole front area of the impeller. This closed clearance (+/- 2-3 mm) chokes the front leakage flow in such a way that it is not any more necessary to use small clearances on the annular ring part of the impeller. This obviates the need for a front wear ring which consists in a major advantage to the users in the chemical industry. As no wear parts are to be fitted in the casing, the latter may remain connected to the piping during all types of maintenace operations. There is no need to bring back the casing to the shop for retro-fitting new wear rings.

Apart from the fact that no wear rings must be stocked, this is felt as a particular advantage to the users in the chemical industry:

- Due to corrosive attack, and particularly "fretting" corrosion, the wear ring/casing interface as well as the set screw threads are very often badly corroded which makes the removal of the old wear ring sometimes very difficult (must be machined thoroughly);

- In the applications where casing jackets and/or special thermal insulation are necessary, the necessity to remove the casing is a strong disadvantage.

The standard values of the annular diametral ring clearance of the Chemstar is 0.9 mm, while most classical process pumps have values between 0.4 and 0.6 mm diametral.

Fig. VIII shows typical values of annular clearances found in process pumps, as well as the minimum clearances recommended by API 610. This makes the CHEMSTAR pump less sensitive to wear as its performances (head, efficiency) will deteriorate at a slower rate than a classical pump.

This has been proven both on field and by many laboratory tests. The following table shows the average mean decrease in efficiency and head on the "Group A" CHEMSTAR pumps.

Clearance (diam.)	Head	Efficiency
Std (0.9 mm)	--	--
2 mm	-0.7%	-1.5%
3 mm	-1.2%	-2.9%

It must be pointed out that 3 mm of diameter clearance is a very severely worn out clearance. The axial thrust exerted on the impeller is towards the driver, as there is more shrouded area on the front of the impeller than on the back (Fig. IX, 1). That is why the front shroud is partially trimmed and has the form of a star.

A major issue in process applications is NPSH. It has been shown that the surface smoothness of investment cast impellers combined with the absence of an obstructing impeller locknut did result in oustandingly low NPSH values.

A phenomenon which is sometimes disregarded by users is the increase in NPSH due to wear. Fig XII shows typical values of the phenomenon. Assuming two impellers having at the beginning the same NPSH, referenced as 100% NPSH. Let us also assume that both the Durco and the classical impeller do have the same global efficiency and also the same volumetric efficiency, which means that the internal leakages are equal in percentage. The capacity output is 100%Q at the discharge flanges, while the capacity in the impeller is 120%Q in both impeller.

The "worn out" condition corresponds to doubled clearances (on the front for Durco, on both rings for the classical impeller). To get the same 100%Q capacity from the pump, one sees that 125% is necessary in the Durco impeller, ans 140%Q in the "classical" one.

If we assume that NPSH varies as the square of the flowrate within the vicinity of BEP, which is generally accepted, one sees that:

Effect of wear on NPSH

DURCO   + 8.5% NPSH

OTHER   + 36% NPSH

This shows the importance in critical applications to closely monitor the rate of internal wear of the pump.

One remark must be said about the suction specific speed parameter.

$$SSS = \frac{Q \; RPM}{NPSH}$$

where Q is capacity flowrate in USPM at BEP and full impeller diam.
NPSH is the required NPSH, in meters of water at BEP and full impeller diam.

The higher the SSS, the lower the NPSH.

Some pump designs have obtained low NPSH values by enlarging impeller eye area. This usually creates instabilities at partial capacity, as it has been shown in reference 4.

Therefore, some users have recommended that a benchmark of SSS=11,000 be set as a maximum value (reference 5) for a stable operation.

It must be noted that this restriction on SSS has less meaning for pumps where the low NPSH is obtained through surface smoothness and obviation of the impeller locknut. Actually, on standardized pumps, the impeller eye diameter is practically imposed by the size of the suction flange, so there is little room for "enlarged" eye areas.

BEARING HOUSING (See Fig. I)

The bearing housing is integral and extends up to the rear cover so that there is no need for an adapter piece. This helps maintaining a good concentricity at the faces of the mechanical seal, as it eliminates one set of tolerances. It facilitates also the dismantling of the pump.

The double-row outboard bearing of the angular type has very low internal clearances limiting shaft end-play to less than 0.03mm. Please note that reference 2 only suggests a maximum of 0.05 mm axial end-play. This feature reduces axial vibrations and limits the pressure fluctuations between mechanical seal faces assuring longer seal life. Actually the axial loadings are so low that a single-row deep groove ball bearing would take the load. The main reason for providing a double-row bearing is the need for reduced end-play.

Appropriate lubrication is obtained by adjusting the oil level to the center of the lower ball. Four longitudinal slots are provided in the bearing carrier to ensure a proper connection between the space behind the bearings and the main sump.

Both outboard and inboard bearings are located against shoulders on the shaft, so there is no need to measure their setting: just heat them and slide them against the shoulder.

The inboard bearing is of the ball-type and the outer cage is free to slide axially versus the bearing housing. This allows for differential thermal expansion and allows an easy dismantling of the shaft assembly (Fig. X, 1).

As the axial thrust is directed towards the driver, it is transmitted by the outboard bearing to a stiff shoulder on the bearing carrier (Fig. I, 5). The conical-type circlips only acts to hold the bearing in place. This is a major difference with most classical designs (Fig. X, 2) where the thrust is directed towards suction, and is taken by a circlips. This by the way is forbidden by ISO 5199 (4.11.9).

Also, when high suction pressures are present, this creates an additional force towards the driver equal to the area under the seal faces times relative suction pressure. In the CHEMSTAR design both hydraulic forces are in the same direction, while on most classical pumps these 2 forces are opposite, so that they can eventually balance each other and provoke detrimental thrust reversal (Fig. IX, 2).

Fig. XI explains the basic difference between the CHEMSTAR concept and most usual designs with regard to the effect of suction pressure. Another area of discussion is the inboard bearing. Many process pumps are equiped with a roller-type inboard bearing. The advantage of this type of bearings is that they can take higher radial loads than ball bearings of the same size (about twice th load). So they might be the solution in pump designs where high radial loadings are applied to the inboard bearing. The reduced overhang of the Chemstar pump as well as the hydraulic design obviate the need for roller bearings, so that the L10 life of the Chemstar bearings is higher than 25,000 hours.

Roller-type bearings have the following disadvantages:

- Dismantling

Fig. X, 2 shows that dismantling a roller bearing is much more time-consuming than a ball bearing. It needs to remove 2 bearing flanges. Many users complain that the inner cage is not always easy to disassemble from the shaft.

- Overheating

When lightly loaded which is the case when operating above minimum continuous flow, roller bearings tend to heat more than ball bearings (+:- 10-15°C more). This is reported by many studies, for instance reference n° 1.

- Pre-load

Roller bearings are usually mounted with a smaller clearance between the outer race and the bearing housing, as the inner race is free to move axially versus the rollers. This makes them more sensitive to loss of internal clearance due to differential thermal expansion of the races.

- Imprint of the races

It has been reported by many users that, when twin-pumps are located on the same concrete slab, the rotating pump transmits its vibrations to the stand-by pump in such a way that the roller imprint both inner and outer races which causes the bearing to fail when the stand-by pump is started. This phenomenon is much more acute with roller bearings than with ball bearings.

For these reasons most API 610 pumps are equiped with inboard ball bearings and not roller bearings.

The CHEMSTAR bearing housing is specially protected against oil contamination (dust, humidity, acid condensations, metallic particles). It is well known that most bearing failures are caused by oil contamination so a special attention has been paid to the design of the CHEMSTAR bearing housing. Double-lip seals isolate (Fig. I, 2) both ends of the bearing housing. A filtered vent makes sure no harmful particle enters the oil sump. A magnetic drain plug (Fig. 1, 3) captures the metallic particles that could be present. Options such as labyrinth-type oil seals and a membrane-type breather are also availabe. Last, the combination of a constant level oiler and a sight glass indicator (Fig.I 4) make sure the oil level and condition are correct.

CASING

Both API 610 and ISO 5199 call for a 3 mm corrosion allowance which is incorporated in the design of the CHEMSTAR pump.

It has been found that to meet the stiffness criterion of ISO 5199 (Annex C) regarding the effect of external forces and moments, it was necessary to increase casing and cover thicknesses in such a way that the pressure parts be designed for 25 bars nominal pressure.

Although ISO 5199 only calls for a PN16 rating at 20°C and using cast iron, it must be recognised that many application in the chemical industry call for temperatures above 100°C, where the material de-rating due to temperature do further limit the pressure capabilities of process pumps. Having a PN25 qualified pump gives extra room that is particularly advantageous in the case of 316 stainless steel whose mechanical resistance decreases very rapidly with temperature.

But the major benefit of the PN25 rating might be found in extra mechanical seal and bearing life: pump components are less distorted by external forces and moments, seal faces remain perpendicular, vibrations are more effectively damped. Last, increased thickness means longer life of components when chemcial corrosion is present.

API 610 OR ISO 5199?

The attached Table I shows a comparison between the requirements of API 610 and

ISO 5199 and their respective applicability to the needs of the chemical industry.

This table shows that for several parameters, ISO 5199 is more conservative than API 610:

* <u>jacket design pressure</u>: the design pressure is higher in ISO 5199.
* <u>securing of axial thrust</u>: snap rings (circlips) cannot be used for transmitting axial forces. This is found in many process pumps.
* <u>need for an external adjustment system</u>

There are several parameters where API 610 is more conservative, and the CHEMSTAR pump was therefore designed to meet those specific API 610 requirements. Let us mention in particular:

1. Seal chamber pressure

The Durco concept minimises and also controls the value of the pressure which does not increase with wear, unlike most classical process pumps.

2. Axial thrust balance

Pumping vanes are not allowed by API 610 because they are susceptible to wear and do not work properly on worn pump which causes uncontrolled increased in seal chamber pressure and axial forces. The CHEMSTAR design minimises and controls axial thrust.

3. Back pull-out design

The uniqueness of the CHEMSTAR design is that there is no need for wear parts to be fitted in the casing, so no need to move the latter away from the piping for maintenance purposes.

It is suggested that ISO 5199 offers an acceptable and competitive alternative to API 610 design for those applications in the process industry which require more reliability than regular "standardized" pumps (to ISO 2858, BS 5257 or DIN 24256).

The term "upgraded medium duty" process pump has been used by some authors to designate those process pump with an increased reliability. It is our opinion that this terminology might be misleading in the sense that "upgraded" means the original design had some weaknesses.

In the case of the CHEMSTAR pump, the original and standard design incorporates upfront all the benefits of increased reliability and operator's safety.

REFERENCES

1. W.D. Anderson, "Five ways to cool bearings", Machine Design, Sept.22, 1983

2. Pump Handbook, 2nd edition, McGraw Hill, pp 9-91

3. J.R. Birk, J.H. Peacock, The Duriron Co., "The Chemical Engineering Guide to Pumps", McGraw Hill

4. Fraser, "Recirculation in Centrifugal Pumps"

5. Hallem "SSS = 11,000 - A benchmark?"

TABLE 1   API 610 VERSUS ISO 5199

No.	PARAMETER	API 610, 6th Edition	ISO 5199	More conservative standard for this parameter	Durco Chemstar Design	DURCO CHEMSTAR MEETS
1.	Seal chamber pressure	Para 2.1.3. For heads above 3.4 bars, the pump should be designed to minimise the pressure on the stuffing boxes.	No recommendation	API 610	Impeller concept is designed to minimise and control seal chamber pressure.	API 610
2.	Axial thrust balance	Para 2.6.1. Pumping vanes shall not be used to establish axial balance.	No recommendation	API 610	Durco impeller is balanced thanks to balancing holes and the controlled setting of the impeller versus the rear cover plate.	API 610
3.	Stable head curves	Para 2.1.7. Continuously rising curves are preferred.	Para 4.1.1. Stable characteristics curves are preferred.	API 610	CHEMSTAR curves are continuously rising towards shut-off.	API 610
4.	Jacket design pressure	Para 2.1.13: 5.2 bar g	Para 4.4.4.3.: 6 bar g at 170°C	ISO 5199	Jackets design for 6 bar g.	ISO 5199
5.	Prevention of oil contamination	Para 2.1.17. Bearing housing shall be designed to minimise oil contamination.	Para 4.12.5. All openings shall be designed to prevent the ingress of contamination.	API 610 = ISO 5199	Double-lip seals; magnetic drain plug, filtered vent, sight glass + oiler	API 610 and ISO 5199
6.	Internal concentricity	Para 2.1.18. Casing and bearing housing shall be designed to ensure accurate alignment on re-assembly.	Nothing mentioned	API 610	Integral bearing housing (no adapter piece), all metal-to-metal contacts with piloting surfaces.	API 610
7.	Back pull-out design	Para 2.2.8. Design shall permit removal of impellers, shaft, bearings without disassembling suction or discharge piping.	Para 4.4.4.1. Back pull-out construction shall be preferred.	API 610	DURCO design obviates the need for case wear rings or wear plates, so the casing may always stay connected on the piping.	> API 610

8.	Studs or capscrews?	Para 2.2.10.5. Studs are preferred to capscrews.	API 610	Only stainless steel studs are used.	>API 610
9.	Effect of external forces and moments	Para 2.4.3. Shaft displacement at coupling to be less than 127 microns when the pump assembly (pump + baseplate) is subject to the forces and moments of table 2.	API 610	Para 4.6. Shaft displacement to be less than 150, 200 or 250 microns when subject to forces F,F,M (Annex C)	ISO 5199
10.	Max. shaft deflection	Para 2.5.12. Max. 51 microns under the worst operating conditions.	API 610	Deflection always less than 50 microns, even at shut-off head.	API 610
11.	Wear surfaces	Para 2.6.2. Wear surfaces to have 50 HBN difference in hardness.	API 610	DURCO impeller is 80 HBN harder than rear cover plate.	>API 610
12.	Min. running clearances	Para 2.6.4.2. Min. clearances are recommended.	API 610	DURCO front impeller annular clearance is 3-4 times min. value of API 610.	>API 610
13.	Allowable vibrations	Para 2.8.2.2. Measures at +/- 10% BEP. Unfiltered vibration: 7.6 mm/sec peak 63.5 microns peak Filtered: 5.1 mm/sec peak	API 610	Para 4.3.2. Measured at rated conditions Unfiltered vibration: <4.5 mm/sec rms (Group A and B) <7.1 mm/sec rms (Gr C)	ISO 5199
14.	Bearings	Para 2.9.1.2. L10 = 25,000 hrs at rated conditions L10 = 16,000 hrs at worst conditions	API 610	Para 4.12.2. L10 = 17,500 hrs within allowable operating range.	API 610
				L10 life of both CHEMSTAR bearings meets API 610 recommendations.	API 610

151

#	Topic				
15.	Securing of thrust bearing	Nothing mentioned		Para 4.11.9. Snap rings (circlips) shall not be used to transmit axial thrust.	ISO 5199
				Axial thrust is transmitted to a shoulder in the bearing carrier.	ISO 5199
16.	Sealing of bearing housing	Para 2.9.1.13. Lip seals are forbidden. Labyrinth seals are mandatory.	API 610	Para 4.12.5. Any sealing device allowed. Oil level indicator or constant level oiler is mandatory.	
				Standard design is double-lip seals with both level indication and constant oil level oiler. Optional labyrinth is available.	STANDARD > ISO 5199 OPTION = API 610
17.	Pressure rating	Para 2.2.2. Thickness to be suitable for max. discharge pressure + allowances.	API 610	Para 4.4.4.1. PN16 at 20°C for cast iron.	
				PN25 at 20°C for cast iron.	> ISO 5199
18.	Seal chamber dimensions	Nothing mentioned		Para 4.13.1. Seal cavity to meet ISO 3069.	ISO 5199
				Seal chamber dimensions exceed ISO 3069.	> ISO 5199
19.	External adjustment system	Nothing mentioned		Para 4.8.3. External means of axial adjustment shall be provided, if necessary.	ISO 5199
				Bearing carrier allows a calibrated adjustment.	ISO 5199

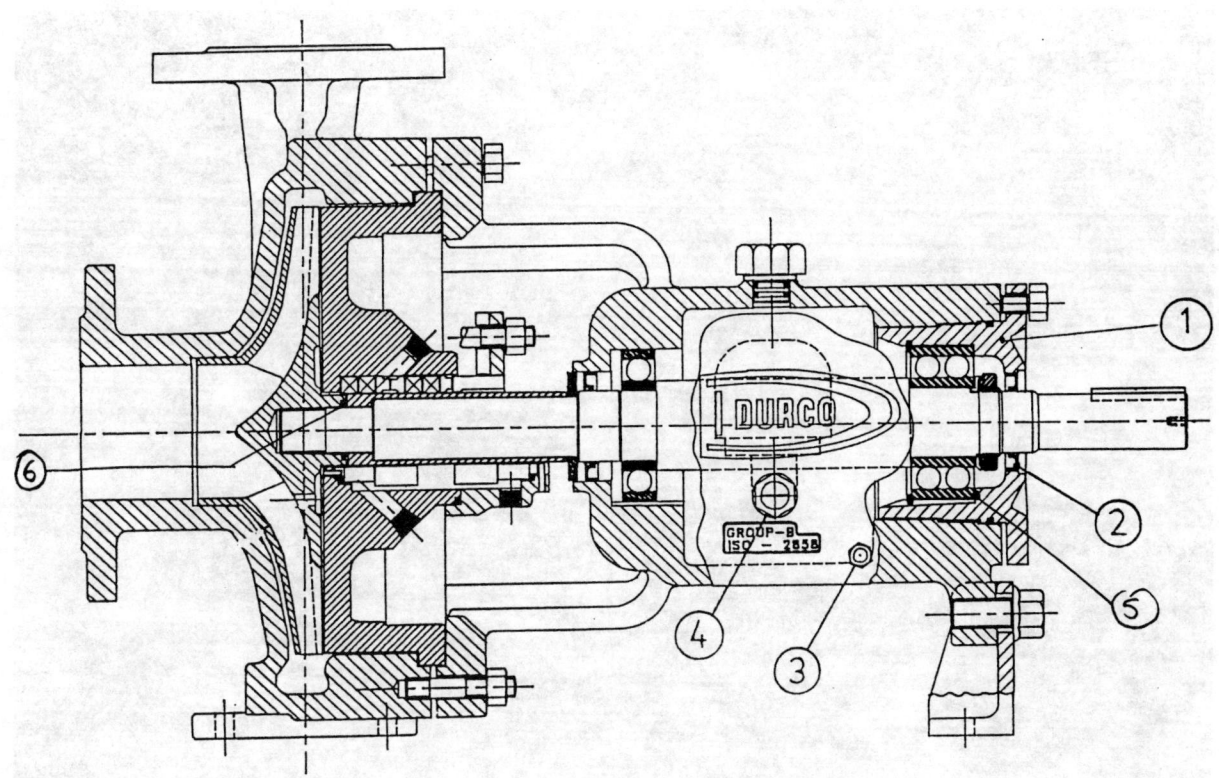

Figure 1. Sectional drawing - Chemstar pump

Figure 2. Cut away view - Chemstar pump

153

Figure 3. Chemstar bare-shaft pump

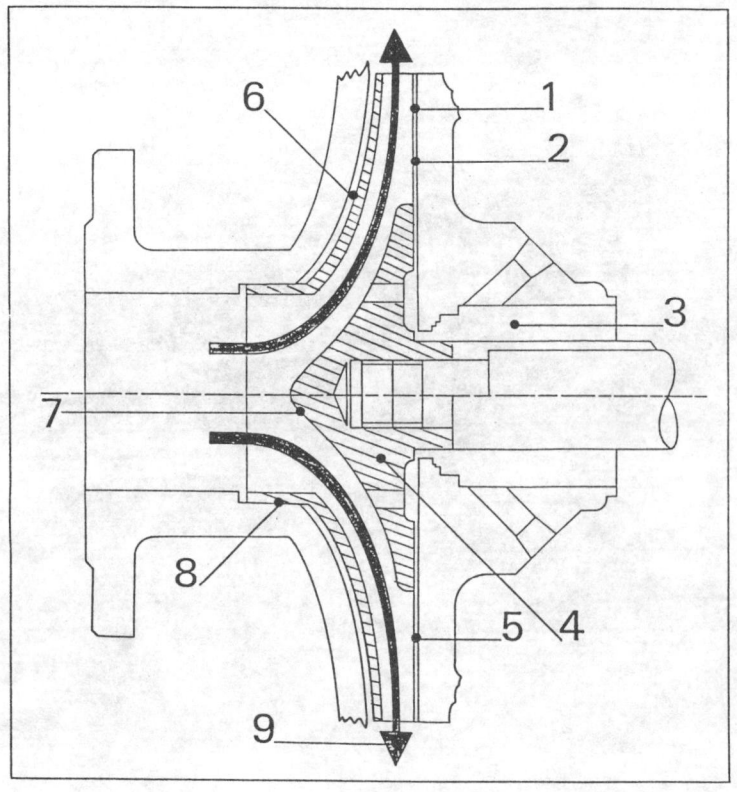

Figure 4. Durco Impeller - principle

154

Figure 5. Chemstar's impeller

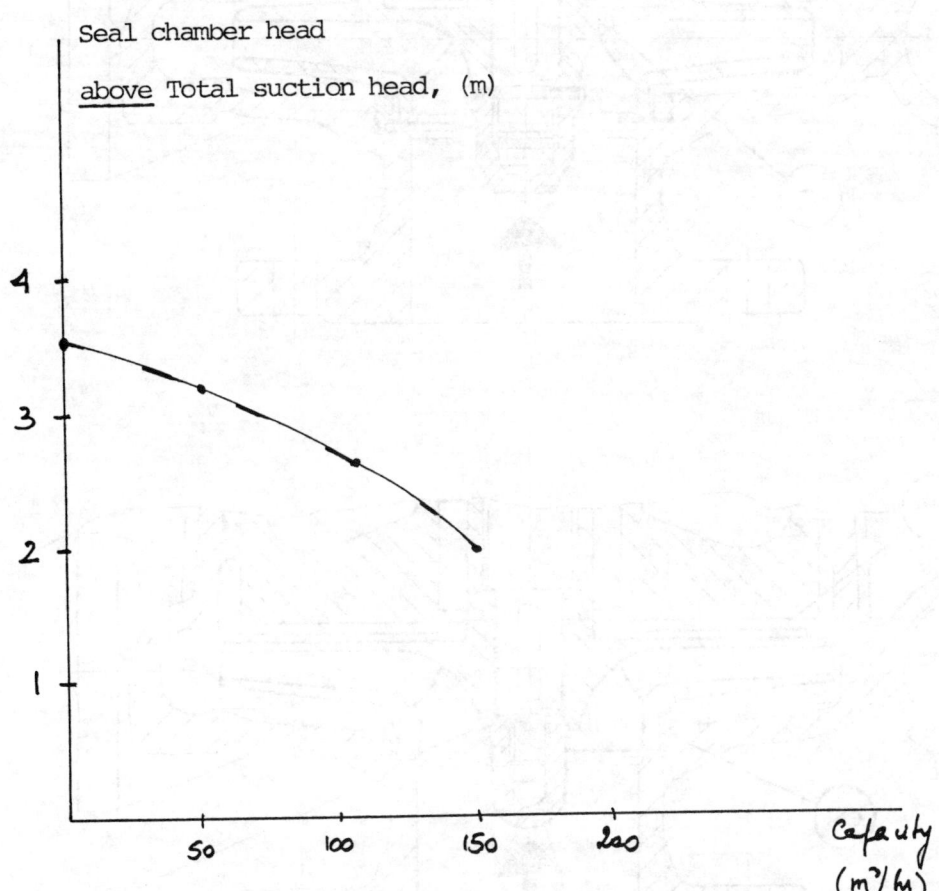

Figure 6. Typical graph of seal chamber pressure vs. capacity

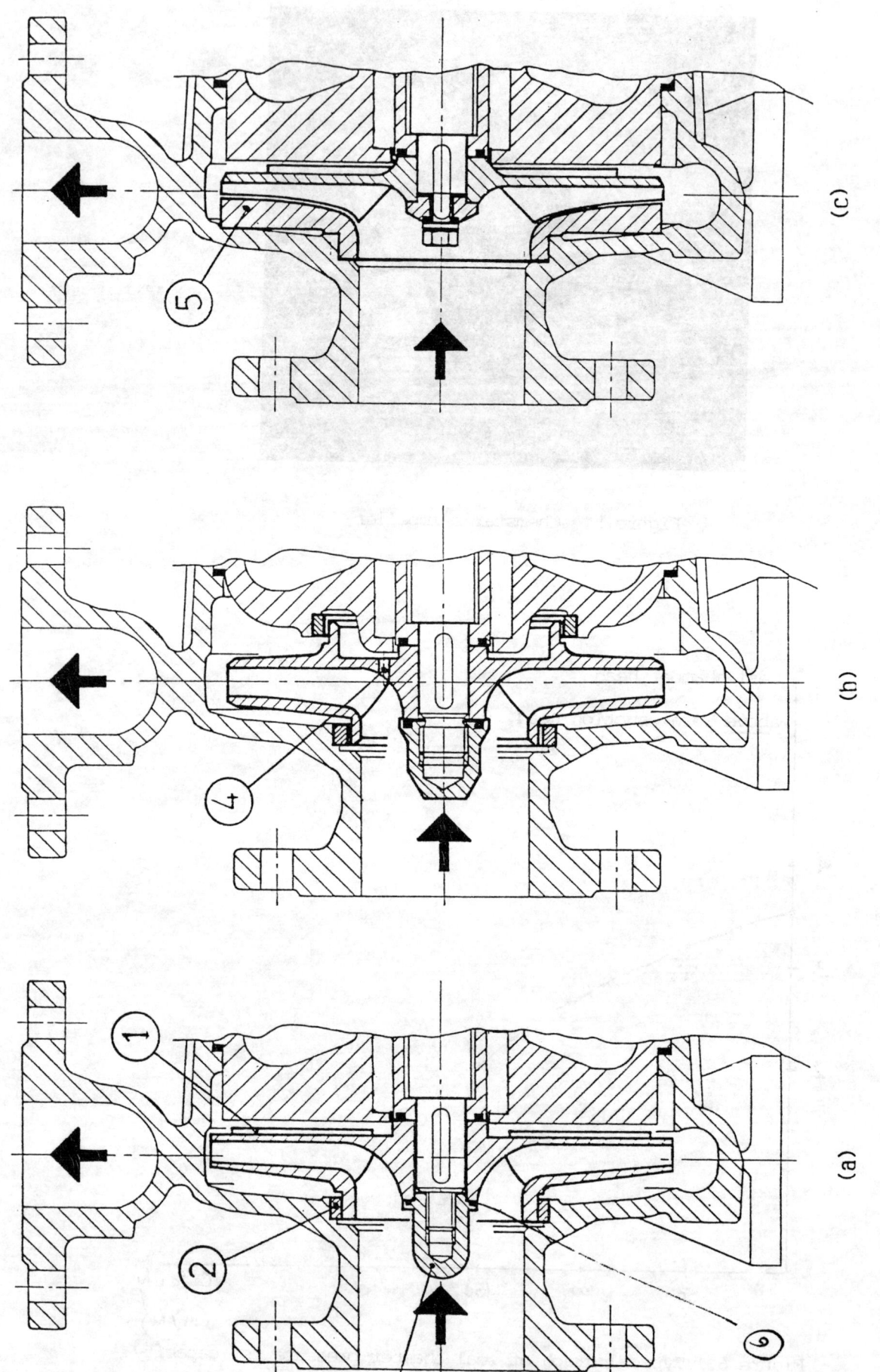

Figure 7. Usual types of impellers

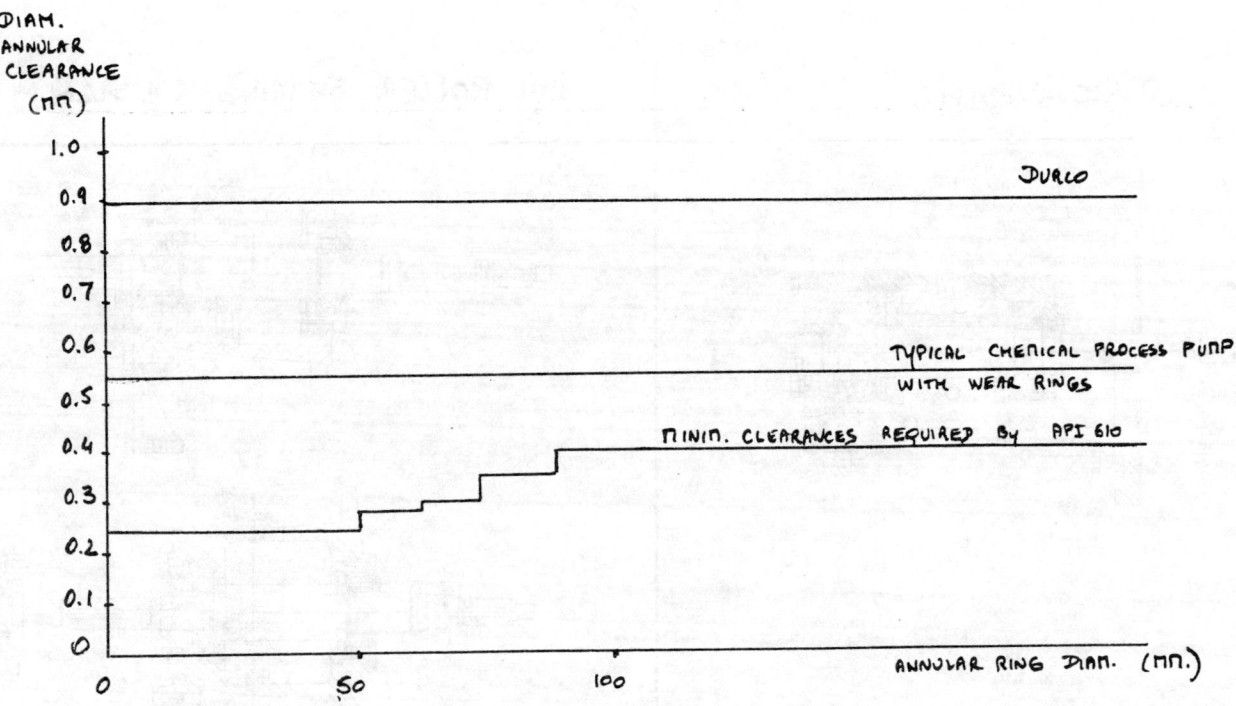

Figure 8   Front annular clearance vs. diameter

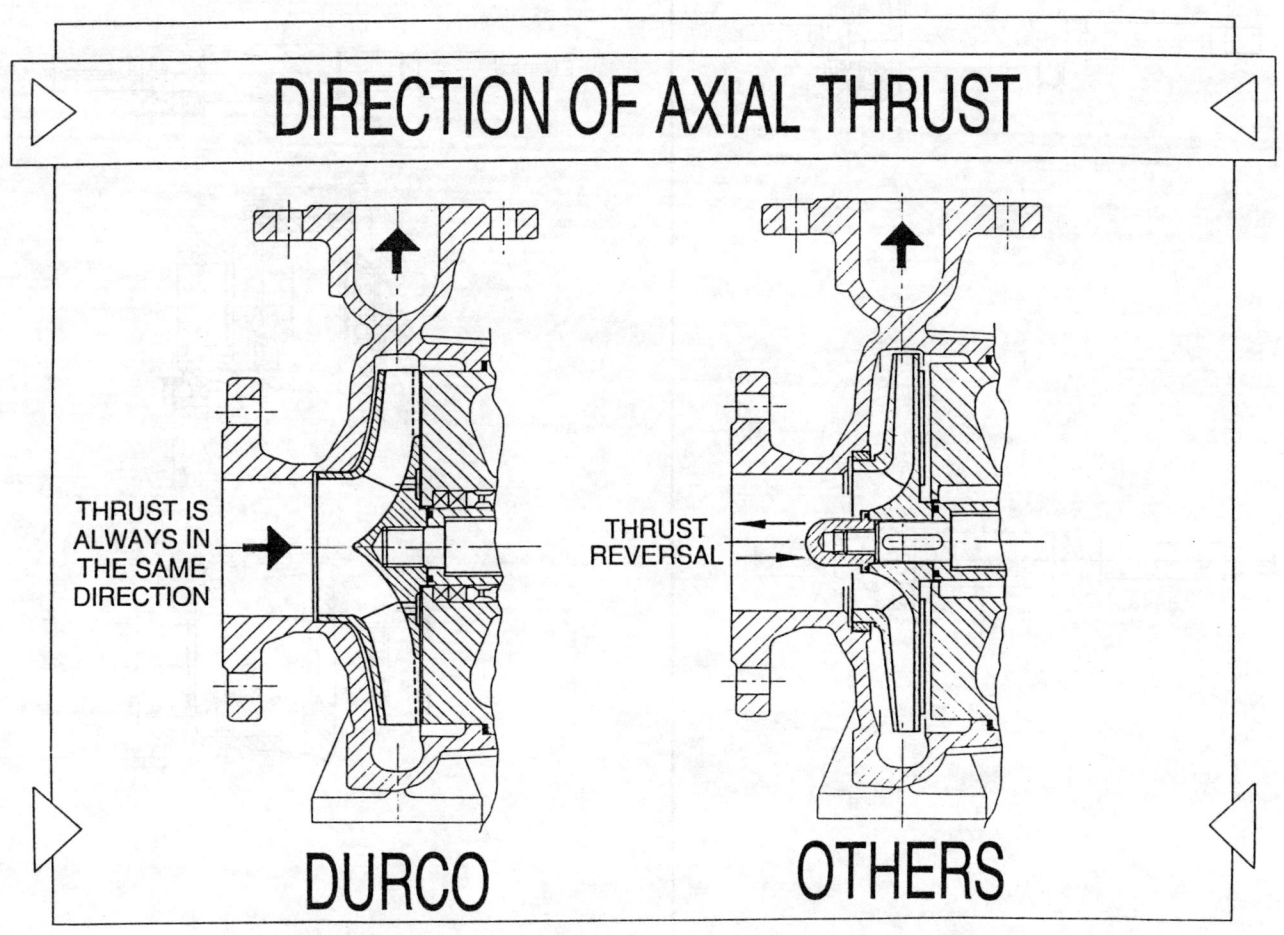

Figure 9.   Direction of axial trust

DURCO DESIGN	CYL. ROLLER BEARING DESIGN

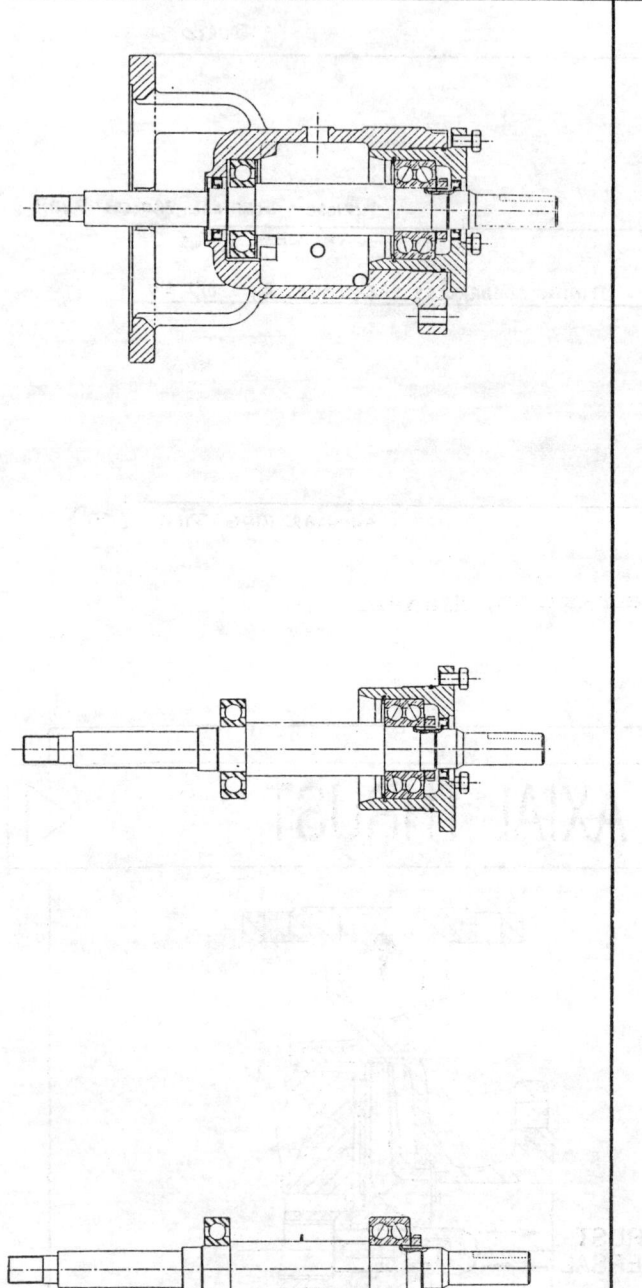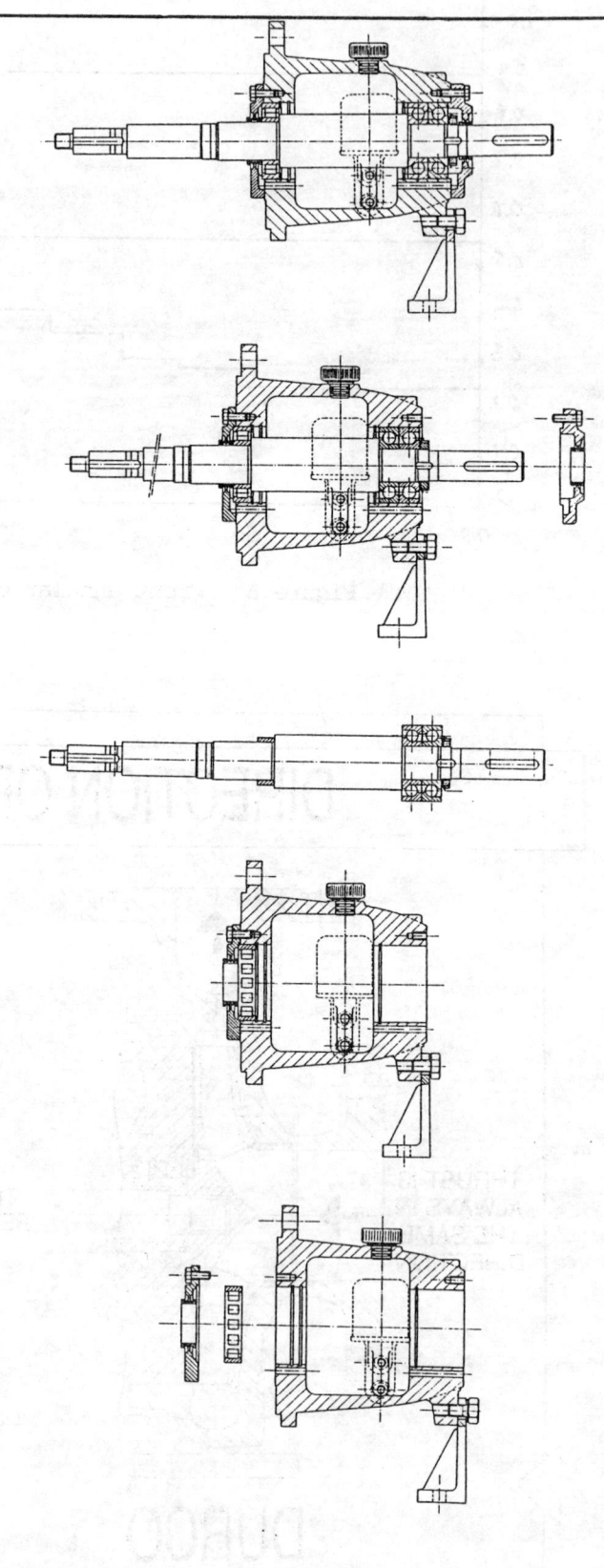

Figure 10. Shaft dismantling

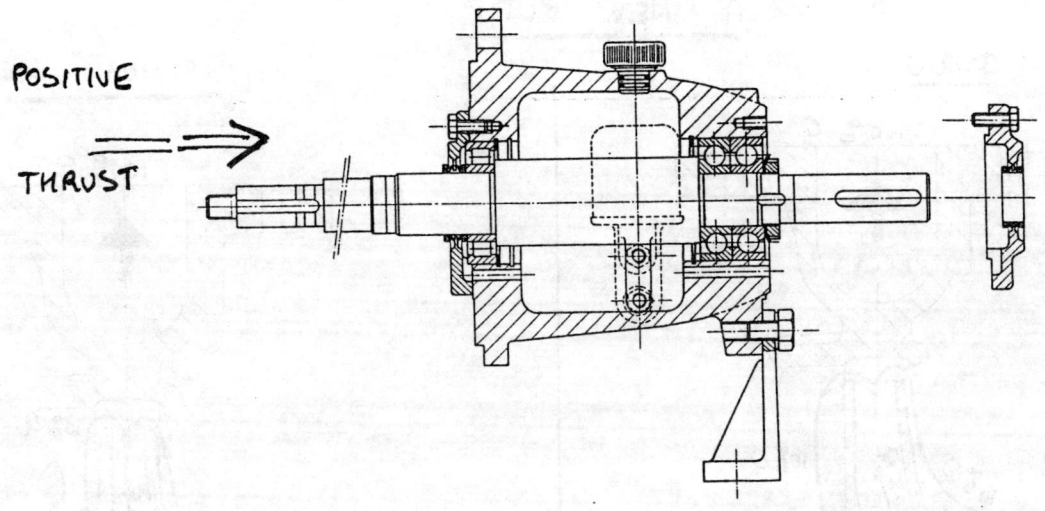

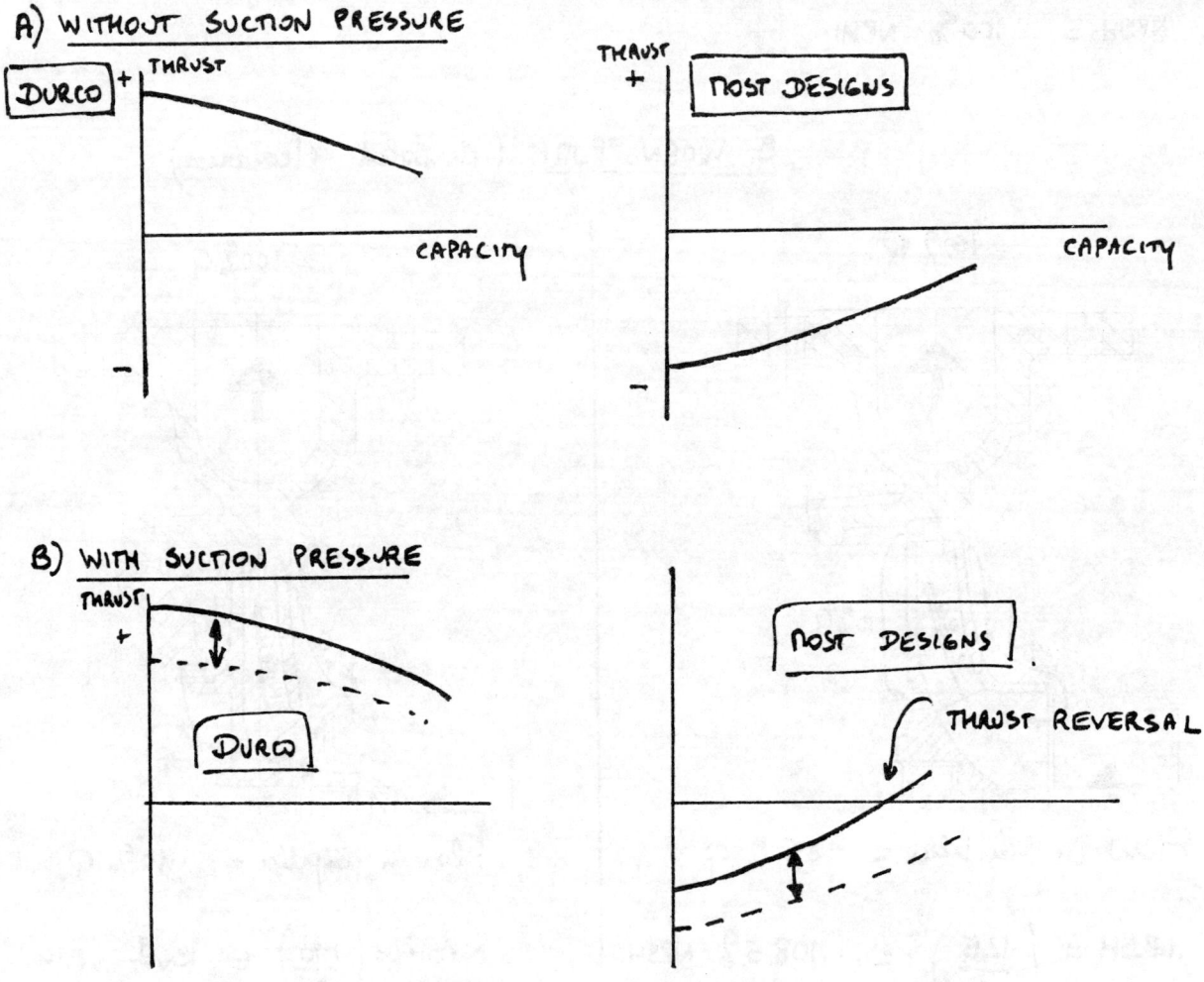

Figure 11. Effect of suction pressure

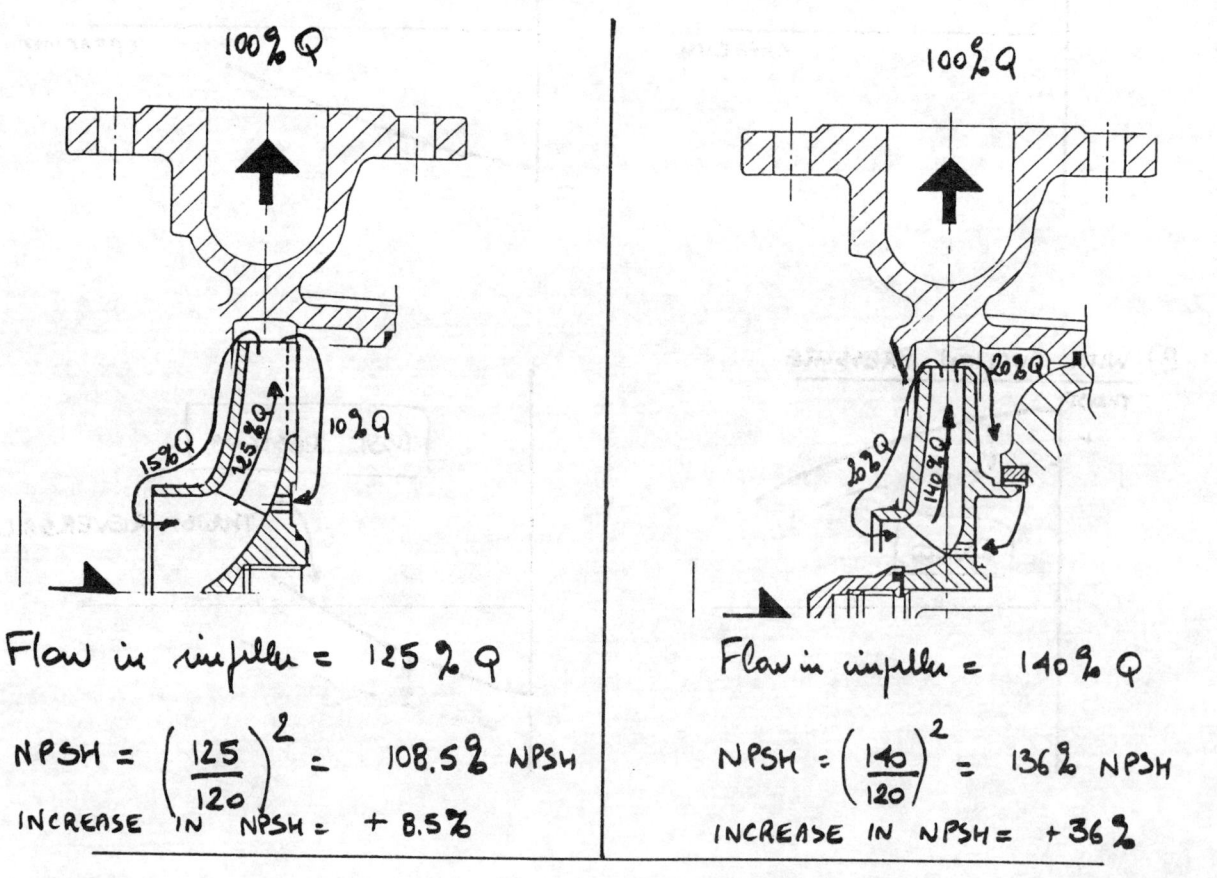

Figure 12. Evolution of NPSH with wear

11th International Conference of the
British Pump Manufacturers' Association

New Challenges – Where Next?

18-20 April, 1989
Churchill College, Cambridge

PAPER 11

DESIGN CONSIDERATIONS FOR PUMPING APPLICATIONS OF VARIABLE FREQUENCY DRIVES

B.P. Zell
Aurora Pump, A Unit of General Signal, North Aurora, Illinois, USA

SUMMARY

This paper addresses electrical, hydraulic, and mechanical design considerations that have been found to have a significant impact on the performance of variable speed pumping systems that utilize variable frequency drives (VFDs).

1.0 INTRODUCTION

The advantages of variable speed pumping have been recognized for decades: lower power cost, increased pump life, better process control. The advent of reliable adjustable frequency drives has simply made it so cost effective that pumps of every type have already been, or soon will be, applied with variable speed. The successful application of this type of equipment requires an understanding of the pump's hydraulic and mechanical characteristics, knowledge of the customer's design and specific operating requirements, and an awareness of the interaction between the electrical, electronic, hydraulic, and mechanical components of the system.

1.1 THE BASIC SYSTEM

The basic variable speed pumping system consists of a parameter sensor, a setpoint controller, a variable frequency drive, a motor, and a pump. Each of the system components will be addressed in turn, with practical recommendations based on experience with operating systems.

2.0 PUMP SELECTION

The pump design point represents a conservative estimate of potential peak head and flow requirements. Constant or variable speed pumps may be selected to provide the highest system efficiency for the range of heads and flows that account for the greatest percentage of operating hours. Most cost effective variable speed pumping applications are characterized by low static head, high dynamic head, many hours of operation at reduced flow, and power ratings above 20 KW.

2.1 DESIGN POINT

Selection of design duty to the left of the best efficiency point (BEP) allows efficient operation at reduced speed when the operating head encountered is substantially less than design. See figure (1) for effects of head less than design on variable speed operation. The constant speed pumping application remedy

Conference organised and sponsored by the British Pump Manufacturers' Association
in conjunction with NEL and BHRA, The Fluid Engineering Centre.
Co-sponsored by the Process Industries Division of the Institution of Mechanical Engineers.

of adding head with a valve is counter to the energy saving intent of variable speed pumping, while trimming the impeller to achieve the needed head eliminates the engineering safety margin and reduces system efficiency. See figure (2) for the beneficial aspects of selecting a larger pump, which provides both efficient operation and a safe operating margin.

2.2 AVOIDING RECIRCULATION

Recirculation at low internal velocities may have a significant impact on system efficiencies, including overloading motors when the KW required exceeds the KW available under reduced cooling air flow conditions. The effect is more prevalent at minimum impeller diameters at flows 50% or less of BEP flow. See figure (3) for an example of a recirculation problem encountered with a secondary chilled water system. The problem was resolved in the field by substitution of a smaller pump with full diameter impeller that provided the required system head and flow at an impeller trim near full diameter. See figure (4) for the performance curve of the replacement pump.

2.3 PARALLEL PUMPING

Multiple identical pumps are commonly used in parallel to efficiently meet system flow and backup requirements. The anticipated system operating area and the individual pump head-capacity characteristics must be carefully compared in order to avoid cavitation when the lead pump is operating alone and approaching full speed. See Figure (5). It may be necessary to program the system controller to add a second pump to operate in parallel with the lead pump to avoid cavitation in some regions of the system operating area. Note the improvement in efficiency achieved by operating two identical pumps in parallel when the system head characteristic is primarily dynamic and there is very little static head. In the example, pump efficiency increases from 75% to 86% when two variable speed pumps are operated in parallel.

2.4 HYDRAULIC CONSIDERATIONS, ONE VFD WITH MULTIPLE PUMPS

It is possible to operate one VFD with two or more pumps by using a technique called "synchronous transfer," but only one pump may be varied in speed at a time, and the remaining pumps are operated at constant speed. Each pump is soft started on the VFD, varied in speed to match demand to full RPM, then transferred to the line by synchronizing the VFD output with the AC power line voltage and frequency and operating contactors. The pump that is operating at constant speed runs out until the real system curve and the pump head/capacity characteristic cross. In those cases where the operating system head is less than the design head, the constant speed pump tends to operate in an inefficient portion of the pump curve at the far right, and the variable speed pump operates in an inefficient area at the far left of the pump curve. The problem becomes evident in Figure (5) if you evaluate pump operation from 390 m3/h to 720 m3/h with one constant speed and one variable speed pump.

2.5 ADVANTAGES OF ONE VFD WITH MULTIPLE PUMPS

The single VFD with logic and contactors is less expensive than the parallel control system with one VFD per pump. If the reduced backup capability, reduced efficiency, and surges associated with pump add and shed cycles are acceptable, then this approach is a viable alternative.

2.6 MIXED SYSTEMS

Multiple pumps of varying sizes may also be used to provide efficient pumping in the face of widely varying demands. Variable speed pump operation becomes inefficient electrically below 50% speed, and may become hydraulically inefficient prior to that, depending on the system operating area. The pumps may include a mix of constant and variable speed pumps, with driver selection based on horsepower, percentage of operating time, and process requirements. Potential savings are reduced by the increased cost of the control logic, and increased by the reduction in energy usage obtained by using smaller capacity constant speed pumps for extended periods of low flow operation. Typical examples include chilled water supplies for facilities with a year round low level cooling demand and a seasonal high level cooling demand.

2.7 HYDROPNEUMATIC TANK/JOCKEY PUMP COMBINATION

Systems that exhibit extended periods of essentially zero flow but require a pressure to be maintained may reduce pumping system power costs significantly by using a hydro-pneumatic tank and low flow (jockey) pump. A constant or variable speed high service pump operates upon demand, and the low service hydro-pneumatic system operates whenever its modest capacity is adequate. Controls may be based on time, flow, pressure, or system KW. Applications include domestic water systems for high rise buildings and small municipal water distribution systems.

3.0 MOTOR SELECTION

Motors for variable speed pump applications should generally be selected totally non-overloading throughout the entire pump system operating range. Exceptions may be made where operation in the high horsepower region will be transitory, if the motor plus variable speed controller combination is capable of generating the required torque. Caution must be exercised with mixed-flow, propeller, and peripheral turbine designs due to the maximum horsepower at no flow characteristics of these pump designs.

3.1 MOTOR LIMITATIONS

Motor insulation system typically permit operation over a 2:1 speed range for constant and variable torque applications without modification. Operation below 50% speed with a constant torque application will probably require an auxiliary blower. Operation below 50% speed with a variable torque application may require an auxiliary blower, depending upon pump, motor, duty cycle, and environment. Constant horsepower operation below base speed will require that the motor and the variable speed controller be increased in size, and may require an auxiliary blower. Consultation with motor and drive manufacturers and an experienced systems application engineer is recommended.

3.2 OPTIONAL MOTOR PROTECTION

Motors need protection from overload conditions when they are operating, and from moisture when they are turned off. Three normally closed thermal switches, one per phase, provide significant protection from gradual motor overloads. The VFD instantaneous electronic trip (IET) and current limit circuits protect the system from rapid motor overloads. Motor heaters prevent moisture from entering the windings when the motor is not operating.

3.3 SPECIAL CASES

Mixed flow, propeller, and peripheral turbine pumps may all exhibit constant horsepower characteristic at reduced speed and flow. Full range performance curves are extremely helpful in selecting the proper motor. In the absence of such data, an extremely conservative course of action should be pursued, up to and including selection of a motor and variable speed controller suitable for continuous operation at no flow.

3.4 HIGH EFFICIENCY MOTORS

High efficiency motors should be used whenever possible. The magnetic characteristics due to the higher quality steel, thinner laminations, greater stack length, tighter control of dimensional tolerances, and improved windings increase the system efficiency by reducing the motor heating associated with the variable frequency drive output waveform. Use of existing or new standard motors as an economy measure fails over the short run due to increased electric power costs, and over the long run due to reduced motor life.

3.5 HIGH EFFICIENCY MOTOR PLUS VFD EFFICIENCY

Questions may arise as to the relative quality of various motors bearing "high efficiency" labeling, and how they will perform with a particular variable frequency drive. Multiplying the motor nameplate efficiency by the nameplate power factor provides a valid numerical quality reference, with "1" being a perfect lossless motor and "0" representing a motor that performs no useful work. Since sine wave motor efficiency and power factor do not, however, accurately represent the motor efficiency that will be obtained when the motor is operating from an adjustable frequency power supply. Testing of the job motor and variable speed drive together is the only effective way to determine actual motor plus drive efficiency. Instrumentation and measurement techniques required for accurate, repeatable results are defined in United States of America IEEE STANDARD 995.2, and may be available in other standards.

3.6 PWM VFD INDUCED MOTOR NOISE

Drive induced motor noise may be a problem at certain speeds and loads with many types of pulse width modulated (PWM) adjustable frequency drives. The extent of the problem varies with the sophistication of the computer program used to develop the PWM waveform, the speed of the switching devices, the filtering provided, and the quality of the motor.

3.7 PWM MOTOR NOISE SOURCE AND CHARACTERISTICS

The low frequency sounds of the typical AC motor are converted to higher frequency sounds of equivalent sound pressure level by the interaction of the adjustable frequency drive current waveform, the motor laminations, and the frame. The noise in question is usually a penetrating 1000 to 2000 hertz tone that varies in amplitude with motor speed, load, and adjustable frequency output waveform.

3.8 SPECIFYING QUIET OPERATION

The motor noise problem may be avoided by not accepting PWM drives, by accepting only those PWM drives with a high frequency carrier, by requiring documentation of the motor/drive noise level prior to approval, or by requiring filters between the drive and the motor on all PWM applications. Noise level may be minimized by specifying high efficiency motors with cast iron frames.

3.9 BALANCING

Motors used with variable frequency drives may run from 3 to 110% of base speed or higher. Every motor for use with variable frequency drives should be balanced over the intended operational speed range. A standard motor is balanced only at the design speed, and may exhibit vibrations at other speeds.

4.0 ADJUSTABLE FREQUENCY DRIVES

Adjustable frequency drives, also known as variable frequency drives, VFDs, and inverters, accept standard frequency and voltage AC power and deliver variable frequency and voltage power to standard three phase AC induction motors so that the motors may be operated at any RPM required to perform work most efficiently. In pumping applications, the working speed range is usually 50 to 100%, although all VFDs have at least a 10 to 100% speed range.

4.1 BASIC VFD THEORY OF OPERATION

Variable speed operation of an AC induction motor with an adjustable frequency drive is achieved by converting the AC input voltage to DC plus an AC ripple, filtering to an acceptable ripple level, then using electronic switches to alternatively connect the positive and negative DC busses to the three leads of an AC motor such that AC power is simulated at the selected frequency. The effect can be demonstrated by connecting any two leads of a three phase AC motor to a DC supply. The motor quickly turns to a position, then stops. Selecting a second pair of leads causes the motor to turn to a second position, and stop. Electronic control of the polarity, duration, and amplitude (optional) of the pulses allows the RPM of the motor to be controlled.

4.2 TYPES OF DRIVES

Three technologies commonly used for adjustable frequency drives are Variable Voltage (VVI), Current Source (CSI), and Pulse Width Modulation (PWM). All three use the conversion, filtering, and switching steps, but each technology has advantages and disadvantages which may affect the successful application of the equipment.

4.3 VARIABLE VOLTAGE INVERTER

The Variable Voltage Inverter (VVI) uses a phase controlled rectifier converter stage which provides a variable DC voltage to the resistor/capacitor filter stage. The DC output is then switched by the inverter stage to provide an output that varies in both frequency and voltage. The phase to phase voltage consists of positive and negative square waves. The line to neutral voltage has a six step staircase appearance on an oscilloscope, and the VVI drive is sometimes called a six step inverter. The VVI drive may use silicon controlled rectifiers (SCRs), transistors, or gate turn off rectifiers (GTOs) as output switching devices. See Figure (6) for voltage and current output waveforms.

4.4 VVI CHARACTERISTICS

The VVI is electronically relatively simple, and has low motor noise. 150% starting torque is commonly available. The transistor version is the most efficient in the 5-150 HP range. The SCR converter stage may create notching in the AC supply voltage waveform if proper impedance is not provided. The VVI causes the motor to cog at low speed, below six hertz. The power factor varies directly with motor RPM, but typically effects facility power factor less than 1% due to the low level of power required at the slow speed. For example, the power factor at half speed would be about .5, but the power required would be 1/8 the amount at full speed.

4.5 CURRENT SOURCE INVERTER

The Current Source Inverter (CSI) uses a phase controlled rectifier converter stage, an inductor filter stage, and an SCR inverter stage. The current is controlled to form a trapezoidal waveform, while the voltage waveform is sinusoidal with spikes at the SCR commutation points. See Figure (7) for voltage and current waveforms.

4.6 CSI CHARACTERISTICS

The CSI, like the VVI, has low motor noise. The regulator is more complex, and the inductive filter is physically large. The SCR converter stage is similar to the VVI. The unit is capable of handling a continuous line to line short without damage, but starting torque is limited to about 110% of design. The CSI is the largest of the adjustable frequency drives for a given KW rating, and is most commonly used for ratings above 110 KW. The CSI drive cogs a motor at low speeds, below 6 hertz. The power factor varies with motor RPM, but typically effects facility power factor less than 2%.

4.7 PULSE WIDTH MODULATION

The Pulse Width Modulated (PWM) inverter uses a diode bridge converter stage, a resistor/capacitor filter stage, and an SCR, transistor, or GTO inverter stage. The voltage output is switched many times per second by a microprocessor in order to create a more nearly sinusoidal motor current waveform. See Figure (8) for voltage and current waveforms.

4.8 PWM CHARACTERISTICS

The PWM is the smallest of the adjustable frequency drives for a given KW rating, and does not create line notching unless the KVA rating of the AC power source is limited, but may cause high frequency power line disturbances when a buss choke is not used inside the VFD. The DC buss level varies with AC power line variations. Some drive manufacturers include circuitry to adjust the PWM modulation scheme to automatically maintain the desired volts per hertz when power line voltages vary, but many do not. There are no motor cogging effects with PWM at low speed. See paragraph 3.6, DRIVE INDUCED MOTOR NOISE, and following for a discussion of PWM motor effects. The power factor remains essentially constant at about .92 to .95.

4.9 DRIVE FEATURES FOR PUMPING APPLICATIONS

The standard VFD is a basic drive with no provision for isolation or control. Naming the following features in a specification ensures that the quoted equipment has basic safety features and may be interfaced readily with control circuitry. A Door Interlocked Disconnect switch provides a safe method of disconnecting power for maintenance purposes. A Hand-Off-Auto switch and Manual Speed potentiometer allow manual control of the system in the event that the automatic control signal fails. An Isolated Process Follower prevents a single variable speed drive failure from affecting the control logic and other connected drives. A ground fault interrupter (GFI) shuts down the VFD without damage if a motor insulation failure should occur while operating. Specification of required input voltage operational limits over a wide range eliminates nuisance shutdowns due to high or low input voltage. A Run light, a Fail light, a Speed meter, and a Running Time meter may be useful additions. Current and voltage meters are commonly specified, but rarely used by the operating staff.

4.10 POWER FACTOR

Variable speed drive manufacturers commonly emphasize power factor if they sell PWM, and don't mention it if they sell VVI or CSI equipment. Variable speed drives of any type have no appreciable impact on the power factor of a facility even when they represent up to 50% of the maximum connected load, because the reduction in power factor is accompanied by a steep reduction in the KW used by the equipment. For example, at half speed a VVI drive will exhibit a .5 power factor, and a CSI drive will have a .25 power factor, but both will draw 1/8 the KW of full speed operation. Power factor correction capacitors must never be used on the input to an adjustable frequency drive, or on a motor controlled by an adjustable frequency drive.

4.11 SERVICE

The single most important specification item is the availability of factory trained field service engineers or technicians and parts support on site or within a few hours travel of the installation. The type of equipment supplied, the mains-to-water efficiency, and the motor noise level become insignificant when the variable frequency drive fails and the owner begins to exhibit a strong desire to have the equipment restored to operation quickly. A modest background in electronic power control systems, basic electronic test equipment, complete documentation, and access to spare parts are normally required to return a faulty VFD to service.

4.12 AVOIDING THE NEED FOR SERVICE

VFDs can run indefinitely if the following protective measures are employed:
- Control the temperature. Increasing the temperature 10 degrees C reduces the equipment life approximately 50%.
- Control the humidity. Condensing moisture on the interior surfaces will reduce the equipment life.
- Keep it clean. The accumulation of dust may allow arcing, or an increase in the operating temperature of a component.
- Protect it from electrical transients. Electrical spikes and surges may reduce equipment life to a fraction of its potential.

5.0 CONTROL LOGIC

An adjustable frequency drive may be controlled by a manual speed potentiometer, by a voltage or current signal through a follower circuit, or by a setpoint controller. Most pumping applications use a parameter sensor and a setpoint controller that compares the parameter sensed with the desired set-point, then changes the pump speed to return the measured parameters to the desired value. The setpoint controller is commonly referred to as a PID controller, for the functions of proportional, integral, and derivative.

5.1 LOGIC LIMITATIONS

Control logic is limited only by the imagination of the specifier and the laws of physics, but the engineering philosophy of keeping the system as simple as possible is highly recommended. A standard control module simplifies submittal, manufacture, startup, and maintenance.

5.2 MINIMUM REQUIREMENTS

The minimum features required for single pump control are process variable display, setpoint, proportional, and integral. Multiple pump controllers require provisions for addition and subtraction of pumps, including minimum run timers and adjustable hysteresis. System gain may have to be modified as a function of the number of pumps operating to avoid instability.

5.3 COMMON CONTROL PANEL OPTIONS

The most common options include RUN and FAULT indicators, and RUNNING TIME meters. LOW SUCTION or HIGH PRESSURE indication with alarm may be required. FLOW or KW indicators may be specified.

5.4 ENERGY SAVING OPTION

Best efficiency control may be incorporated in multiple pump control systems when local power costs are such that energy savings offset the increased complexity. The logic controller monitors system head, flow, and input KW. The mechanical work in KW is calculated from the head and flow data, then divided by the input KW to develop the system efficiency. The system efficiency is then compared to the range of system efficiencies attainable for the current head and flow conditions with different numbers of pumps operating. The number of operating pumps is then controlled to optimize the system efficiency. Minimum run timers are incorporated to preclude excessive cycling. The cost effectiveness of this option depends on the local power rate and the KW rating of the pump motors.

5.5 PROCESS OPTIMIZATION

The end product of a process may be of sufficient value that the major goal of the entire logic system will be to maximize the production of that product. The logic controller for this type of application will commonly be a PLC (programmable logic controller) with a number of parameter inputs that have been programmed for dedication to a specific product line. For example, production of a long chain polymer may require that sheer conditions be avoided at all times. Use of variable speed pumps instead of control valves may minimize sheer stress, and may increase the percentage of useable product from a given volume of feed stock.

6.0 SENSOR BASICS

The sensor measures the process variable and sends an analog or digital signal to the control logic. The quality and cost of the sensor are functions of accuracy, repeatability, ability to withstand environmental conditions, durability, signal interface provided, and mechanical installation features provided. The location of the sensor in the system affects system response and efficiency. The parameter measured must be related in a direct way to pump operation.

6.1. SENSOR QUALITY

An industrial quality sensor provides years of trouble free operation at an extremely modest cost per year. Commercial grade sensors are not recommended, as the reliability and durability tend to be far less than the equivalent industrial devices. The industrial pressure sensors are rated for +/- .25% accuracy with overpressures to 7,000 KPa or more, while the commercial devices are commonly rated +/- 1% with overpressure ratings of 1.5 x design pressure. When the transmitter is sensing 55 KPa differential pressure and the system static pressure exceeds 900 KPa, the potential for damage to the commercial unit during installation or maintenance becomes apparent.

6.2 SENSOR LOCATION

A guage pressure sensor should be mounted at the highest and most remote point possible to minimize power consumption, with a hydro-pneumatic tank in the immediate vicinity to moderate pressure surges. A differential pressure sensor is commonly used for control of closed loop heating and cooling, and one should be mounted at the most remote point on each loop in the system, typically between the supply and returns lines at the last fan-coil unit or air handler.

6.3 TEMPERATURE SENSING

Controlling the speed of a pump based on water temperature or differential temperature is not recommended. Controlling differential pressure between supply and return to minimize system energy, and then using a separate control loop for temperature is effective. Attempts to control pump speed for heating and cooling based on differential temperature only have resulted in the temperature sensors controlling accurately, but a majority of the facility served reporting no temperature control because providing water to only the first air handler satisfies the loop differential temperature control without providing adequate pressure to serve the remainder of the facility.

7.0 COUPLINGS AND SHAFTS

The mechanical elements that couple the variable speed motor to the pump are an important part of the system. The wide range of operational speeds available increases the potential for damaging resonant conditions. Rigid couplings transmit vibrations originating in the non-sinusoidal current waveform directly to the pump, and occasionally transfer them to the water and throughout the building, while elastomeric inserts of incorrect durometer may actually amplify vibration at certain RPMs and loads. Flexible shafting exhibits the normal critical speed problem, but also exhibits a torsional vibration phenomenon related to the pump WR2 and number of impeller blades that extends over a broad RPM range. The solutions include dynamic balancing with a variable speed balancer, torsional analysis, and careful selection of elastomeric materials.

8.0 AC POWER SUPPLIES

The mains represent a significant portion of the system. The generating station, sub-station, distribution transformer, power lines, power factor correction equipment, and emergency generator may all interact with the variable frequency drive. Other equipment installed on the same power distribution system may affect the variable speed drives, or be affected by them, even if line chokes or isolation transformers are provided.

8.1 SOURCE KVA EFFECTS

An AC power source must have adequate KVA for the drives installed, but care must be taken to ensure that the supply KVA and the input impedance of the variable speed drive are compatible. Variable speed drives draw current from the mains in pulses from zero to about half load, and then may go into continuous conduction. The magnitude of the current pulse is a function of the variable speed drive input impedance, the line voltage, the internal impedance of the source, and the impedance of the power line. A low impedance source with a short power line to the variable frequency drive may blow fuses or damage input devices due to the magnitude of the current pulses developed. For example, a 1000 KVA supply transformer with a 5 KW variable speed drive connected by 50 feet of power cable typically requires line chokes or an isolation transformer to prevent damage to the drive input circuitry, while a 140 KVA transformer with a 5 KW drive connected probably needs no additional protection. Consult the drive manufacturer for specific recommendations.

8.2 HIGH LINE VOLTAGE

Variations in the electrical load cause wide variations in the AC line voltage, particularly in industrial areas that do not operate three shifts seven days. The source voltage and internal impedance of the generator remain essentially constant, but the line voltage drop due to current flow decreases as the square of the reduction in load. Night and weekend voltages may rise 20% or more, and may cause VFDs to trip off on over-voltage or blow input fuses. Switching of generators and power factor correction capacitors during this low load period may also generate more significant electrical transients which may damage equipment. The use of an isolation transformer with taps to allow selection of a lower secondary voltage is an effective way to deal with this problem.

8.3 THE ENGINE POWERED GENERATOR AS A CHALLENGE

The application of an engine powered generator as a source of AC power to static power conversion devices, including variable frequency drives, uninterruptible power supplies, and solid state welders, must be approached with caution due to the pulse nature of the current draw under lightly loaded conditions. If the intent is to power a variable frequency drive as the sole load on the generator, the ability of the regulator to handle sub-cycle current events becomes significant. Many generator regulators cannot effectively control the engine in the absence of a base sinusoidal current. A constant speed motor or a resistive load bank may be used to establish a base current in order to stabilize the generator. The limited KVA available and the high source impedance allow significant voltage notches to develop from the operation of the conversion stage of the variable frequency drive, which may cause the drive to set its low line voltage monitor and shut down.

8.4 MUTUAL INTERFERENCE WITH DRIVES

Variable frequency drives connect to a common AC power supply may interact and affect the firing of phase control rectifiers in the converter stage of VVI and CSI drives, or the conversion efficiency of PWM drives, due to line notching with limited KVA sources.

8.5 LIMITING PEAK CURRENTS AND INTERFERENCE

Both peak currents and mutual interference may be controlled by the addition of impedance between the AC power source and the variable frequency drive. This may be accomplished by using a line choke or an isolation transformer at the input to the VFD. Many variable speed drive manufacturers provide line chokes and bus chokes as standard equipment that limit current peaks and conducted interference, and are also suitable for high source KVA applications. Where line notching must be held to 10% or less to protect transient sensitive equipment, an isolation transformer may be used in conjunction with a line choke.

8.6 PROTECTING COMPUTERS

Computing devices are adversely affected by electrical transients from any source, including variable frequency drives. Although occasionally requirements to protect 1 KW computers by filtering 100 + KW drives have been encountered, the more cost-effective solution is to provide a UPS (uninterruptible power supply) for the sensitive low KW device. The UPS isolates the computer from all electrical transients regardless of source by rectifying the AC power and using it to charge a battery. The battery power is then inverted and filtered to provide a protected source of standard frequency AC power. The battery power also provides a few minutes of operation after AC line power loss to allow a computer to store files and shut down in an orderly fashion.

8.7 PROTECTING VARIABLE FREQUENCY DRIVES

The primary cause of failure for variable frequency drives is an electrical voltage transient. The source of the transient may be an electrical storm, the power company, or the normal operation or failure of other electrical equipment. There are four levels of protection that are generally available.

1. Resistive/Capacitive (RC) network
2. Metal Oxide Varistor (MOV)
3. Gas Discharge Tube
4. Hybrid Series/Parallel

8.8 PROTECTIVE DEVICE CAPABILITIES

The hybrid series/parallel device is capable of intercepting lightning related transients, 50,000 amps and 10,000 joules, and includes a low resistance parallel path to earth (gas discharge tube) for up to 200,000 amp surges. The series element is an inductor that limits rate of rise to the secondary shunt devices, which provide a parallel path to the AC power neutral connection. The gas discharge tube acts as a parallel device only and is similar in rating, 50,000 amps at 10,000 joules, but provides less isolation. It is adequate when there are sufficient users on the power line to limit the total energy delivered to the protective device. Metal oxide varistors are solid state parallel devices available for over 100 joules, but are a consumeable device that will handle 1 large surge in the 2000 amp range or millions of smaller surges prior to failing short. The resistive/capacitive network acts as a parallel filter that limits the rate of rise of transient waveforms. It is suitable for short, fast rise time transient of limited amplitude, but fails to regulate serious disturbances.

7.5.2 PROTECTION DEVICE COSTS

Resistive/capacitive networks or line-to-line metal oxide varistors (MOVs) are commonly provided as standard equipment on the input circuits of variable frequency drives. Line to neutral MOVs are available as an option for about £200. The parallel gas discharge device is available separately for about £800. The hybrid series/parallel device is used primarily at the electric service entrance to a facility, and varies in cost from £2,000 to £20,000 depending upon system amperage.

9.0 SYSTEM RESPONSIBILITY

The successful application of variable frequency drives to pumps is a multifaceted combination of art and science that approximates a symphony orchestra. Like a symphony, the best results are achieved when an experienced director is leading. The individual or firm to be charged with system responsibility should be defined clearly in the specifications. Qualifications based on years of experience, number of similar operational systems, availability of parts and service, and financial responsibility should be used to limit the potential bidders.

10.0 CONCLUSION

Thousands of variable frequency drives are currently in operation on pump systems throughout the world, but most pumps are still operating constant speed. Every low static high dynamic head pump application over 20 KW represents a significant opportunity to improve performance and save energy. Those opportunities can be made into reliable operating systems by referring to the above information and using is as a chart to guide you around the bars and snags and into the safe harbor of effective conservative VFD pump systems design.

Figure 1

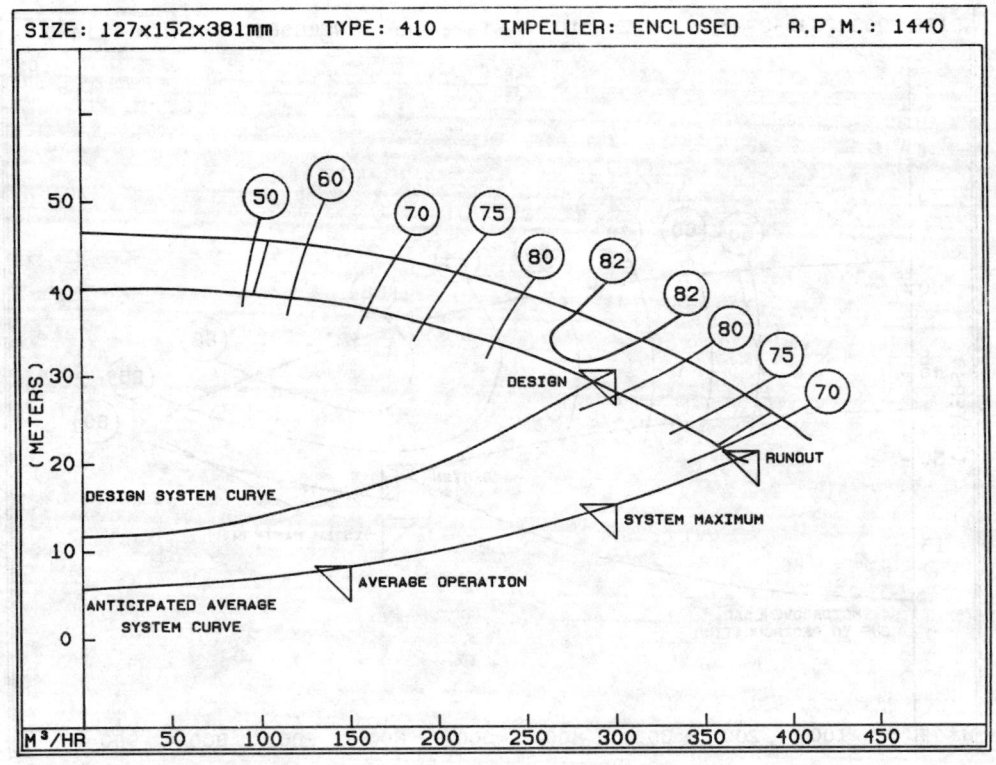

Figure 2

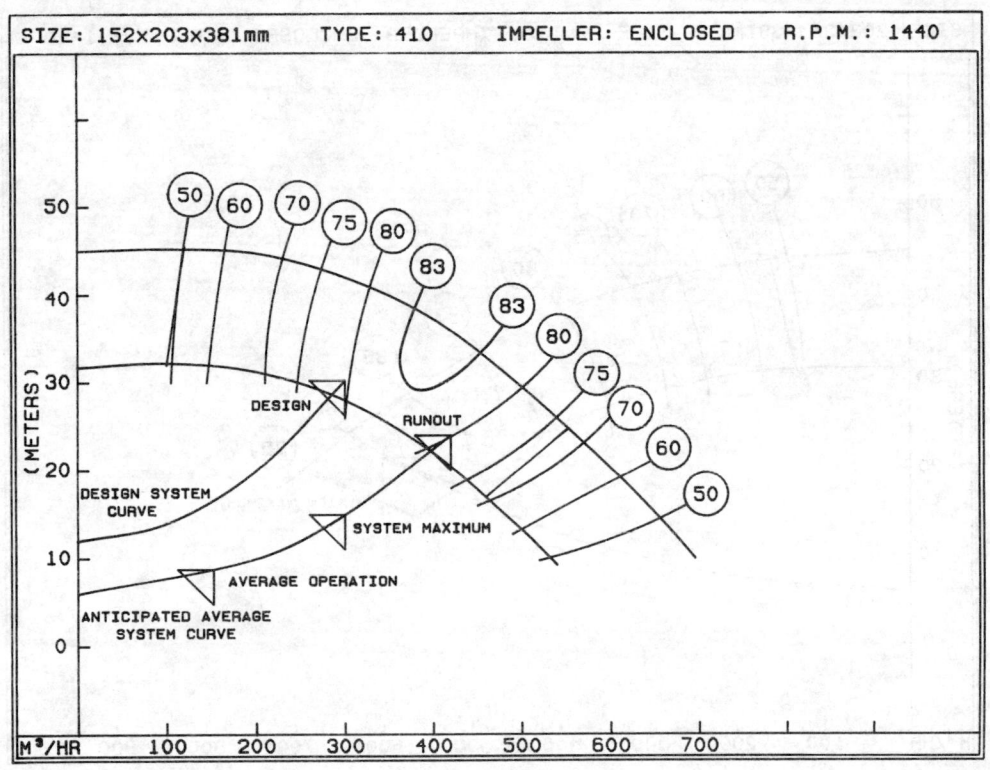

Figure 3

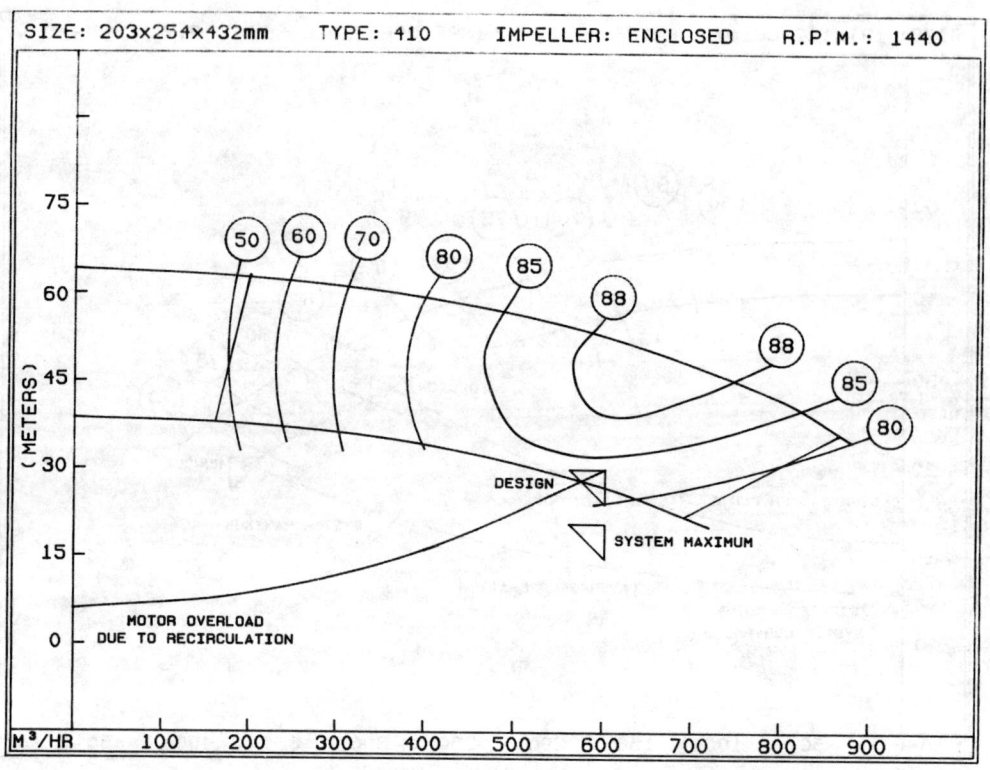

Figure 4

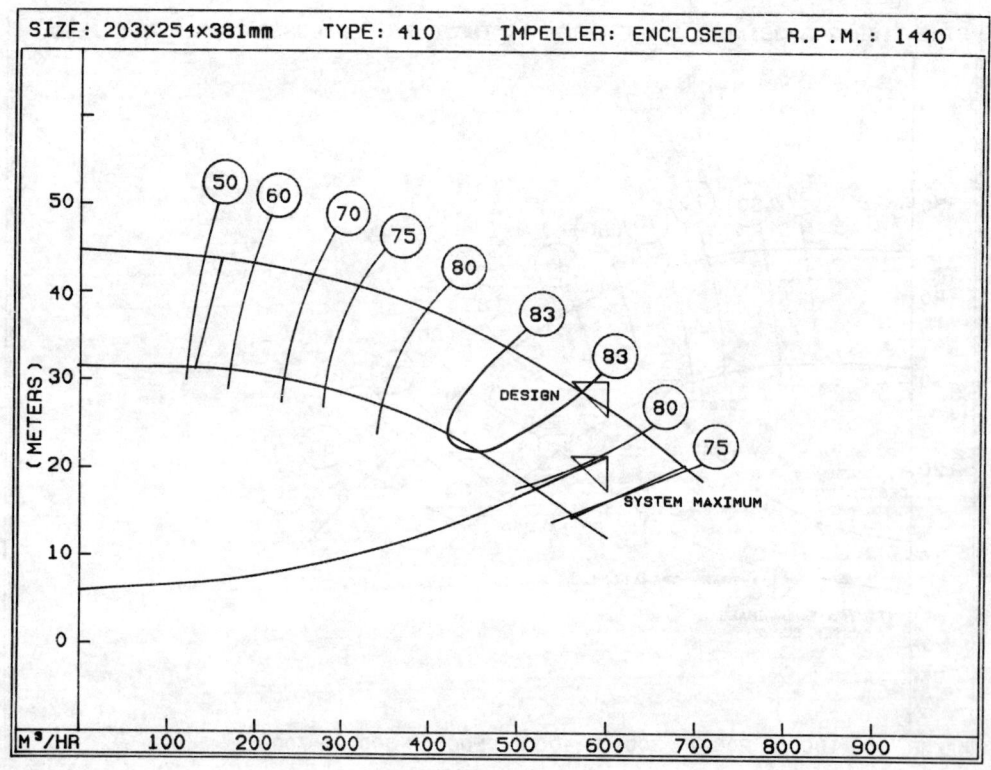

Figure 5

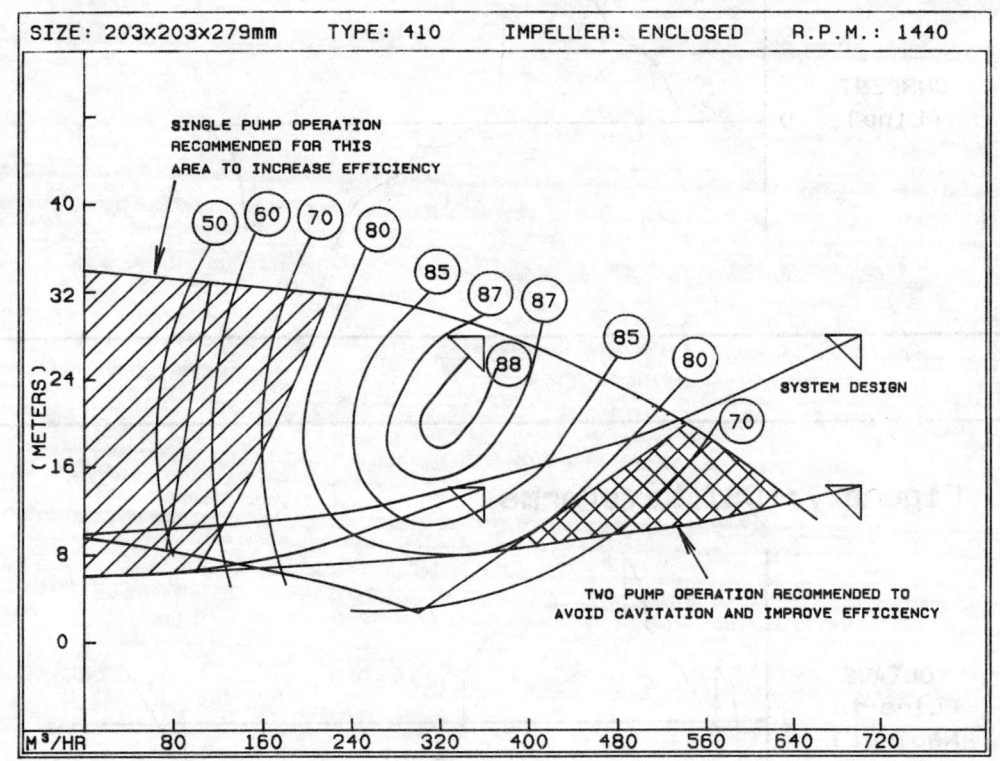

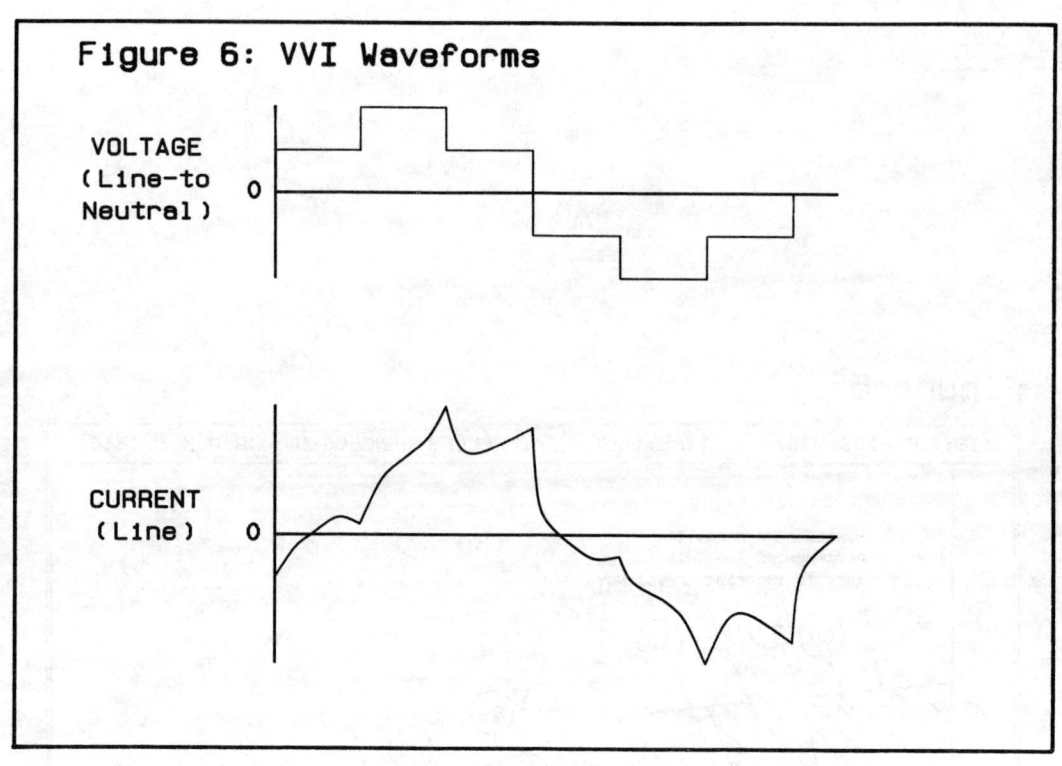

Figure 6: VVI Waveforms

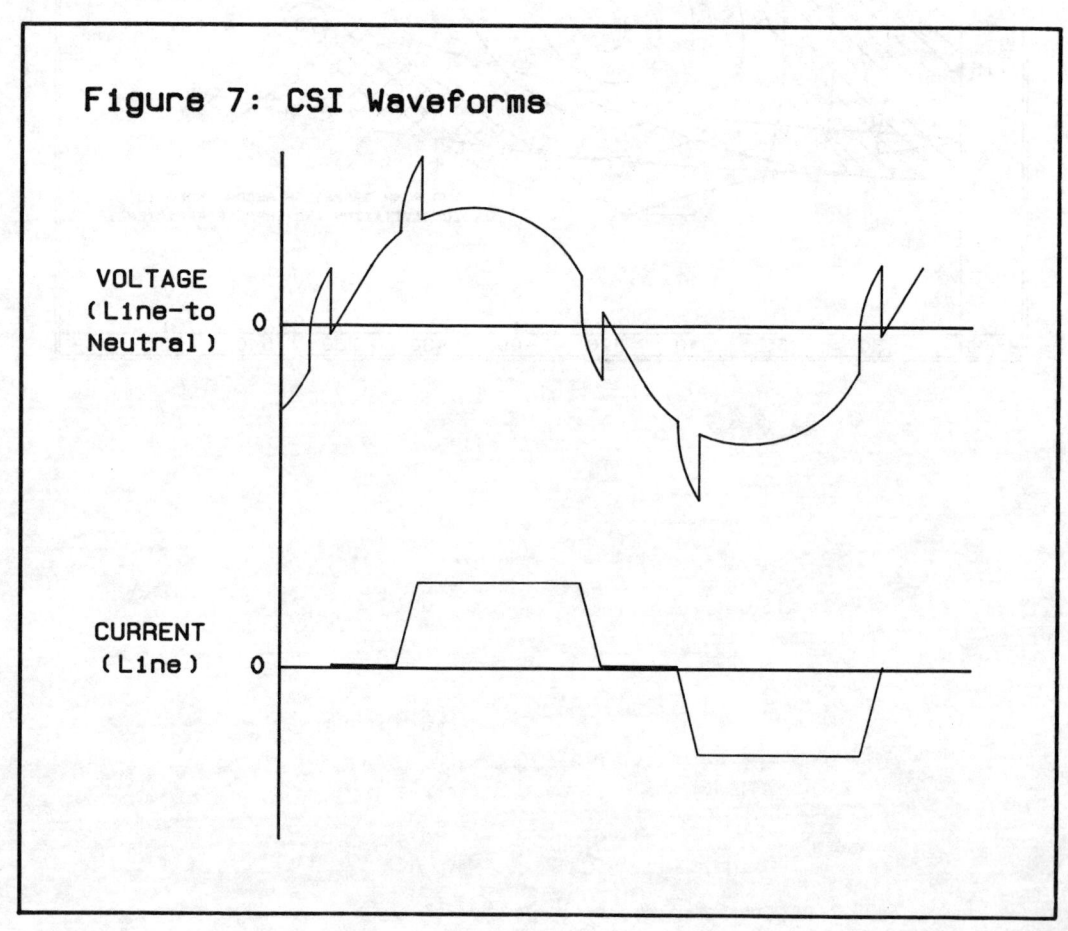

Figure 7: CSI Waveforms

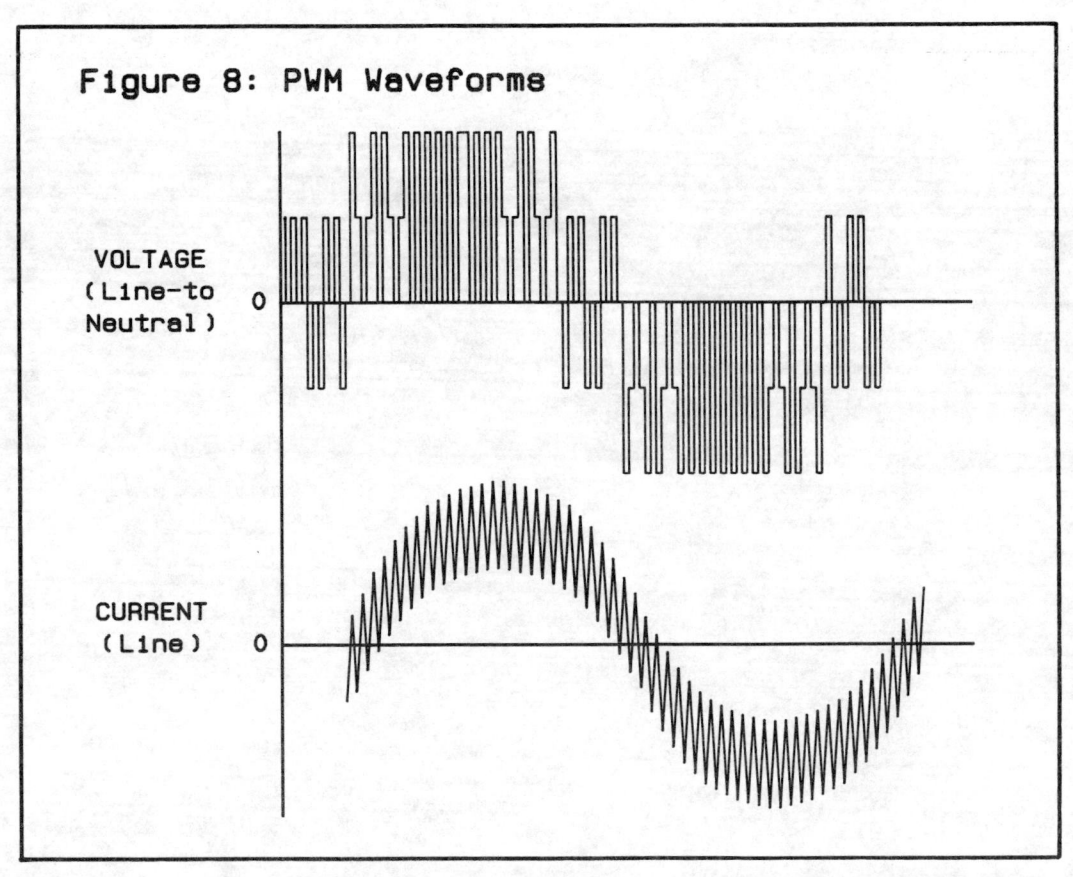

Figure 8: PWM Waveforms

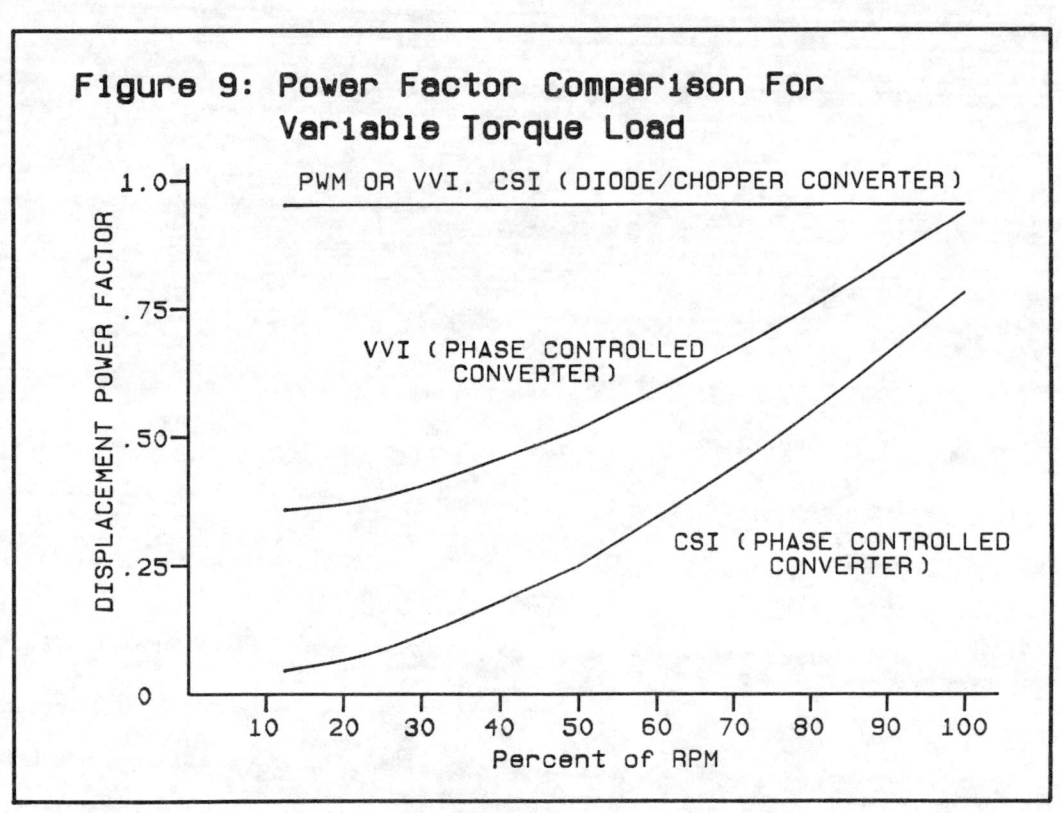

Figure 9: Power Factor Comparison For Variable Torque Load

11th International Conference of the
British Pump Manufacturers' Association

New Challenges – Where Next?

18-20 April, 1989
Churchill College, Cambridge

PAPER 12

ON A PILOT PUMP USING THE WEIS-FOGH MECHANISM

Michihisa Tsutahara and Takeyoshi Kimura
Department of Mechanical Engineering
Kobe University, Rokko, Nada, Kobe 657 JAPAN

Summary

A pilot pump using an efficient lift-generating mechanism found in hovering flights of small insects, which is called the Weis-Fogh mechanism, was built and its characteristics were studied experimentally and it was shown that this mechanism works well for pumps. The maximum efficiency was larger than 45%. A flow visualization about the wing was performed and clusters of vortices were found to be produced periodically in the wake. Some possibilities for raising the efficiency were discussed.

1. Introduction

The Weis-Fogh mechanism, which was found by observing the hovering flight of a small insect called Encarsia Formosa, is a novel and efficient mechanism of lift generation. Some investigations on this mechanism itself have been done as a problem of fluid mechanics(1)-(4), but only few for its engineering applications. Ffowcs Williams have applied this mechanism to an axial pump(5), and Tsutahara and Kimura have used its two-dimensional model for the propulsion of ships(6). We have applied the same model to a pump and we shall describe here the characteristics of a pilot pump.

The model mentioned above is shown in Fig.1. Two wings move in a square channel as shown from (a) to (h), and the direction of the flow is downward in the figure. The detailed wing motion is as follows. A point p located between the leading edge and the trailing edge (closer to the leading edge) of each wing reciprocates in the direction perpendicular to the flow. In (a) the two wings rotate up to some angle α about the common trailing edge, which slides upstream as the point p moves rightward. In (b) the wing translates with the angle α being fixed, and in (c) the wing rotates about the leading edge. The motions in (e) to (h) are the same as those in (a) to (d), except that the direction of the motion is opposite. Considering the images due to the side walls of the channel, the circulation is generated immediately around the wing in (a) and (e), by which the wing works very efficiently.

Conference organised and sponsored by the British Pump Manufacturers' Association
in conjunction with NEL and BHRA, The Fluid Engineering Centre.
Co-sponsored by the Process Industries Division of the Institution of Mechanical Engineers.

2. Experimental apparatus and method

2.1 The pilot pump

The apparatus of the experiment is shown in Fig.2. The frame of the pump is made by plywood of 21 mm thick, and consists of a water supply tank, a pump section, and a head tank. Width of the channel is uniform and 70 mm. It is difficult to keep airtight inside the equipment, therefore the water is pumped up from a pit by another pump to the water supply tank. The water level in the tank is kept constant by a weir inside it. The pump section is 60mm in height and 550mm in length, and there is the wing moving section in its center. Above the wing moving section, an acryl plate is placed and the flow about the wing can be observed. The head tank is for keeping the head of the pump constant. At the end of the channel is an orifice, by which the flow rate is measured and controlled.

A wooden wing of chord length 55mm, span 55mm, and thickness 5mm, is set in the pump section. The motion of the wing as shown in Fig.1 is performed as follows. A fly wheel of diameter 100mm is attached to a DC motor(14W), and a cam follower is located at a point 30mm from the center on the fly wheel as shown in Fig.3. This cam follower moves inside the grooved cam and a slide unit connected with the cam is given a reciprocal motion along the rail laid in the direction perpendicular to the channel. A push rod of 4mm diameter is also connected with the slide unit and it drives the wing. The motor is controlled by a controller to keep constant speeds even when the load changes.

On the other hand, a thin hole is opened in the span direction of the wing at the point 15mm distant from the leading edge, which corresponds to the point p in Fig.1. A 2mm diameter pin of stainless steel is put through the hole and the wing is freely rotatable about the pin. The pin is connected with the push rod stated above. When the push rod moves the wing rotates and opens up to a certain angle by the moment about the pin acted by the fluid. The opening angle is determined by the length of the hole which is cut as shown in Fig.4. When the wing approaches to the channel wall the wing rotates and closes because the leading edge touches the wall. The motion of the wing is shown in Fig.5, and the same is true in the motion of the opposite direction.

2.2 Experimental procedure

The experiment was performed for two opening angles 30 deg and 45 deg. The reciprocating speed of the wing motion was chosen from 3 to 5/sec (which means that the wing reciprocates the channel three to five times per second). The procedure of the experiment was as follows. The rotating speed of the motor was set to a certain reciprocating speed of the wing. The width of the orifice was varied from 6mm to 22mm and the head was read from the difference between the water levels of the head tank and the water supply tank. After this process was over, the rotating speed of the motor was changed and the same process was continued.

The efficiency of the pump η is defined as

$$\eta = L_w/W \times 100 \; (\%)$$

where L_w is the water horsepower, W is the brake horsepower or the power input to the wing. The water horsepower is also defined as

$$L_w = g\rho Q H$$

where g is the gravity acceleration, ρ is the density of water, Q is the flow rate or the discharge, and H is the total head. The power W is obtained by measuring the voltage and the electric current into the motor and multiplied by the efficiency of the motor and the mechanical loss is subtracted.

2.3 Flow visualization

A flow visualization about the wing was also performed using sawdust (small wooden particles) as tracers. The tracers were thrown into the water supply tank and the flow patterns were taken by a video camera.

3. Results

3.1 Characteristic of pump

The characteristics of the pump for the opening angle 45 deg and the reciprocating speed of 3, 4, and 5/sec are shown, respectively, in (a), (b), and (c) of Fig.6, where the flow rate is taken in the axis of abscissa and the total head, the power to the pump, and the efficiency are taken in the axis of ordinate.

In all cases, the total head and the power decrease as the flow rate increases. On the other hand the efficiency increases up to some maximum value and decreases as shown in (b) and (c) of Fig.6. In (c), the maximum value was not detected because of limit of this equipment, but it would surely exist and would be higher than 45%. The experiment using another wing with a larger cut (7) shows that the peak of the efficiency curve moves towards larger flow rate as the reciprocating speed increases. The tendency will be the same for the present case.

Although the results are not presented here, the experiments are performed for the opening angle 30 deg. The efficiency for 30 deg is less than that for 45 deg in all the cases.

At the same flow rate, as the reciprocating speed increases the total head and the power increase also, but the efficiency decreases. This means, especially as to the efficiency, that this type of pump has high efficiency when the flow rate is large and the reciprocating speed is rather slow and inevitably the total head is small. The extreme case is the previously reported ship's propulsion mechanism in which the maximum efficiency is up to 80%. Of course the difference between the pump and the ship's propulsion is whether the pressure raise exists or not and this reflects the ratio of the moving speed of the wing to the flow velocity V/U at which the maximum efficiency takes place. For the propulsion mechanism, this value is less than 0.5 (6), on the other hand for the pump in the present case this is around 1.3. The higher value of this ratio will lead strongly disturbed flow in the wake of the wing. This will be considered briefly next with this ratio as a parameter.

3.2 Flows around the wing

Some sketches of the flows past the wing are shown in Fig.7 and Fig.8. The flow shown in Fig.7 is for the case of the ratio V/U = 0.3, and the flow is smooth. This type of flow, however, is not realized in the pump but in the ship's propulsion mechanism, and this is for the sake of comparison. In Fig.9, the flow for V/U = 1.3 is shown. Clusters of vortices are found to be produced periodically in the wake of the wing and the flow is meandering strongly. The clusters move downstream when the ratio is not so large, but they do not move smoothly until the next cluster is produced when the ratio is larger than 1.5. This corresponds to the low efficiency for the case of small flow rate.

4. Discussions

Even when the efficiency is the highest, the flow has strong disturbances unlike that in the ship's propulsion mechanism. But the disturbances are rather periodic and synchronized with the wing motion. Therefore the disturbances can be weakened and the efficiency can be raised by setting two or more wings in series in the channel and controlling their motion.

As to the opening angle, for the propulsion mechanism the efficiency is high for small opening angle, say around 15 deg. But for pump the efficiency is higher in 45 deg than in 39 deg. Optimum opening angles correspond to the speeds ratio and they are rather different between the pump and the ship's propulsion equipment.

This pump is a kind of tangential pump, but the mechanism is completely different from any existing pump. The weakest point of this pump is that the wing has a reciprocating motion. We are now considering a rotating type, not reciprocating type, using the same principle of the Weis-Fogh mechanism.

5. Conclusions

A pilot pump using the Weis-Fogh mechanism is made and the characteristics of the pump is studied. The conclusions obtained are as follows.
(1) The complicated wing motion is realized by a rather simple mechanism and it is shown that the Weis-Fogh mechanism is applicable to pumps.
(2) The moving speed of the wing when the efficiency is maximum is about 1.3 times flow velocity.
(3) The maximum efficiency was higher than 40%, but a full scale model will, of course, have higher efficiency.
(4) There are clusters of vortices produced periodically in the wake.
(5) Performance of this pump will be increased by equipping several wings in series.

6. References

1. Weis-Fogh,T.: "Quick estimates of flight fitness in hovering animals, including novel mechanism for lift generation". Journal of Experimental Biology, 59, 1973, pp.169-231.
2. Lighthill,M.J. : "On the Weis-Fogh mechanism of lift generation". Journal of Fluid Mechanics, 60, 1973, pp.1-17.
3. Maxworthy, T.:"Experiments on the Weis-Fogh mechanism of lift generation by insects in hovering flight. Part 1. Dynamics of the 'Fling'". Journal of Fluid Mechanics, 93, 1979, pp.47-63.
4. Edwards,R.H. and Cheng,H.K.:"The separation vortex in the Weis-Fogh circulation-generation mechanism". Journal of Fluid Mechanics, 120, 1982, pp.463-473.
5. Furber, S. B. and Ffowcs Williams, J. E.:"Is the Weis-Fogh principle exploitable in turbomachinery?" Journal of Fluid Mechanics, 94, 1979, pp.519-540.

6. Tsutahara,M. and Kimura,T.:"An application of the Weis-Fogh mechanism to ship propulsion". Journal of Fluids Engineering, Transactions of the ASME, 109, 2, 1987, pp.107-113.
7. Tsutahara, M. and Kimura, T.:"A pilot pump using the Weis-Fogh Mechanism and its Characteristics". Transactions of the JSME, 498,1988, pp.393-397. (In Japanese).

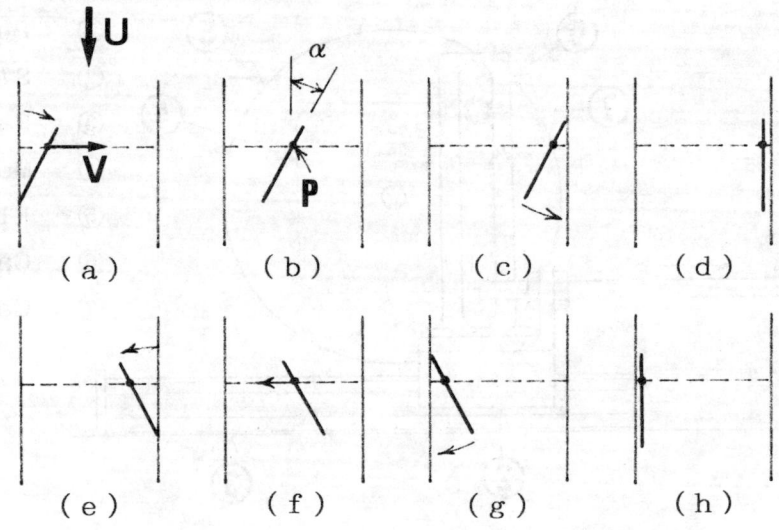

Figure 1:   Model of wing motion

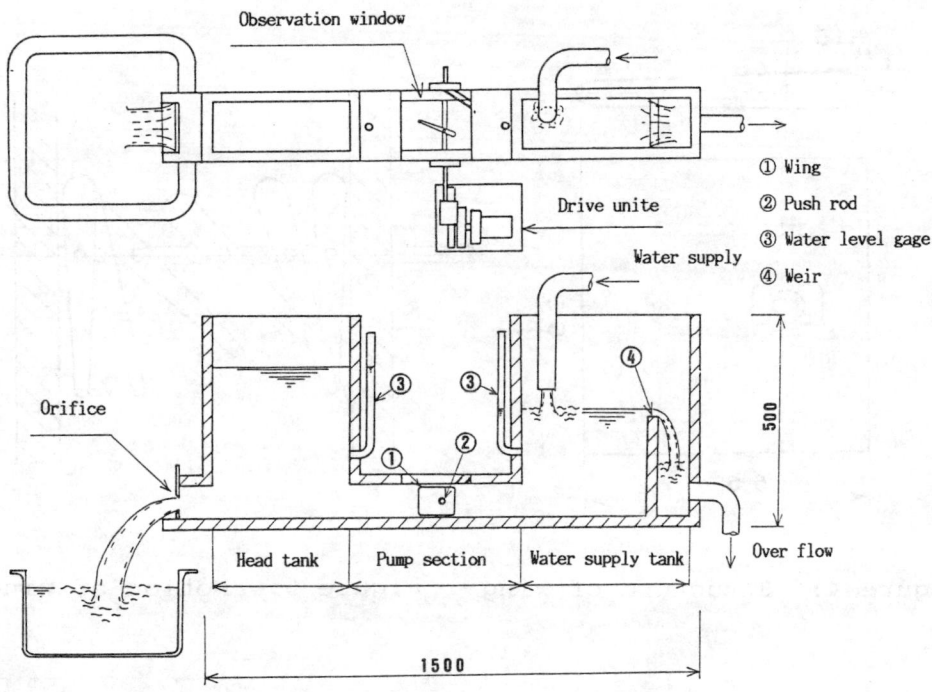

Figure 2:   Schematic of experimental apparatus

181

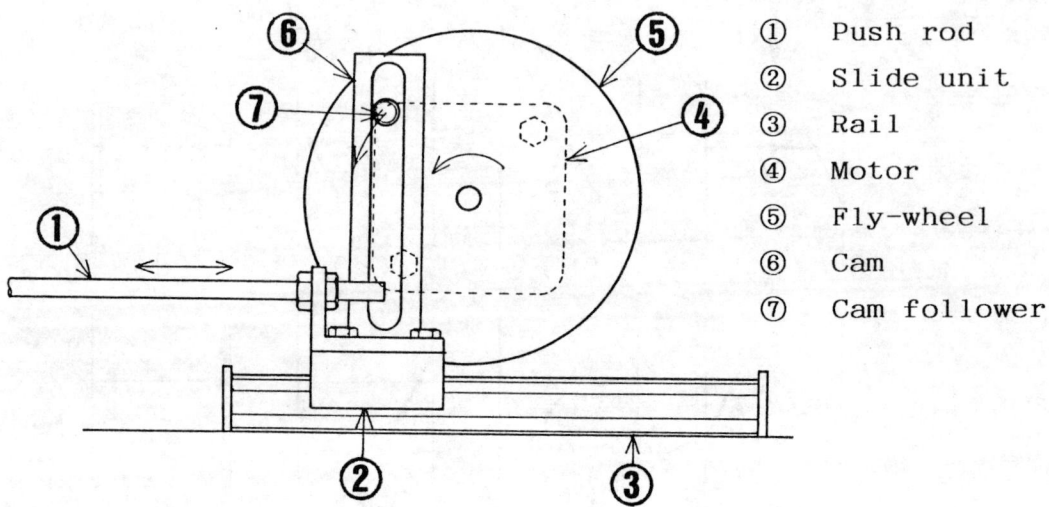

Figure 3: Wing drive unit

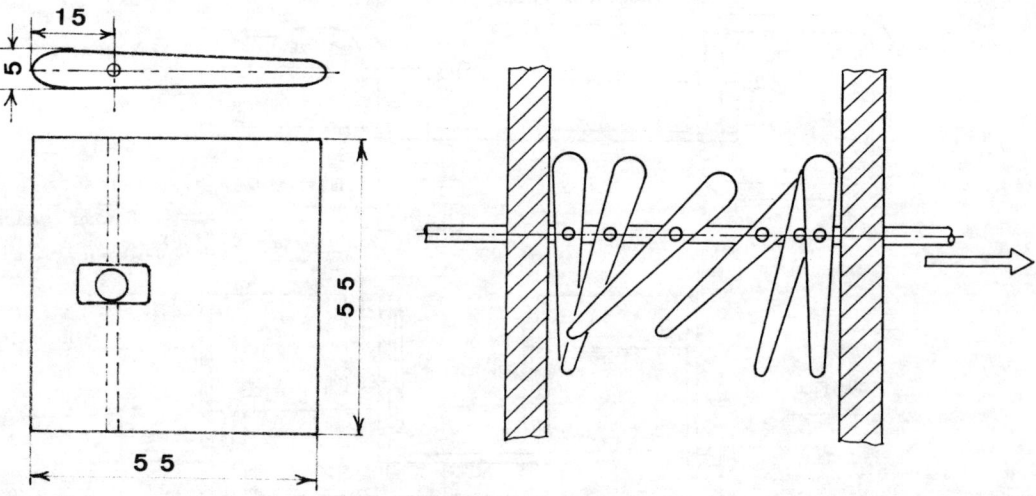

Figure 4: Structure of wing   Figure 5: Motion of wing

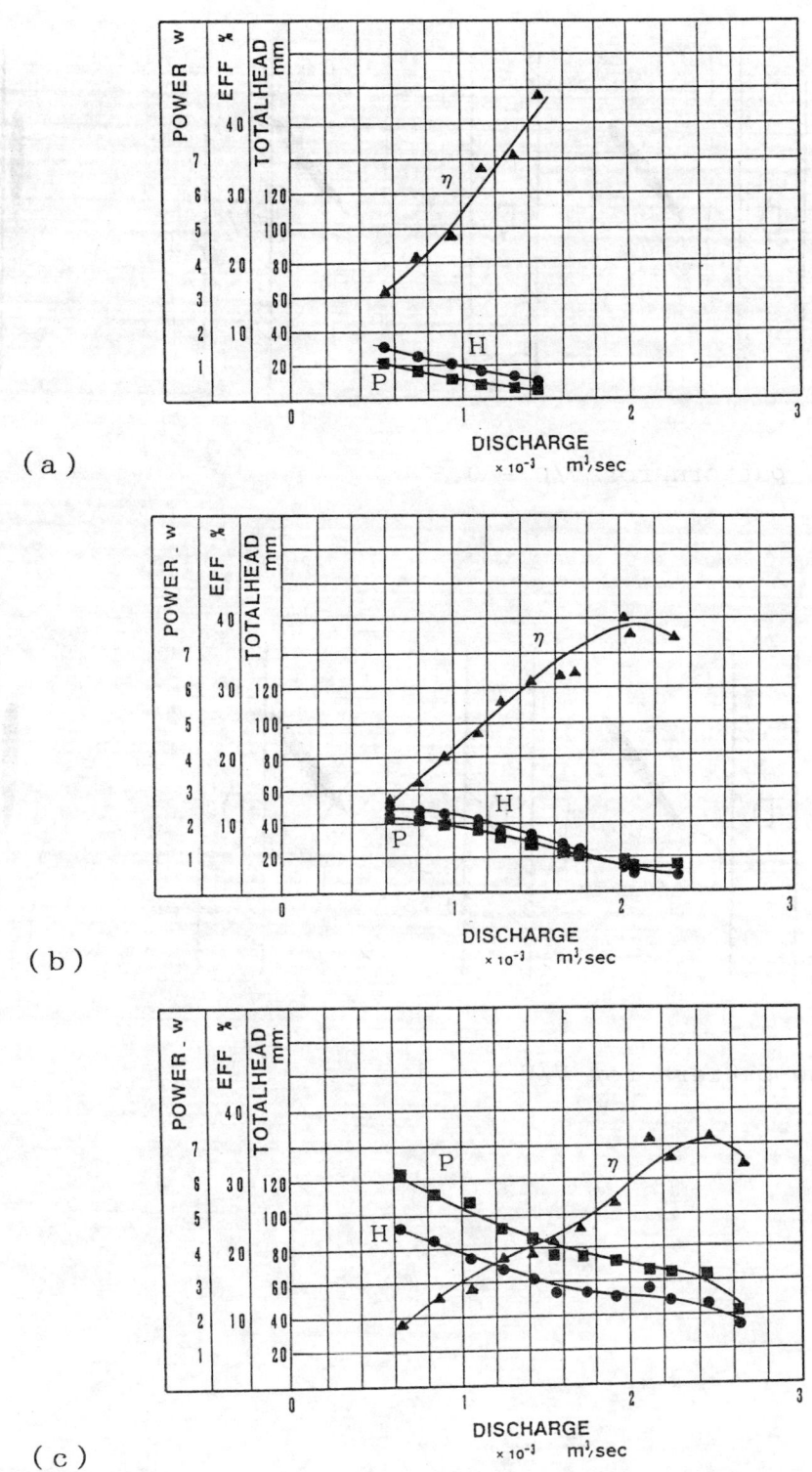

Figure 6: Characteristic curves for opening angle 45 deg
(a) Reciprocating speed 3/sec (b) Reciprocating speed 4/sec
(c) Reciprocating speed 5/sec

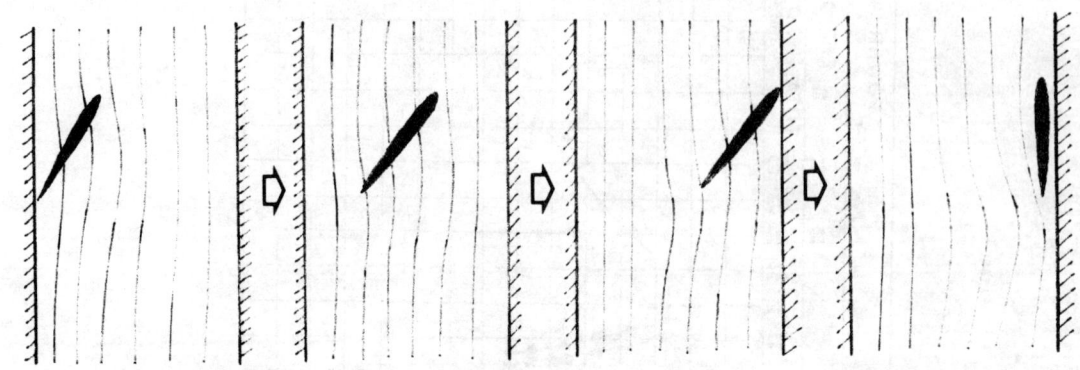

Figure 7:   Flow pattern for V/U = 0.5

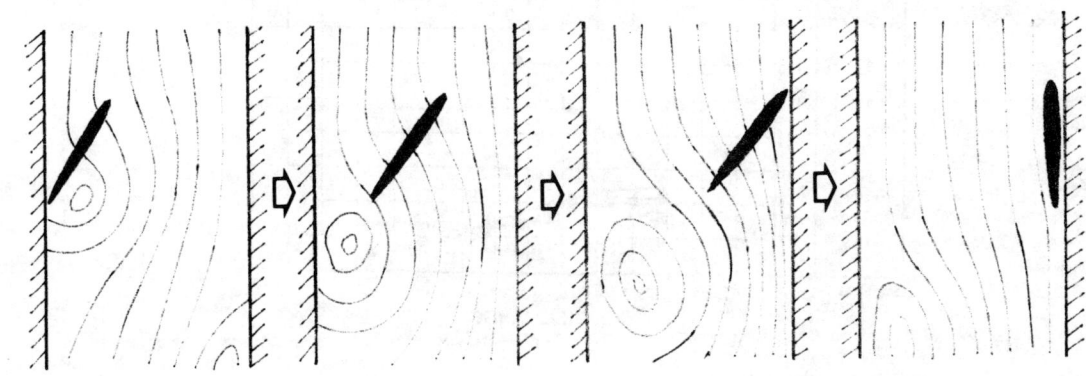

Figure 8:   Flow pattern for V/U = 1.3

11th International Conference of the
British Pump Manufacturers' Association

New Challenges – Where Next?

18-20 April, 1989
Churchill College, Cambridge

PAPER 13

HRC – A NEW SEALING CONCEPT
FOR STANDARDIZED PUMPS IN THE CHEMICAL INDUSTRY

U. Reinfrank and H. Nowak,
FEODOR BURGMANN Dichtungswerke GmbH & Co., F.R. Germany

Summary

This report demonstrates that the presented seal concept is the logical consequence of

- The development history of the "Norm" mechanical seal

- Experience obtained with the "Norm" mechanical seal

- The current requirement definition for a new "Norm" Seal

- Experience obtained with pumps with seal optimized stuffing boxes

- Subsequent development and testing

The evident advantages of this concept and it's ready acceptance by reputable end users recommends it's further use and adoption by more original equipment manufacturers. (O.E.M.)

Historical development of the standard (norm) mechanical seal

The use of this machine element, known since the 1920's was at first limited by narrow constraints.

The development and application of a wide range of seal face materials, together with demands for lower leakage rates, gave rise in the sixtees to the breakthrough of the substantially maintenance free mechanical seal.

As a result of the increased use of mechanical seals there came (firstly from the end user, and then from the original equipment manufacturer) a desire to standardize dimensions for installation and connection.

Defined objectives

- Optional application of stuffing box packings or mechanical seals

- Interchangeability between seals from different manufacturers without modification of the device being sealed.

These objectives resulted, within West-Germany in the distribution of the VDMA 24 960 recommendations of March 1970. These recommendation were readily accepted and applied as a standard.

The publication in November 1976, of the official standard DIN 24 960 included the content of VDMA 24 960 as standard version N (= normal) and added the optional K (= short) seal version with reduced axial length; all other dimensions remaining unchanged.

The revised DIN 24 960 edition of June 1980 contained additionally the dimensions for a axial lock of the stationary seat. The stuffing box dimensions for a double seal installation with two single seals in "back to back" arrangement.

In most West-European countries there exists either a standard corresponding to DIN 24 960 or one of the standards is adopted and treated as a European "Norm".

The incorporation of DIN 24 960 into other standards e.g. ISO 2858 and correspondingly DIN 24 256, it's distribution and significance has been further increased.

Consequences of mechanical seal standardization

The rapid acceptance and widespread use of the "Norm Seal" has brought indisputable advantages.

- Interchangeability of different brand mechanical seals

- Ready availability

- Reduction in the number and variety of seal housings adaptors and connections

- Price reduction in spite of improvements in design and materials of construction, through simultaneous expansion of the operating limits for the mechanical seals.

Equally indisputable however is, that the existing standards prevented (at least within "Norm Pumps") the development of the optimum machine/seal combination.

It may be remembered that during the seventies, the trend towards lighter pumps and simultaneously more demanding duties led to many instances of premature seal failure. The principle couse of these failures was immediately recognisable as excessive shaft deflection in the region of the mechanical seal, and the resultant oscillation of the seal components.

Some of the pump manufacturers tried to eliminate these deficiencies by the use of a more robust shaft and bearing construction. The implementation of these design improvements however was substantially prevented by the restraints imposed by prevailing standards.

Instead of the elimination of these seal failures causes, the market demanded "better" mechanical seals capable of compensating for them.

The fulfillment of this last requirement has not proven practicable although it's pursuit has resulted in many detailed improvements by the leading seal manufacturers, in particular among the birotational seals preferred within the chemical industry.

In parallel with the further development of pusher type seals there has been increased use of elastomer and metal bellows seals. These seals are less vulnerable to shaft vibration and their further development is being pursued today.

In spite of ever possible detail improvement it can be said that the replacement of shaft packings by small diameter, inboard mounted mechanical seals with rotating face loading springs has reached or has almost reached it's ultimate limit of development.

The high level of performance achieved certainly assures this well proven and cost effective component a wide range of applications for many years to come.

Within the dimensional restrictions imposed by DIN 24 960 however, the requirements for resistance to media either aggressive and/or containing high percentages of solids cannot be met to the extent demanded.

Demands on the new mechanical seal

Investigation of almost countless numbers of failed mechanical seals and pumps continually indicate the same reasons for failure, and the necessary measures for their elimination. Particularly relevant examples are provided by the EXXON study "Evaluation of cost effective design features for centrifugal pumps" (1) and the "VCI - Forderungskatalog" (2). Both studies, in spite of their differing starting points and objectives, come to the same conclusions regarding the requirements necessary for mechanical seals.

Main requirements:

EXXON	VCI
"1. Pump incorporates complete engineered package seal system or, alternatively: "Pump incorporates oversized stuffing box."	- Utilization of the centrifugal effect within the stuffing box - Arrangement of the seal within either a generous radial clearance or a positive conical chamber immediately behind the pump impeller - Prevention of compacted solids rotation.
"4. Shaft system design: Low overhung distance, large shaft diameter."	- Optimize distance between the point of action of the radial impeller forces and the drive side rollerbearing. - Reduction of radial shaftdeflection to within total Q-H region (= 50 µm)
2., 3., 5., 6. and 8.: Reinforced bearing with improved lubrication and sealing.	Reduction of shaft-vibration
	Non-rotating load springs not-contacted by pumped medium.

Most of these requirements have since been inducted in ISO 5199.

Additional demands made by potential users of the new chemical "Norm" seal:

- Interchangeable installation of the following seal options without changes to shaft sleeve or housing.
  - Single Mechanical seal.
  - Single Mechanical seal with pressureless (fluid) quench.
  - Double Mechanical seal.
- Maintenance of the sealing function in the event of buffer fluid pressure loss when running double seals.

Experience obtained with pumps with seal optimized stuffing boxes

e. g. pumps in Flue Gas Desulphurization Plants (FGD):

Mechanical seals of this design are presently operating in a range of shaft diameters from 20 mm to 260 mm.

Characteristics of FGD-media:

solid content	up to 50 %
ph-value	1 ... 14
chloride content	up to 200 000 ppm
pressures to be sealed	up to 6 bar
sliding velocity	up to 15 m/s

Nearly all flue gas desulfurization plants in the Federal Republic of Germany are equipped with single seals of this design.
In this application they have proven their performance.

Further development and trials

Experience gained using the illustrated (fig 1) seal concept within slurries and corrosive media, combined with the established user requirements, led in several development stages, to a modular construction comprised of a minimum number of components. The form of these components was further optimize with the aid of "Finite Element Methods" (FEM).

With the object of achieving a long and fault-free operational life, the highest priority was placed on optimum dissipation of friction induced heat from the seal faces, and simultaneously to prevent stress concentrations within the ceramic seal face materials.

This led to the

   H   R   C (Norm-) Mechanical seal
```
|   |   |
|   |   > for  c hemical industry
|   > with  r otating counter face
> h ydraulically balanced
```

The FE plots show the favourable stress and temperature distribution within the seal faces for the most commonly encountered pump size (35 mm shaft dia measured under the shaft sleeve) and the following operating conditions.

- Pressure to be sealed   p1 = 25 bar
- Media temperature   t1 = 160 °C
- Shaft speed   n = 3600 rpm
- Sliding speed   v = 14,5 m/s
- Seal face pairing   SiC/SiC

The following test program was undertaken using the FEM optimized Mechanical seal.

Program 1: Design and Material trials, determination of application limits of the single acting seal HRC/58-G11.

Seal face material pairing SiC/SiC

Test conditions:

Temperature	(°C)	= 60/90/120
Shaft speed	(1/min)	= 1500/3000/...6000 (n = 500)
Sliding speed	(m/s)	= 6/12/...24
Sealed pressure	(bar)	= 5 ... 50 (p = 5)
Running time/Test period		= 15 min.

Program 2: Repeat of program 1 after incorporation of detail improvements derived from evaluation of initial (prog.1) results.

Main results

At n = 3000 rpm and t = 120 °C, uniform, fault-free running up to p = 50 bar; at n = 6000 rpm and t = 120 °C up to p = 35 bar.

Program 3:	1500 hour endurance test of the single acting seal.
Medium:	demineralized and deaerated water, 60 °C
Pressure:	12 bar
Shaft speed:	1500 rpm
Result:	fault-free running
Wear rate:	0,018 µm/hour

Program 4:	as progam 3 but with double acting seal
Sealed pressure:	10 bar
Buffer pressure:	12 bar
Result:	fault-free running
Wear rate:	0,016 µm/hour

Further testing is presently being carried out with the object of determining actual application limits.

Results obtained from tests concluded to date indicate that the defined application limits will be exceeded.

Operating limits

- pressure   0 ... 25 bar
- temperature   -10 ... +160 °C
- velocity   ... 20 m/s

Materials

- face materials
  standard: SiC versus SiC
  special : SiC versus carbon

- O-rings (with enlarged cross section)
  elastomers or PTFE-coated elastomers

- metal parts
  standard: 1.4571 (AISI 316 Ti), 1.4462 or similar
  special : e.g. Hastelloy C-4

Common features

- modular system: all versions fit into the same seal cover

- grouped shaft diameters

- conical seal-cavern

- independent of direction of rotation

- hydraulically balanced

- allowable axial shaft movement +/- 1mm

- rotating mating face (reversible),
  fully flowed around by the medium,
  o-ring mounted, with shrink bandage

  ### principal advantages of rotating mating faces

  - excellent heat dissipation:
    the mating face is fully washed around by the medium

  - open space design in face area reduces gas bubble concentration

  - reversible mating face

  - elastic torque transmission reduces damages by impacts and vibrations

  - shrink bandage prevents excessive leakage in case of broken face

Economical benefits

- standardized design materials

- in abrasive media single arrangement, instead of previously necessary double

- no flushing and buffer medium required

- therefore low investment- and operating costs

Technical benefits

- cartridge design

- easy installation

- robust design

- extended lifetime

- independent of direction of rotation

- HRC3: no medium leakage to atmosphere
        in case of buffer fluid failure

REFERENCES

(1) EVALUATION OF COST-EFFECTIVE DESIGN FEATURES FOR
    CENTRIFUGAL PUMPS, Heinz P. Bloch, Exxon Chemical Company,
    Central Engineering Devision, CED DOCUMENT 84CE 1 0114

(2) Schwachstellen der Abdichtungspartie und Forderungen an
    die modifizierte Gleitringdichtung und Pumpe,
    VCI (Verband der Chemischen Industrie), 4936/03-er, 09.02.1987

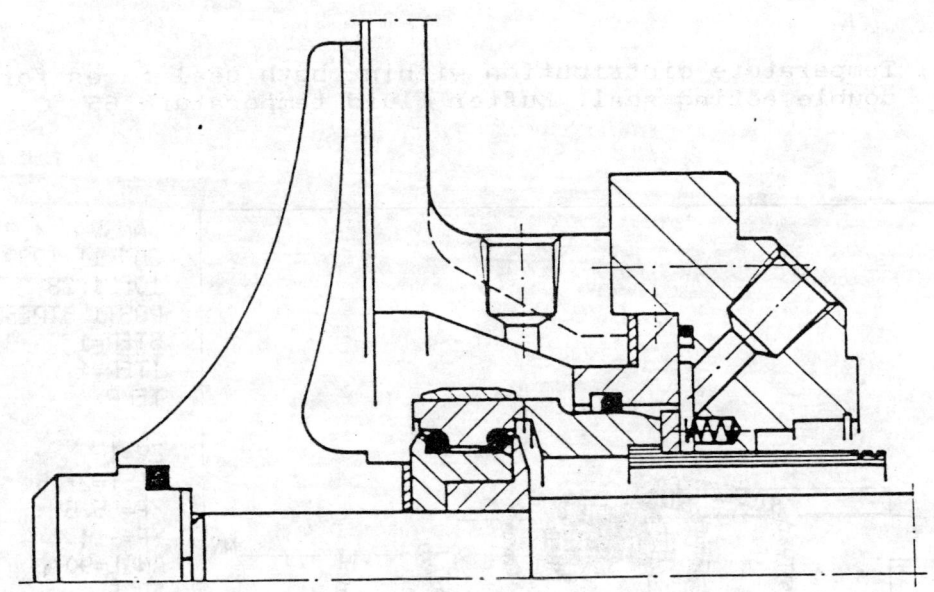

figure 1: typical mechanical seal for FGD-pumps

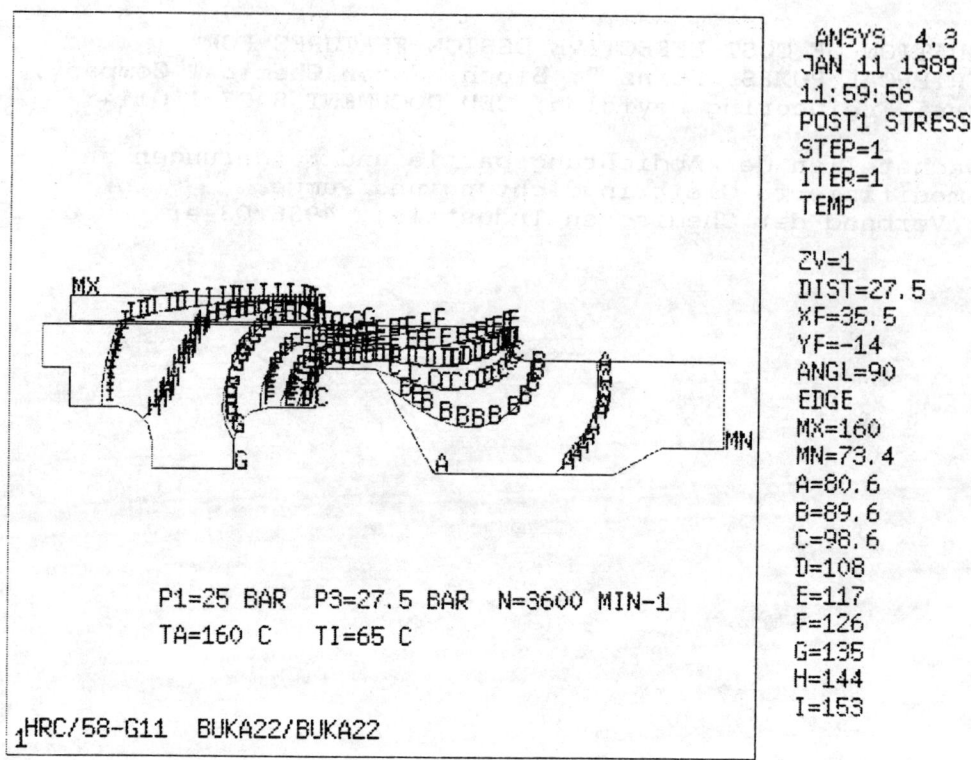

figure 2: Temperature distribution within both seal faces for a double acting seal. Buffer fluid temperature 65 °C.

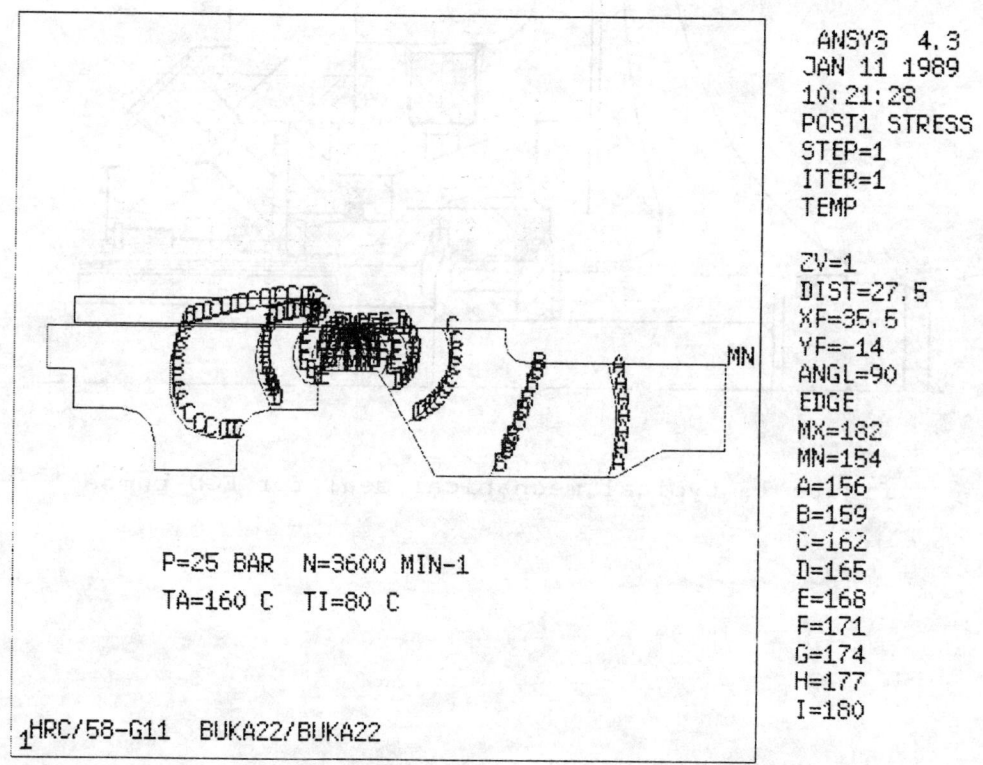

figure 3: Temperature distribution within both seal faces for a single acting seal.

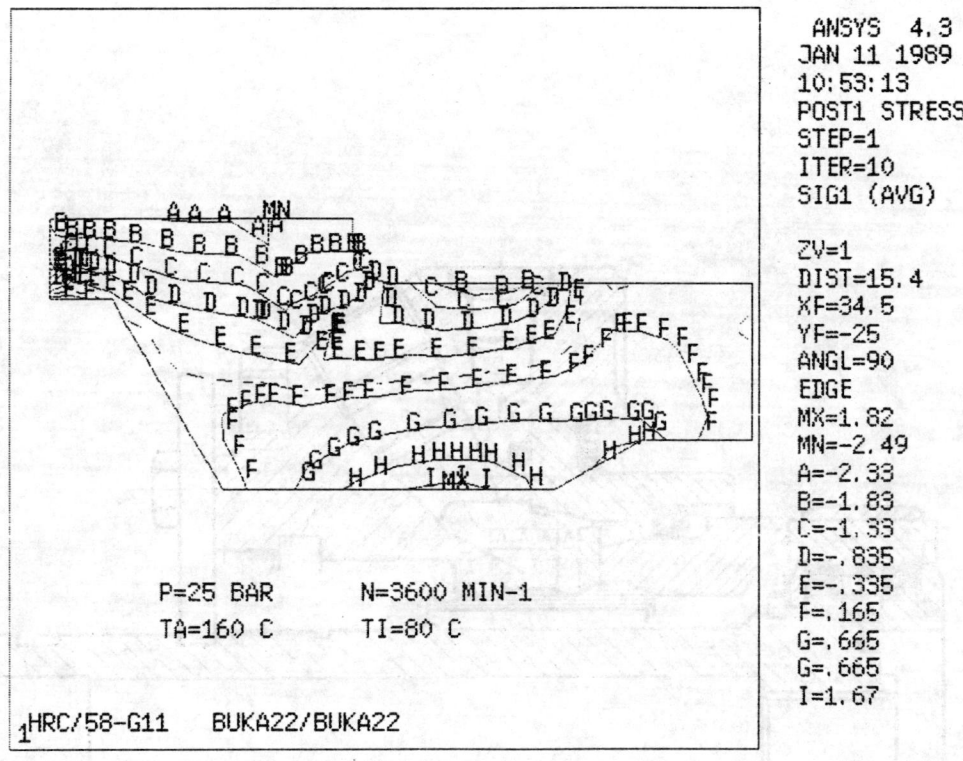

figure 4: Stress distribution within the seal face for a single acting seal.

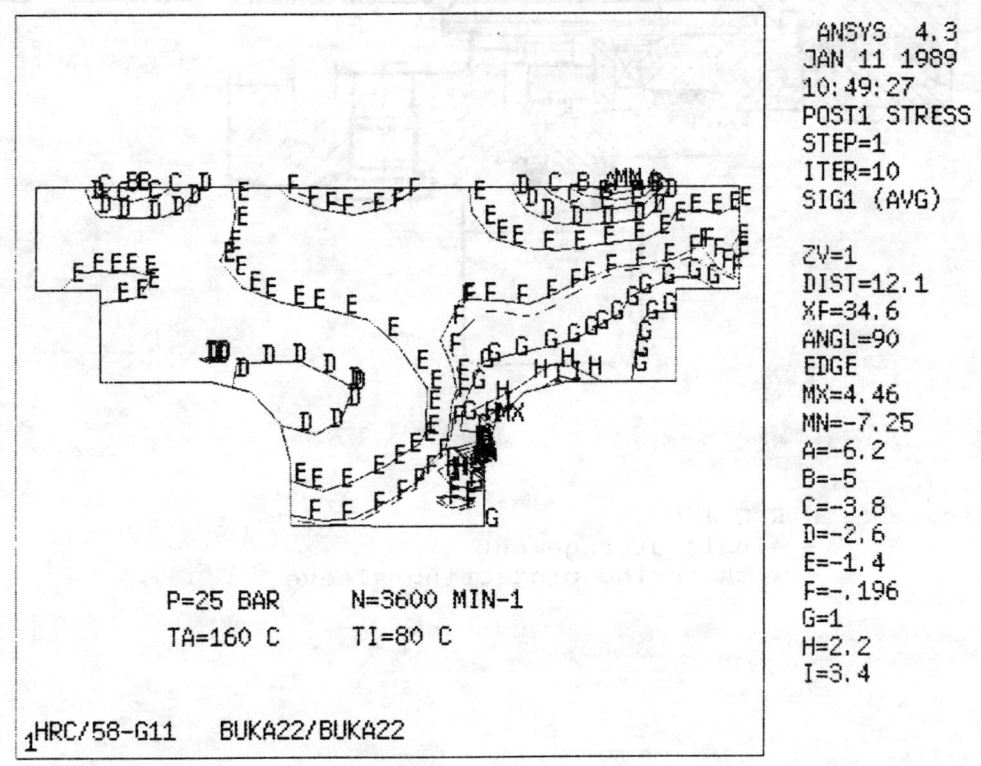

figure 5: Stress distribution within the seal counter-face for a single acting seal.

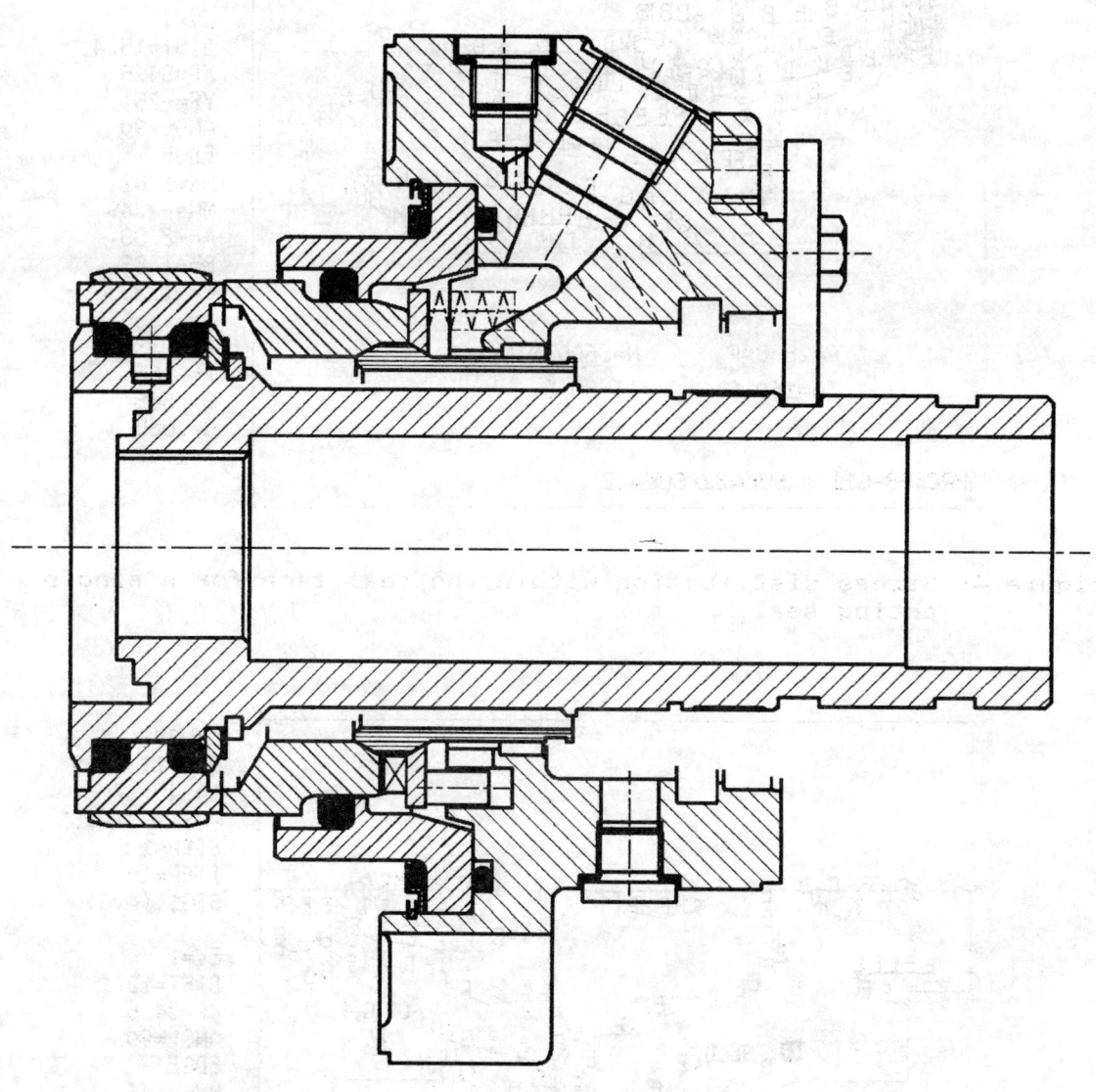

figure 6: H R C 1
- single arrangement
- with spring protecting sleeve

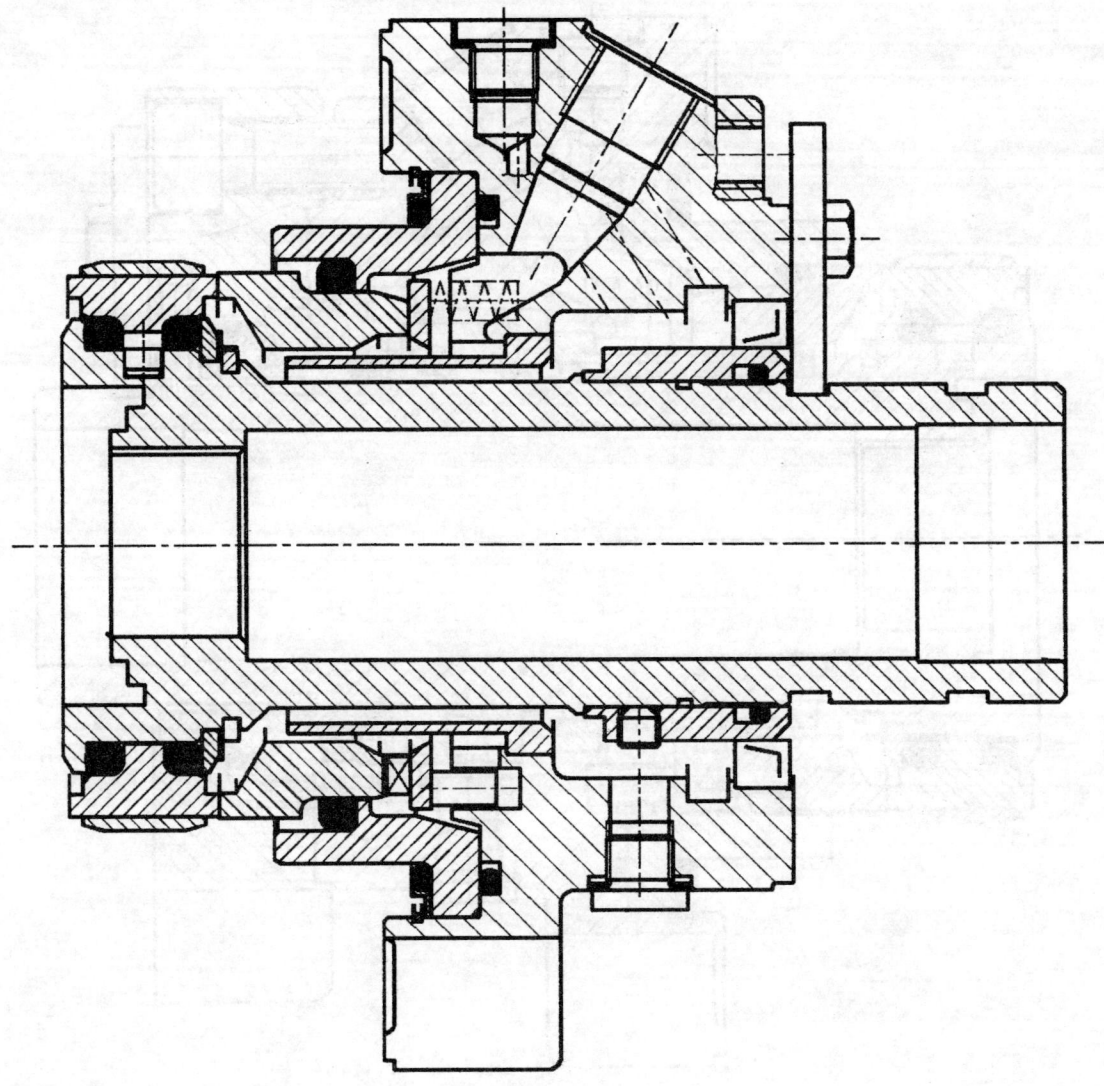

figure 7: H R C 2
- single arrangement with pressureless quench
- flow guide for quench fluid

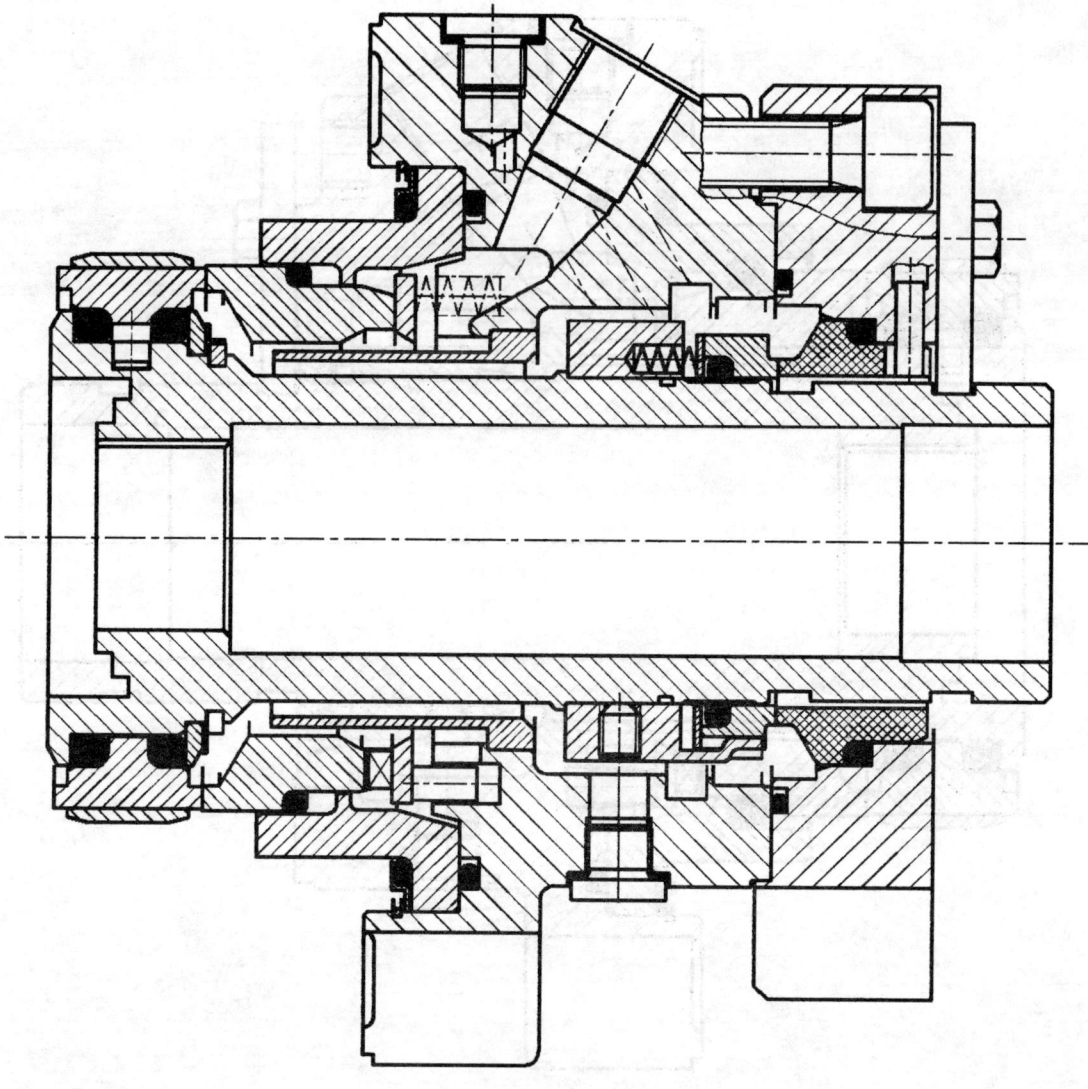

figure 8: H R C 3
- double arrangement
- self-closing, does not open in case of buffer pressure failure

11th International Conference of the
British Pump Manufacturers' Association

New Challenges – Where Next?

18-20 April, 1989
Churchill College, Cambridge

PAPER 14

THE DEVELOPMENT OF SPECIAL SEALS FOR MULTI PHASE PUMPING

N.M.Wallace C.Eng. BSc F.I.Mech.E
Flexibox Limited, Manchester, England

Summary

Multi phase pumping offers great advantages in oil production and a number of pumps are being developed for that purpose.

This has created a requirement for new generations of mechanical seals which must meet the difficult duty requirements and give reliable service for long periods whilst simultaneously being very compact to fit into restricted space envelopes. This Paper describes the design and development of such seals for a positive displacement screw pump.

It further describes initial field experiences and some of the problems encountered, including a most unusual mechanism which had a dramatic effect on seal life but which was eventually traced and corrected.

The Paper concludes that mechanical seals are quite capable of meeting the service requirements and makes a number of operational recommendations for getting the best out of them.

1. Introduction

Progress in offshore exploration and production has triggered many new technical challenges for the design engineer.

This is particularly true in the area of pumping and shaft sealing where difficult new duty conditions have been encountered.

Over recent years, seals for conventional main oil line pumping service have been developed to the point where lifetimes measured in years are being achieved regularly.

Such a seal is illustrated in Fig 1. This is a single mechanical with a simple PTFE lip seal for diverting any leakage to a safe drain.

Conference organised and sponsored by the British Pump Manufacturers' Association
in conjunction with NEL and BHRA, The Fluid Engineering Centre.
Co-sponsored by the Process Industries Division of the Institution of Mechanical Engineers.

Typical duty conditions are:-

Table 1   Typical Main Oil Line Pump Seal Duty Conditions

Seal Size	:	95mm
Speed	:	3600 rev/min
Sealed Pressure		
(Series pumping) Pump 1	:	15/50 bar g
Pump 2	:	85/105 bar g
Sealed Fluid	:	Crude oil/gas/sand/water/wax
Pumping Temperature	:	80°C
Fluid Vapour Pressure	:	10 bar at pumping temperature

Seals of the type illustrated often employ two hard faces (typically silicon carbide versus tungsten carbide) which are very good at resisting seal face vaporisation and its effects and can handle the abrasion of sand and salt in the crude oil mixture.

Such seals in single or tandem configuration (for extra safety) operate well with a guaranteed supply of fluid to the faces at a positive pressure. They have, however, very limited tolerance to the highly variable conditions encountered in 'mixed phase' pumping where pumps may operate at high pressures on the gas phase alone for substantial periods of time.

Whilst it is important to minimise the complexity of the mechanical seals for mixed phase pumping, the duty conditions make it essential to use double seals which remain lubricated on such occasions.

Whilst double seals are intrinsically safe and very resistant to changing duty conditions in the pump and they have been used for main oil line service in the past, they do require a barrier fluid system and can create problems of their own as will be illustrated later.

This Paper describes the design and development of a special double seal for mixed phase pumping. It outlines two problems that occurred and were resolved in the field evaluation phase and concludes that reliable seals are now available and that close co-operation with the OEM and the user is vitally important in developments of this kind.

During 1983/84 we took part in a small scale evaluation of mixed phase pumping in a series of experiments commissioned by a consortium of oil Companies on Stothert and Pitt PLC who were engaged in extensive development work on positive displacement mixed phase pumping.

(Stothert and Pitt have now formed Multiphase Systems PLC to specialise in multiphase pumps).

The seals used at that time were simple DIN standard double seals and attention was focused on the viability of mixed phase pumping rather than mechanical seals. The results of the trials were very encouraging and resulted in full scale trials being considered.

We obtained the contract for the seals for those trials and our first task was to review the operational requirements and come up with a suitable seal design which would then be evaluated in the laboratory and then in the field.

2.   Mechanical Seal Requirements

The mechanical seal requirements comprised operating conditions, space available, special design and performance requirements and technical data which would be used in the design of the pump and associated systems.

2.1 Operating Conditions

Parameter	Specific Applicational Values	Design & test values
Pumped Fluid	Mixture of crude oil, gas, water, Nacl, H_2S	
Fluid Temperature °C Normal/max/min.	75/130/10	
Viscosity (centiPoise)	0.4/20	
Forward Dynamic Pressure bar max/min	10/0	Normal (7/28) Max. 150(15 mins)
Reverse pressure bar max/min	99.3 vacuum	150
Shaft speed - Rev/min. Normal/max/min	3000/3600/1200	Up to 4000

2.2 Space Available

Typical shaft diameter	130/140 mm
Radial Space Available	26.5 mm
Length available box	124
Outside box	160
Total	284mm

Barrier Pressure normal	88 bar
Barrier Pressure maximum	155 bar
Barrier Fluid	Light mineral oil

2.3 Special Design and Performance Requirements

2.3.1 The pumped fluid may be a mixture of liquid and gas in any proportion. The pump must be able to run dry at the start up for a minimum of 15 minutes and with pressures up to the maximum level.

2.3.2 The seal must tolerate pressure rises/falls at the rate of 27 bar/minute under start up conditions.

2.3.3 The seals must allow for axial shaft movements of up to 3mm.

2.3.4 Cartridge seals are preferred.

2.3.5 Special materials are required for wetted parts.

2.3.6 The seals must be capable of containing a box pressure of up to 150 bar with zero barrier fluid pressure.

2.4 Technical Data Requirements

2.4.1 Maximum Seal Leakage Rate to be determined

2.4.2 Break out torque per seal

2.4.3 Maximum absorbed power per seal

3. Design

3.1 Choice of Seal Type

All of the mechanical seal requirements outlined in Section 2 could be met by a single or tandem seal as outlined in Section 1 <u>except</u> for 2.3.1., the dry running requirement and 2.2 the space available.

Double seals had been developed for use on very high pressures for offshore operation and special features for enhanced safety had also been devised.

Figure 2 illustrates a double seal for high pressure NGL service. It has a specially developed "anti-blowout" inner seal which is capable of taking the full box pressure in the unlikely event of loss of barrier fluid pressure.

General duty conditions as follows:-

Table 2.   NGL Condensate Pump Seals: Duty Conditions

Seal Size	: 95mm
Speed	: 3700 rev/min
Product	: NGL condensate
Product Pressure	: 100 bar
Barrier Fluid Pressure	: 110 bar
Barrier Fluid	: Mineral Oil (46 cS at 20°C)
Oil Flow Rate	: 23 L/min
Reverse Pressure Capability:	Tested to 150 bar

The NGL double seal design is thus capable of meeting the technical requirements for mixed phase pumping but is basically too large and complicated for the space available.

The twin rotor screw pump for mixed phase pumping requires four double mechanical seals to seal the four shaft ends.

There is only 53mm of space between the shafts in way of the seal and so a radially very compact seal was required.

Alternative seal configurations had been developed for another project for seals for high speed charge pumps for down-hole pumping. These seals were very much simpler and more compact and variants had been developed for use on high speed compressors and also for high pressure MOL service.

Figure 3 illustrates the general design principles.

For compactness, the seal has a single monobloc central rotary seal ring. The ring is manufactured from high grade reaction bonded silicon carbide and has a duplex alloy band shrunk onto its outer periphery. The band serves to put the silicon carbide material under compressive stress and is capable of containing the centrifugal forces which would result if the material cracked under maloperation. This is only possible with low density silicon carbide at high speed. Tungsten carbide is too dense.

The simple stationary seal ring construction contains a 'built-in' double balance line which ensures that the inboard seal seals even if the barrier oil pressure is lost.

The principle of this feature is illustrated in Figure 3.

This type of seal had also been fully tested for the MOL service as outlined in table 1 and had been subjected to reverse pressures of up to 100 bar.

Other versions with built in pumping rings are in service on variable speed pumps in the North Sea and have been tested under reverse pressures of 289 bar (Fig 4).

It was decided to follow these design prinicples for the multi phase seal development.

3.2 Special Features

The general design is illustrated in (Fig 5). It will be immediately clear how little space was available for two seals with such close shaft centres.

In consequence the seal cartridge is of a very novel design.

Each cartridge contains two double seals and there are two cartridges per pump. On the initial design, the seals ran in a common chamber and the rotary seal rings contra rotated within a few millimetres of each other.

The seal specification requirements which had a profound influence on design were:-

* Maximum barrier fluid pressure 155 bar
* Reverse pressure capability of 150 bar
* Normal barrier fluid pressure of 35 bar
* Minimal starting torque and running power absorption required

From the general arrangement drawing (Figure 5) it can be seen that the monobloc central rotary seal ring design is very stable from a pressure distortion standpoint. It is essentially balanced hydraulically except for a small nett thrust towards the atmospheric side from the product pressure acting on the small area between the balance line and the rotary seal ring/shaft 'O' seal. This thrust keeps the faces square to the axis of rotation.

In order to seal under forward <u>and</u> reverse pressures, the stationary seal ring features a double balance line. In effect, the seal selects one of two balance lines automatically depending on the direction of pressure.

In this design the seal uses the lower balance diameter for normal forward pressure and the seal is hydraulically 'balanced' to suit the range of operating pressures from 35 to 150 bar.

In the reverse direction the seal selects the upper balance diameter with the 'O' ring moving to the opposite end of the groove. In this case, the seal is <u>un-balanced</u> to give high sealing integrity although this limits the ability to run under reverse pressure to short periods which are, however, adequate for 'shut-down' purposes.

The stationary seal ring itself was very carefully designed to give minimal distortion under forward and reverse pressures so that good sealing integrity could be attained without increasing absorbed power to unacceptable levels. In fact, the two requirements are mutually contradictory and, since low leakage and power consumption were prime requirements, adjustments were made to this component during development (see section 4) for optimum performance.

The stationary seal ring is a composite of Duplex alloy and high grade carbon. Figure 6 a) illustrates three designs which were considered for this project.

The composite seal ring must be capable of staying together under all conditions of temperature and pressure.

With design 1, there is a temperature at which the interference fit between the carbon and its housing is lost due to differential expansion. There is also a pressure at which the fluid pressure is equal to the carbon/housing interface pressure due to the shrink fit and where the carbon can be 'floated' out of its housing. Both of these values can readily be calculated and are shown on Figure 6 b) in the form of an operating envelope.

In the reverse pressure mode, there is a hydraulic ejection force on the insert which acts on an area bounded by the outer balance diameter and the outside diameter of the insert itself. An envelope can also be drawn for reverse pressure operation and this is similarly affected by differential expansion as in the forward pressure direction. The operating envelope for design 1 is shown on Figure 6 b) and it can be seen that the maximum duty points were well outside and a better design was required.

Design 2 utilises an inner steel band which is shrunk into place to increase the shrink pressures and to help exclude high pressure fluid from the rear of the insert.

The effect is dramatic and the operating envelope is expanded. Note that the drop out temperature becomes irrelevant with this design because of the inner retaining ring. The value which is used is the temperature at which the carbon ring goes into tension and this is lower than the previous drop-out temperature because of the initial expansion of the carbon during fitting of the inner ring.

Rings of this design were made and were expected to work on the basis of the internal pressure directly increasing the interface pressure at the inner ring/carbon ring and carbon ring/housing interfaces and assuming a coefficient of friction of 0.15 (carbon/steel, dry).

In the event, the carbon ring was ejected at around 120 bar and it was necessary to find a way of increasing the interface coefficient of friction to a value of around 0.25.

Design 3 shows how this was achieved by using sharp edged grooves which mechanically lock into the carbon surface. This design also increases the forward pressure capability.

Design 3 was used in the final seal design after some adjustment to optimise running performance.

It was previously mentioned that this positive displacement screw pump uses four double seals, one at each shaft end. With eight seal faces, it is extremely important to minimise faceloading, torque and power consumption to avoid excessive pump losses and difficulty in starting.

The following equations show how face loading, torque and power are affected by the basic seal parameters:-

Closing Force

Outer seal face: $F_s + p_b \cdot B \cdot A_f$

Inner seal face: $F_s + p_b \cdot B \cdot A_f + p_s \cdot (1-B) A_f$

Total, $F_t = 2(F_s + p_b \cdot B \cdot A_f) + p_s (1-B) A_f$ [1]

F_t = closing force (total) N
F_s = spring load N
p_b = barrier pressure N/mm^2
B = balance ratio
A_f = seal face area mm^2
p_s = stuffing box pressure N/mm^2

Torque

$$T = F_t \frac{Dm}{2} \mu \quad Nm \quad [2]$$

Dm = mean face dia m
μ = face coefficient of friction

Absorbed Power

$$Q = \frac{T.N.}{9.549 \times 10^3} \quad kW \quad [3]$$

N = speed rev/min
Q = absorbed power - kW

The use of narrow faced seals with optimum balance ratios minimises F_t and hence T and Q by minimising A_f and B. Low values of friction coefficient μ are obtained by careful control of face flatness under all conditions of running so as to create the right lubrication regime at the face.

Values of µ less than 0.01 were created at high pressure as will be illustrated later.

A most interesting performance area is 'starting torque'.

If a seal is pressurised with barrier fluid only and the starting torque established, then one would assume that the introduction of a pressure on the pump side of the inner seal would reduce the starting torque by reducing the pressure differential on the inner seal.

Surprisingly, this is not true and the reverse has been seen in the laboratory and in the field.

Trials offshore on a high pressure NGL seal gave a starting torque of 52 Nm with a barrier pressure of 103 bar and 72 Nm when a pump pressure of 76 bar was established!

Figure 7 a) shows how this effect has been confirmed in the laboratory.

It can be seen from the graph that, for a given barrier pressure, the starting or 'break-out' torque increases with increasing pump or product pressure.

Figure 7 b) illustrates how the nett face load in a seal, pressurised on both sides, can vary according to the hydrostatic model assumed.

For all four possible models the nett closing Force F_N for that model with <u>no</u> pump pressure is shown in parentheses.

For model 1, the nett force, and hence the torque reduces with increasing pump pressure.

For model 2, the parallel film, there is no difference whether there is pump pressure or not and only models 3 and 4 show an increase in torque for an increase in pump pressure.

Since model 4 is potentially unstable and would leak if the pressure pb was high enough, it is presumed that model 3 is more representative of the truth - an interesting effect! As before, narrow seal faces and carefully chosen balance ratios are important in minimising break-out or start-up torque.

4. Development and Test

4.1 Basic Testing

The test capsule shown in Figure 8 was used to evaluate the seal on a variable speed test rig with thyrister speed control.

The capsule allowed testing with various combinations of barrier pressure and pump pressure and under static reverse pressure conditions up to the maximum requirement of 150 bar.

<u>Running under barrier pressure alone:</u>

A critical requirement was that the seal should run with a barrier pressure of up to 155 bar for periods of a quarter to half an hour.

After some adjustment to the stationary seal ring distortion characteristic, a stable seal was achieved with a leakage of the order of 1cc/min at maximum duty condition.

Attempts to reduce the leakage to zero were successful but with a penalty of very high friction values at the maximum barrier pressure of 155 bar.

Extended normal running gave a pattern of reducing leakage with time as the seal faces 'bedded-in'.

The frictional heat at the seal faces was measured and is shown on Figure 9 for pressures between the normal running value of 34.5 bar and the maximum of 155 bar. Also shown are derived effective coefficients of friction at the face. (The effective coefficient of friction is derived from the <u>measured</u> power absorption, at the seal and the theoretical value based on the seal geometry and the operating conditions) reference 1.

Values of 0.01 and less were measured for pressures in excess of 100 bar.

<u>Reverse Pressure Testing</u>

The essence of this work was covered in Section 3.2; suffice to say that the carbon insert was ejected on occasions and lost its flatness on others after reverse pressure testing at 150 bar.

When the final modification of the grooved recess was introduced the seal proved to be totally stable under all operating conditions and the face flatness remained good.

5. In-field Testing

Whilst the laboratory testing had been relatively uneventful, two problems occurred during testing in the pump, one at the pump makers and the other in the field.

5.1 'O' Ring Compression Problems during Pump Test

After a number of hours of pump test on water some leakage of water from within the pump occurred to atmosphere. The leakages were always seen after the pump had cooled down after a run and on the application of water pressure.

It can be seen from Figure 10 that the only logical route for a leakage that appeared from between the sleeve and sealplate was via the rotary seal ring/sleeve 'O' ring.

This Viton 'O' ring was found to have taken a permanent set with a total loss of compression when cold.

At elevated temperatures the Viton 'O' ring expands more than the housing it is contained in. Additionally the duplex alloy sleeve expands more than the silicon carbide insert and thus the radial groove depth reduces with temperature.

'O' rings with poor temperature compression properties take on a 'set' at the elevated temperature and subsequently leak when the temperatures are reduced and the rubber contracts.

The 'O' ring in question was originally designed as a 3mm section metric 'O' ring. Because metric 'O' rings are not easy to obtain in the UK it was sourced in Germany and was eventually identified as Viton A and not Viton E60C as specified.

The 'O' seal was re-designed as a 3.53mm section and an investigation into various alternative materials was started and followed by laboratory tests.

a) Review of manufacturers Data

A number of materials were reviewed and the following table summarises the most important properties:-

Table 3   'O' Ring Material Properties

Material	Manufacturer	Hardness Shore A	Max. temp. Continuous/ extreme °C	Compression set at 200°C in air	
				Number of hrs	% set
Viton A	Du Pont	60-95	204/260	70	20-70
Viton E60C	Du Pont	60-95	204/260	70	10-20
Kalrez 1058	Du Pont		288/316	70	82
Kalrez 3018	Du Pont		288/316	70	63
Fluoraz 797	Green Tweed	75	260/288	22	27
Fluoraz 799	Green Tweed	93	260/288	22	25.6

From the above data it was decided to evaluate Viton E60C using Viton A as a control.

b)   Laboratory test

A test capsule was manufactured to take 'O' rings of 145mm inside diameter (Figure 11).

The inner sleeve was made from AISI 316 stainless steel with a coefficient of expansion of $16 \times 10^{-6}/°C$ and the outer sleeve of AISI 413 stainless steel with a coefficient of expansion of $12 \times 10^{-6}/°C$.

The use of dissimilar materials simulated, to a degree, the service situation.

The rig was designed to take a 3mm section 'O' ring at one end and a 3.53mm section one at the other.

The following test procedure was then carried out:-

Test Procedure

i)   Pressurise the seal with water at 34.5 bar

ii)   Place in oven at 175° for 70 hours

iii) Remove capsule and allow to cool

iv)   Reduce pressure in steps, leaving for 5 minutes at each step - check for leakage and record.

v)   Strip capsule and examine 'O' rings measuring for compression set and checking for hardening.

NB.   A shadow graph with 50x magnification was used to display and measure the 'O' rings before and after test.

Results

On re-application of pressure after the capsule cooled to 34°, leakage occurred from the 3mm Viton A 'O' ring but not from the 3.53mm section E60C one.

Pressure	Leakage (Viton A) cc/min.
34.5	50
27.5	45
21	40
14	20
7	12
0	0

A number of measurements were then made and compared with theoretical groove dimensions, hot and cold.

Table 4   Summary of Results

'O' Ring Material	Initial nip% (cold)	Initial nip% (hot)	Nip % after test (cold)	Compression set %
Viton A (3mm Section)	9.9	11.5	(- 2.5)	100+
Viton E60C (3.53mm Section)	11.7	13.0	5.2	50.2

These results are most easily represented in Figure 12 as shadow graph outlines.

As a result of the tests, the 3.53mm section Viton E60C was used and no more service problems with this seal were experienced.

5.2   Problems of High Inboard Stationary Seal Ring Wear on Site

During production tests of the pump in the field it was found that the inboard stationary seal rings were suffering from excessive wear although the outboard seals were in perfect, unworn condition (and with the highest pressure differential!)

With a barrier pressure of 34.5 bar and a pump pressure of only 7 bar, the duties were nominally very similar. One seal showed 0.09mm face wear after 100 hours opration which equates to 2,200 hours life!

Another seal that had run (apparently very well) for 3,500 hours had worn down completely at the face.

It is no exaggeration to say that dozens of possible mechanisms were considered as the cause of the problem until finally a most unusual effect was identified and subsequently proved by test.

The stationary seal ring was fitted with two PTFE anti-extrusion washers, one each side of the 'O' ring seal to cater for forward and reverse pressures.

Apart from the heavy face wear, the anti-extrusion washer on the barrier fluid side of the 'O' ring showed strong signs of extrusion towards the barrier oil and indeed had extruded on one occasion (Figure 13).

With a reverse pressure of only 7 bar, this was surprising to say the least!

We learned that the operators had allowed the pump seal to sit under reverse pressure for long periods when the pump was idle and the barrier pressure off.

The pump was subsequently re-started and run normally. In fact, the reverse pressure feature in the seal is really there for emergencies only since even a small seal leak could allow contamination of the barrier fluid after a long period of time.

That apart, this practise, gave us the idea for a mechanism which would explain the high wear.

Mechanism

The possible mechanism for high wear can be explained in conjunction with Figure 14a).

Under the action of a steady reverse pressure, the right hand anti-extrusion washer 'forms' to the end of the 'O' ring groove to the extent that it makes a perfect seal.

When the barrier pressure is applied, the anti-extrusion washer lifts slightly and allows oil into the cavity which moves the 'O' ring to the other end of the groove.

As the pump starts to operate and get warm, the contents of the 'O' ring groove (oil, rubber and PTFE) expand relative to the groove itself and the right hand anti-extrusion washer re-establishes a seal which seals off the cavity completely.

Differential expansion then establishes a pressure in the cavity which creates more load on the seal which creates more heat which creates more pressure etc. etc. and caused heavy face wear.

A theoretical evaluation produced values as shown on Figure 14b). The first part of the slope is where the 'O' ring is becoming compressed elastically and the steep second slope treats the 'O' ring like liquid with a high bulk modulus.

It was decided to try to verify this possible mechanism in the following way.

i) Demonstrate that the inner anti-extrusion washer could form an effective seal against the housing after a period under a reverse pressure and not otherwise.

ii) Show that an artificially induced pressure in the cavity would increase as the seal was run, due to thermal effects.

iii) Show that the seal could generate its own pressure and increase the face loading **without** artificially introducing a pressure into the cavity.

A fine drilling was made into the 'O' ring cavity so that pressures could be introduced and monitored.

Test 1

The seal was subjected to a reverse pressure of 7 bar for 15½ hours with zero leakage.

The seal was then subjected to a barrier pressure of 3.5 bar and a pressure of 7 bar introduced into the cavity between the 'O' ring and the inner anti-extrusion washer.

The pressure held without decay overnight indicating an effective seal.

Test 2

The pressure in the gap was set at 33 bar with a barrier pressure of 21 bar and the seal run at 3000 rpm. The gap pressure rose to 40 bar within 5 minutes proving that the pressure increasing mechanism existed.

At this point, the barrier pressure was reduced to 3.5 bar by accident and the gap dropped only to 27.5 bar.

The seal face temperature rose from 115° to 140° almost immediately and smoke began to eminate from the seal indicating high face loading and imminent breakdown of the lubricating film.

A second experiment with an initial gap pressure of 16.5 bar gave a pressure increase to 18.6 bar over 100 minutes (See Figure 15).

Up until this point, the pressure measurement system added considerably to the gap volume and hence limited the pressure rises.

We removed the pressure measurement system to conduct the final test to see if the seal could generate its own pressure and monitored seal face temperatures only.

The results of this test showed that the inboard seal was 21°C hotter than the outboard seal _after_ a period under reverse pressure.

The results confirmed that the mechanism existed and the solution was easy - remove the barrier fluid side anti-extrusion washer, adjust the clearances and 'O' ring hardness to avoid extrusion problems.

The final design is shown on Fig. 16.

Conclusion

Whilst the problems described in this Paper were experienced during development, the seals otherwise worked very well with very low friction and wear and minimal leakage.

Fortunately, the solution to the problems was easy (which is not true of the diagnosis and the confirmation by demonstration to the customer).

The high inner seal wear problem highlighted the practise of using the reverse pressure feature as a convenience during shutdown conditions. This is not good practise because barrier fluid contamination is possible via any small face defect or leak path. Over a long period this could have a significant effect. The same small leak from barrier fluid to pump under normal conditions would be acceptable, however.

The reverse pressure feature is for emergency conditions and other provision in the systems should be made to avoid its regular use as a convenience.

At the end of the day, very effective seals for multi-phase pumping were established and this was largely due to close co-operation between the seal maker, pump maker and user at the specification, design development and test stages and in establishing production standards.

References

1. 'Mechanical Seal Practice for Improved Performance'. I Mech E reference ISBN 0 85298 671 8 Chapter 1 Section 1.2.3.

Appendix

Notes on Derivation of Equations (Figure 7b)

F_C = Seal Closing Force
F_O = Seal Face Opening Force
F_N = Nett closing Force = $F_C - F_O$

Model 1

$$F_C = F_s + p_b \cdot B \cdot A_f + p_s A_f (1-B)$$
$$F_O = p_s A_f$$
$$F_N = [F_s + p_b \cdot B \cdot A_f] - p_s \cdot A_f \cdot B$$

Model 2

$$F_C = F_s + p_b \cdot B \cdot A_f + p_s \cdot A_f (1-B)$$
$$F_O = p_s A_f + \tfrac{1}{2}(p_b - p_c) \cdot A_f$$
$$= p_s A_f + \tfrac{1}{2} p \cdot A_f$$

(where $p = p_B - p_S$)

$$F_N = F_s + p_b \cdot B \cdot A_f + p_s A_f (1-B)$$
$$\quad - p_s A_f - \tfrac{1}{2} p \cdot A_f$$
$$= F_s + A_f \cdot B \cdot (p_b - p_s) - \tfrac{1}{2} p \cdot A_f$$
$$F_N = [F_s + p \cdot A_f (B - \tfrac{1}{2})]$$

Model 3

$$F_C = F_s + p_b \cdot B \cdot A_f + p_s \cdot A_f (1-B)$$
$$F_O = 0$$
$$F_N = [F_s + p_b \cdot B \cdot A_f] + p_s (1-B) A_f$$

Model 4

$$F_C = F_s + p_b \cdot B \cdot A_f + p_s A_f (1-B)$$
$$F_O = p_b A_f$$
$$F_N = [F_s + p_b (B-1) A_f] + p_s A_f (1-B)$$

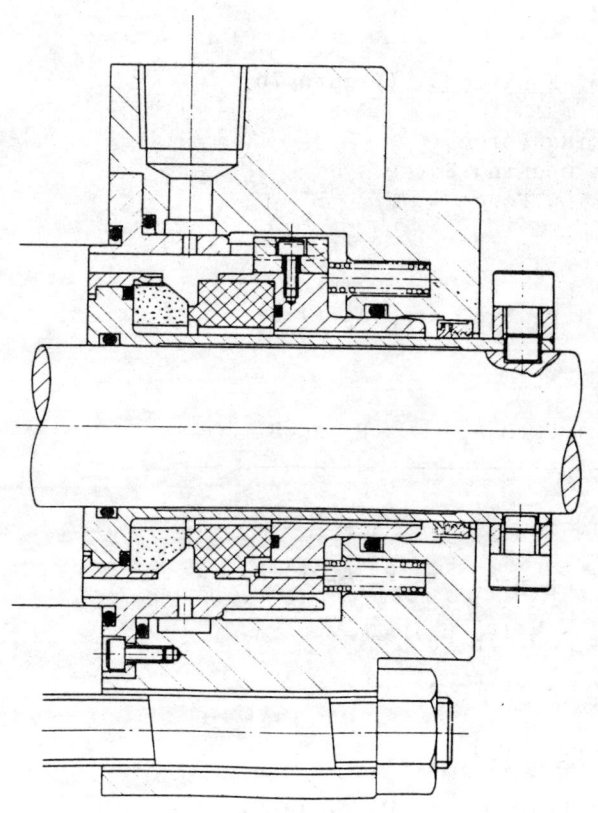

Figure 1 High Duty Single Seal

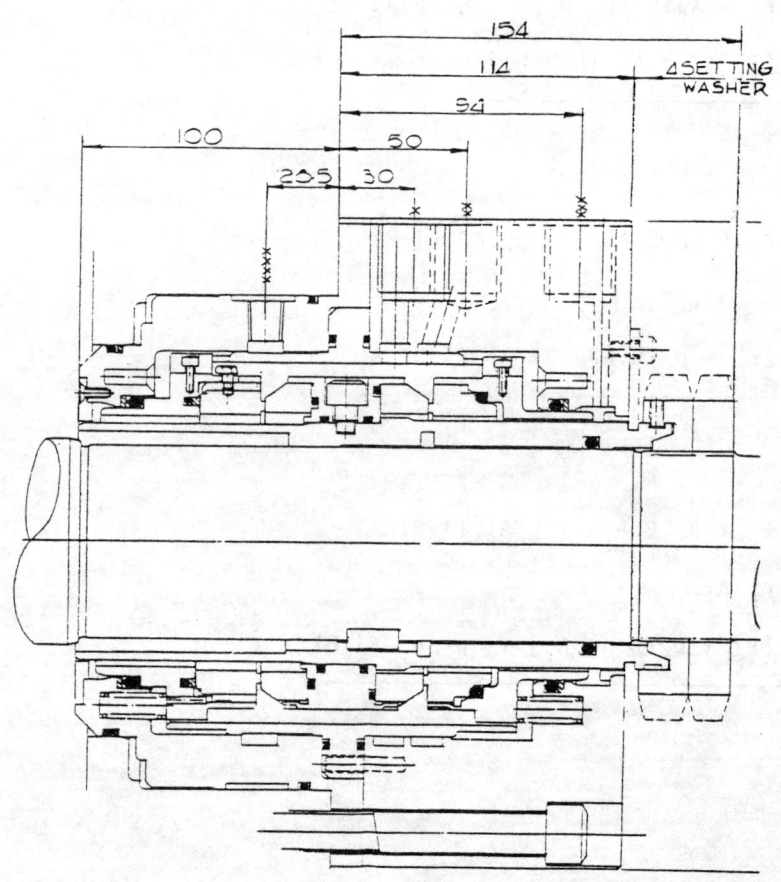

Figure 2 High Pressure Double Seal

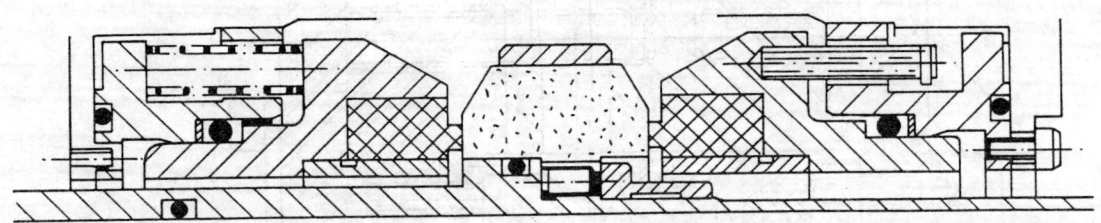

DOUBLE BALANCE LINE

a) UNDER BARRIER
   PRESSURE P_B.

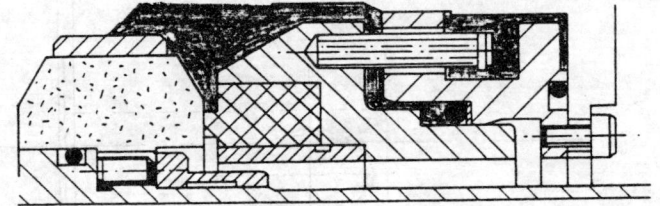

b) UNDER PRODUCT
   PRESSURE P_S.

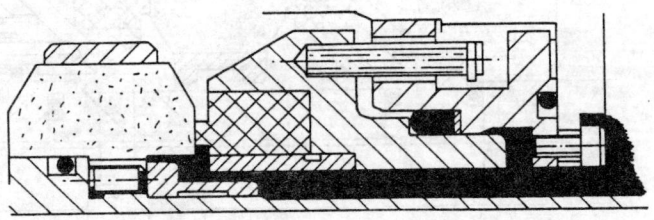

Figure 3   High Speed Compressor Seal

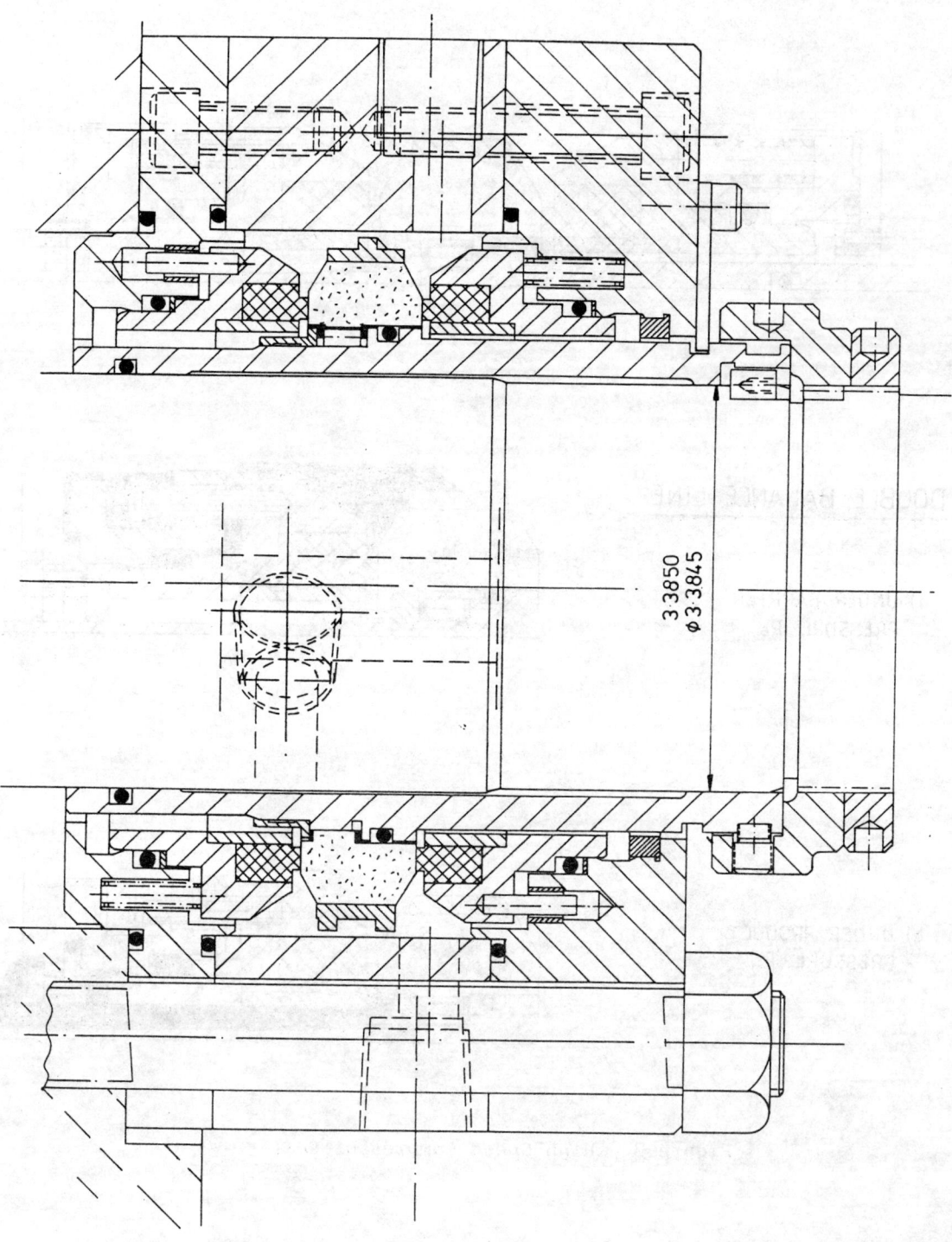

Figure 4 Main Oil Line Double Seal

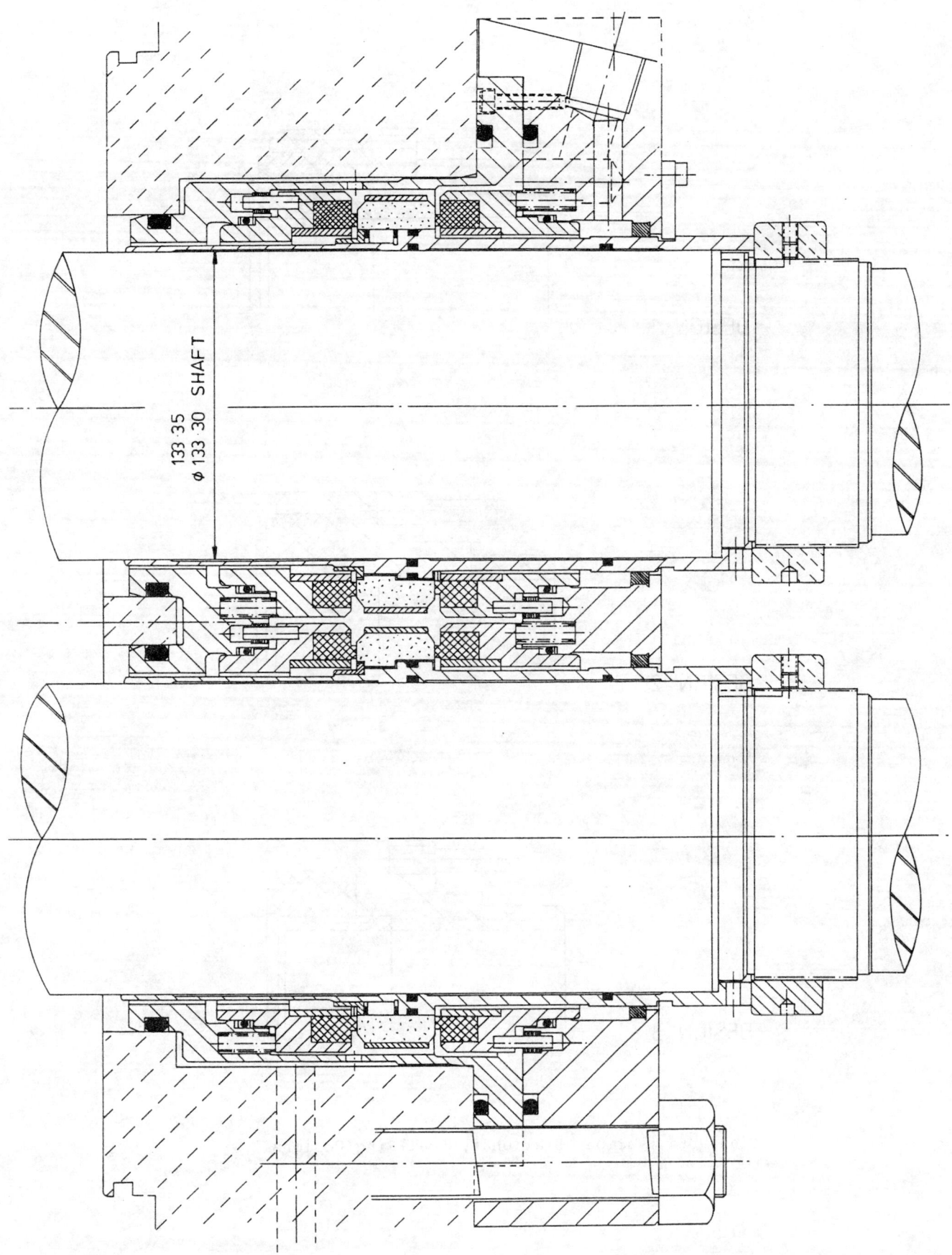

Figure 5   Multi-phase Pump Seal - initial design

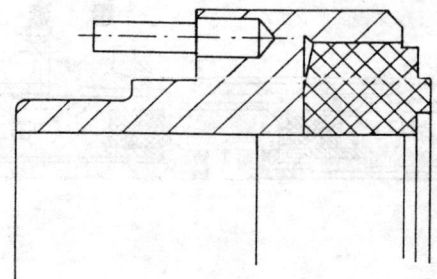

DESIGN 1

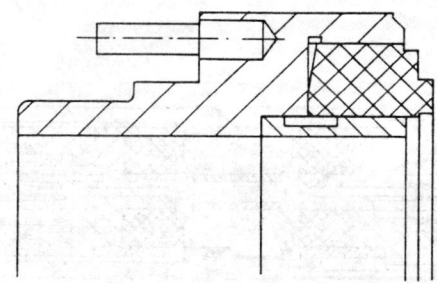

DESIGN 2

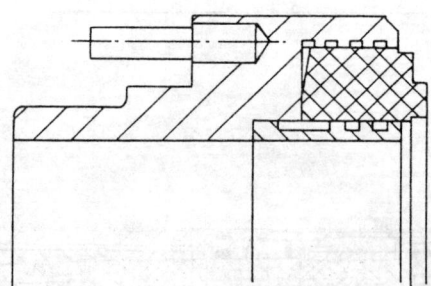

DESIGN 3

Figure 6a Various Stationary Seal Ring Designs

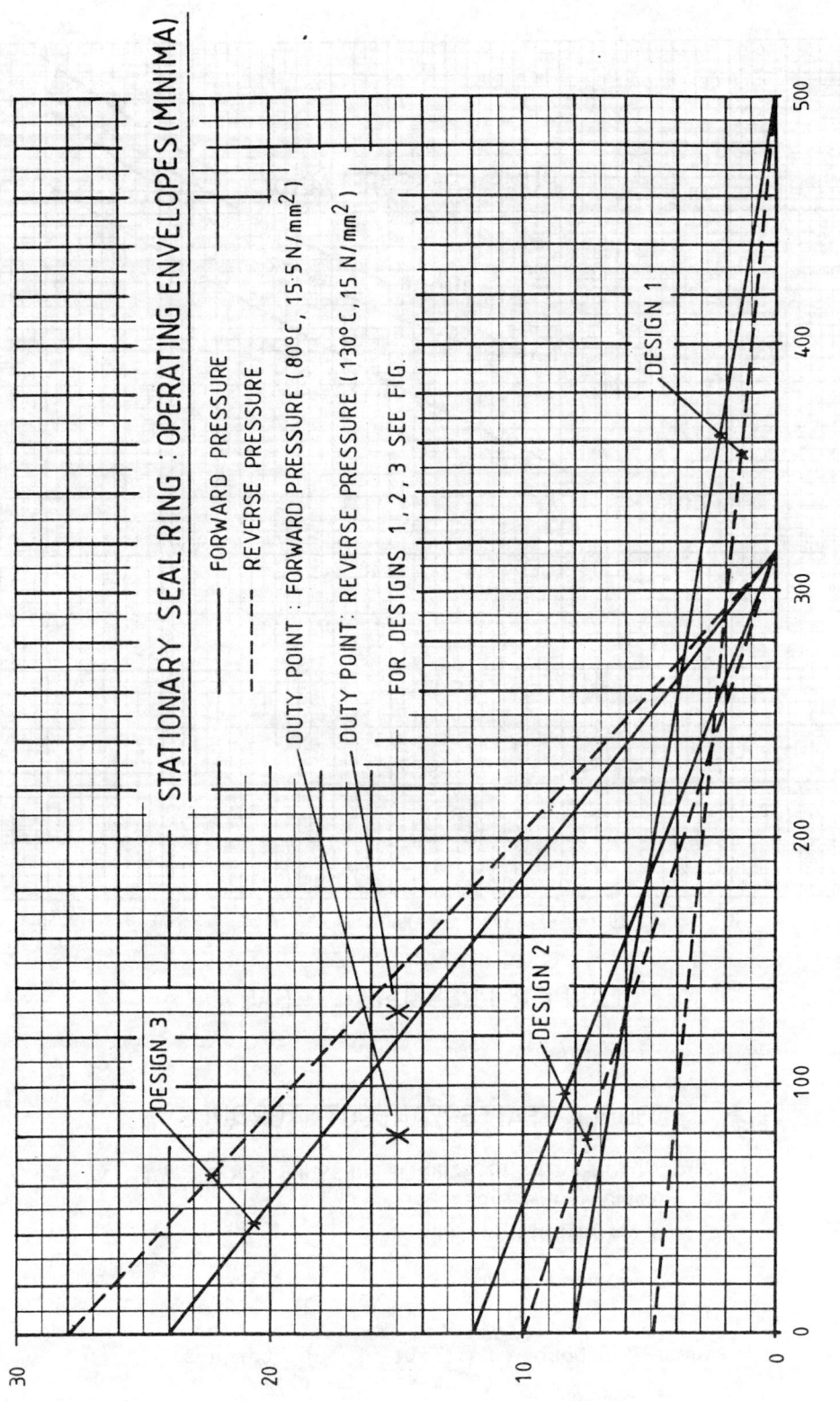

Figure 6b Stationary Seal Ring Operating Envelopes

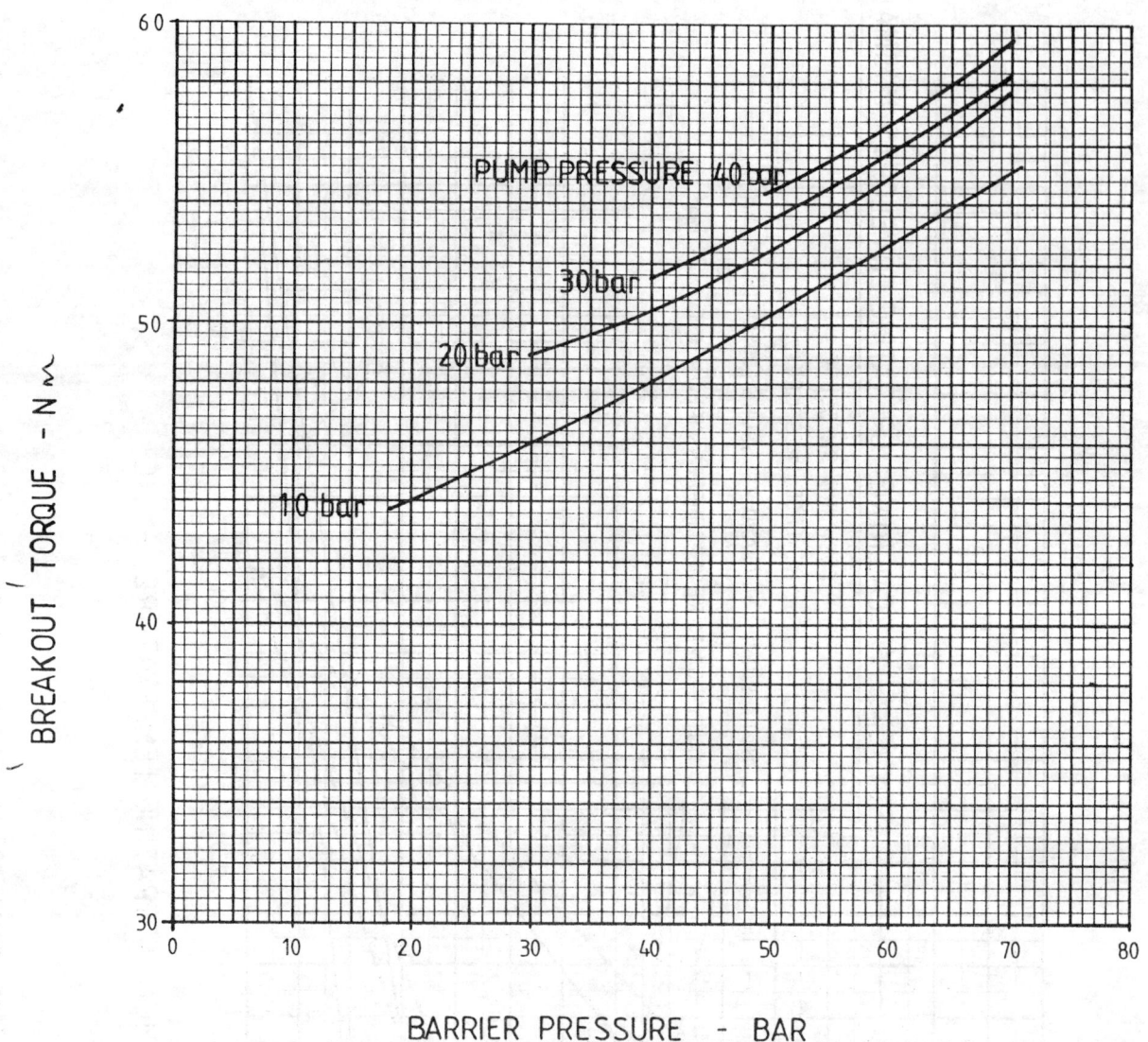

DOUBLE SEAL 'BREAK-OUT' TORQUES

TORQUE VERSUS BARRIER PRESSURE FOR
VARIOUS PUMP PRESSURES
EXPERIMENTAL VALUES

Figure 7a Double Seal 'break-out' torques
(67mm size seal)

SEE APPENDIX FOR DERIVATION OF EQUATIONS.

1. MODEL 1 – O/D CONTACT

FACE OPENING FORCE F_o
NETT CLOSING FORCE F_n

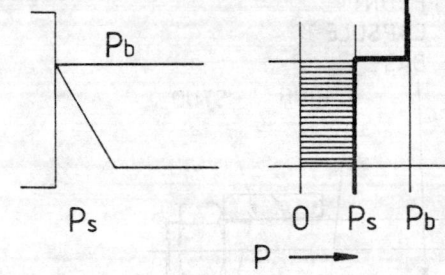

$F_o = P_s \cdot A_f$

$F_n = [F_s + P_b \cdot B \cdot A_f] - P_s \cdot A_f \cdot B$

2. MODEL 2 – PARALLEL FILM

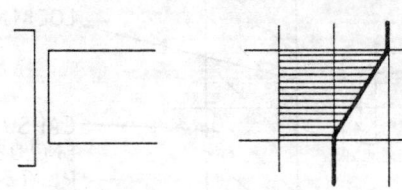

$F_o = \left(\dfrac{P_b + P_s}{2}\right) A_f$

$F_n = [F_s + \Delta p \cdot A_f (B - \tfrac{1}{2})]$

$\Delta p = (P_b - P_s)$

3. MODEL 3 – NO FILM

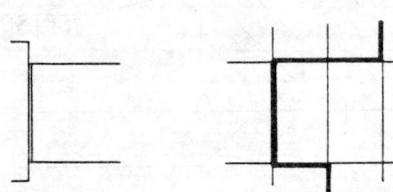

$F_o = 0$

$F_n = [F_s + P_b \cdot B \cdot A_f] + P_s (1 - B) A_f$

4. MODEL 4 – I/D CONTACT

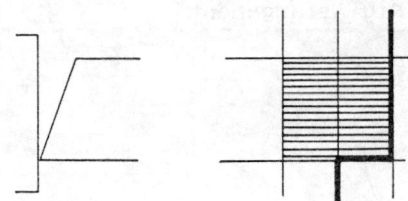

$F_o = P_b \cdot A_f$

$F_n = [F_s + P_b (B - 1) A_f] + P_s \cdot A_f (1 - B)$

Figure 7b Hydrostatic Conditions at Inner Seal Face

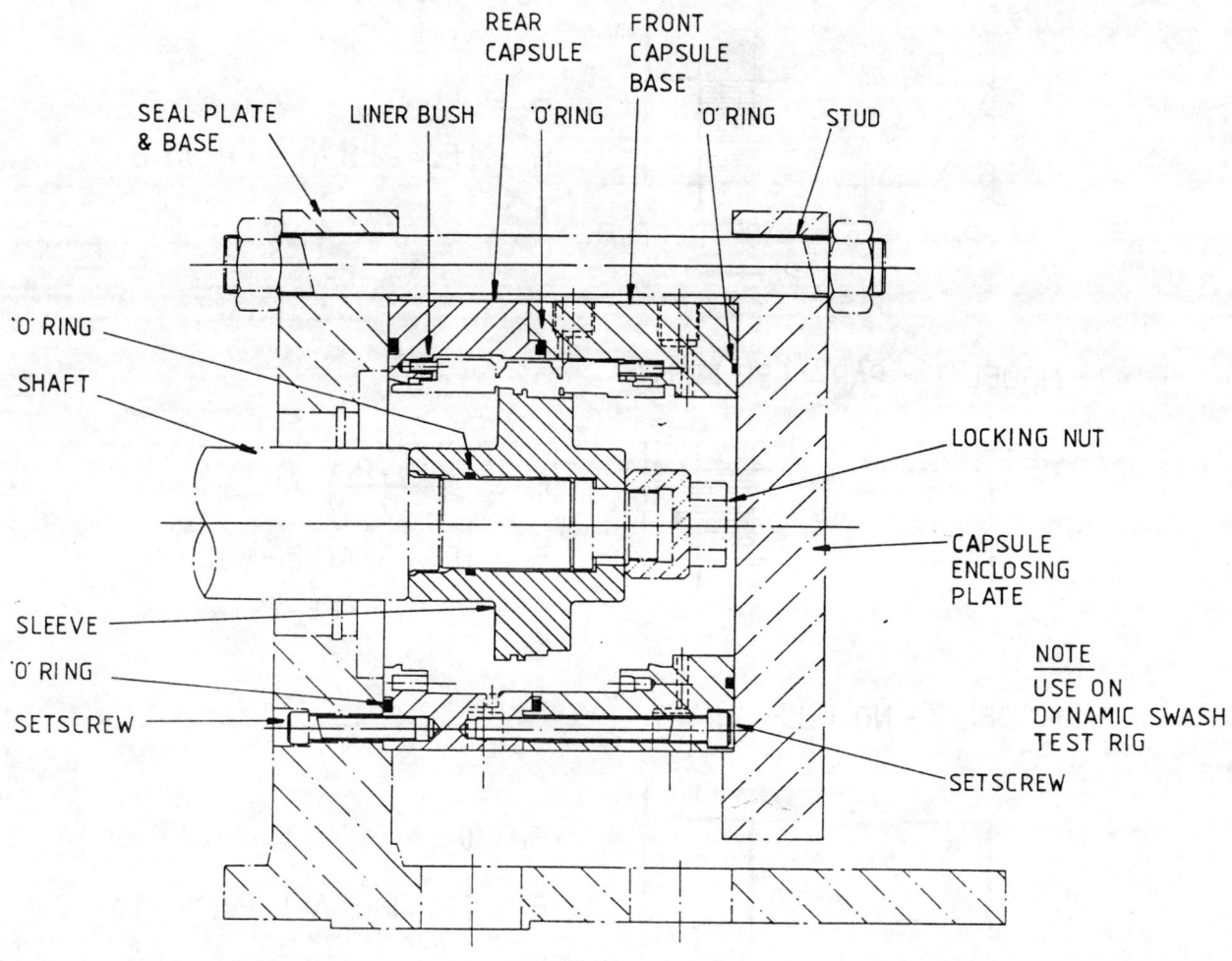

Figure 8   Seal Test Rig Arrangement

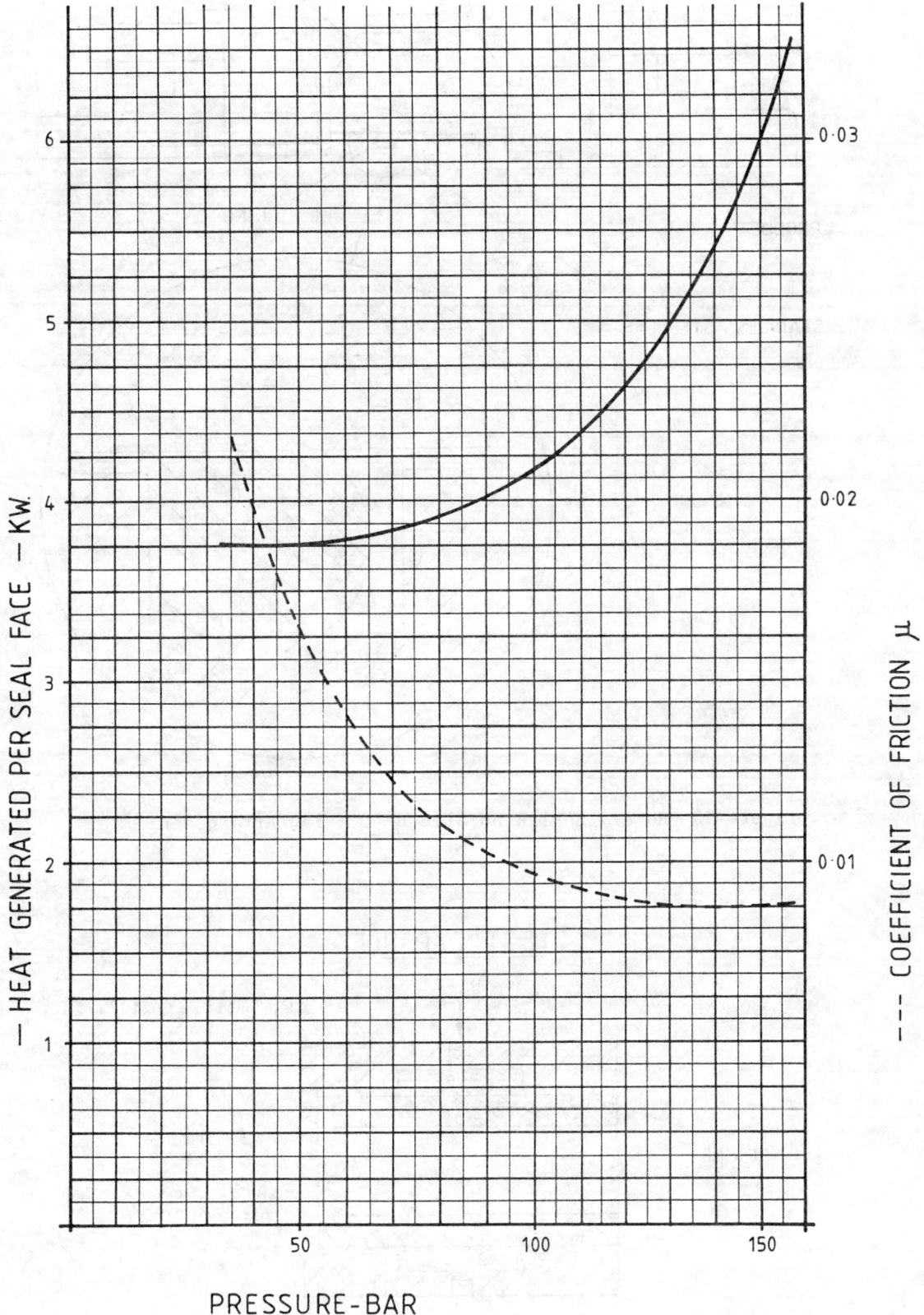

Figure 9   Heat Generated and Friction

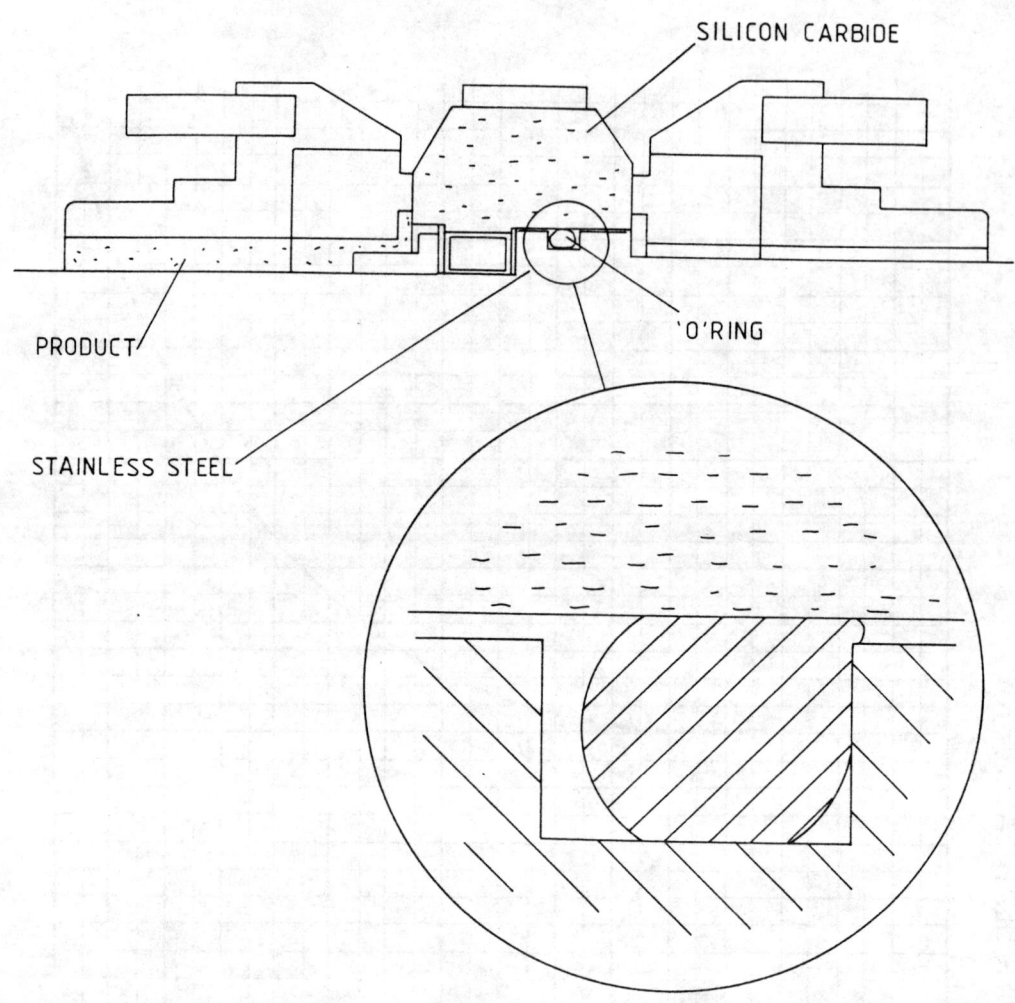

Figure 10   Rotary Seal Ring/Sleeve 'O' Ring Arrangement

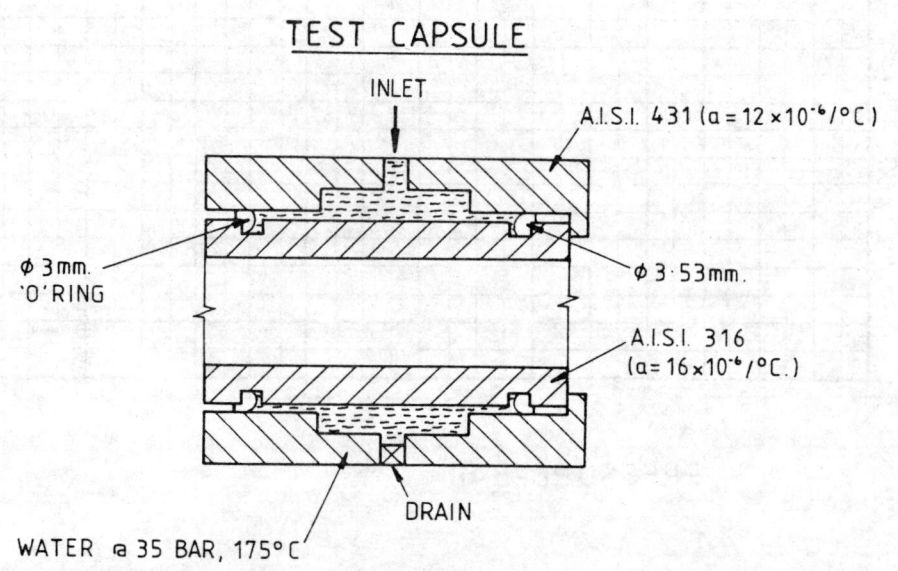

Figure 11   'O' Ring Test Capsule

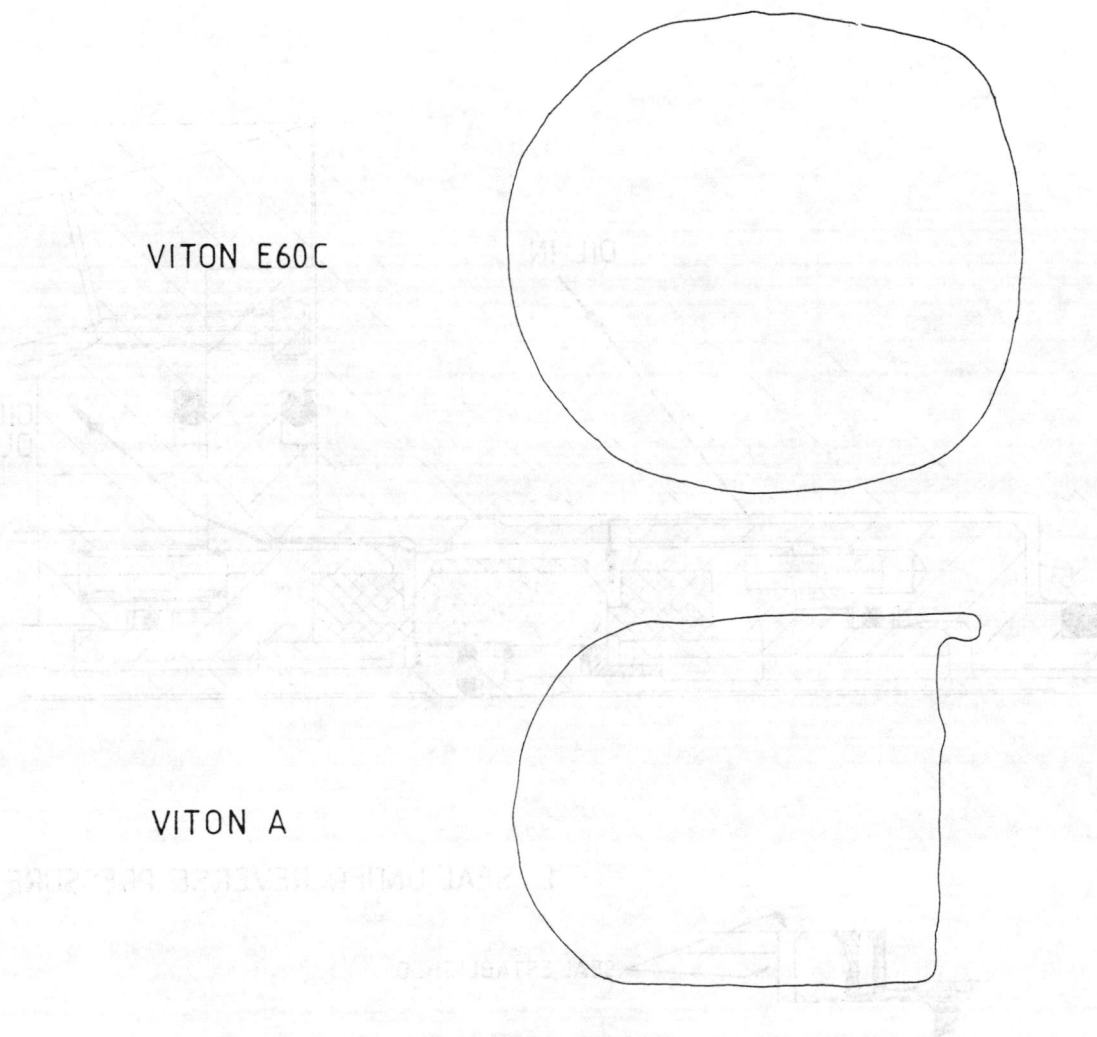

Figure 12 'O' Ring Profiles After Test

Figure 13 Inboard Stationary Seal Ring Showing Anti-Extrusion Washer As Dismantled

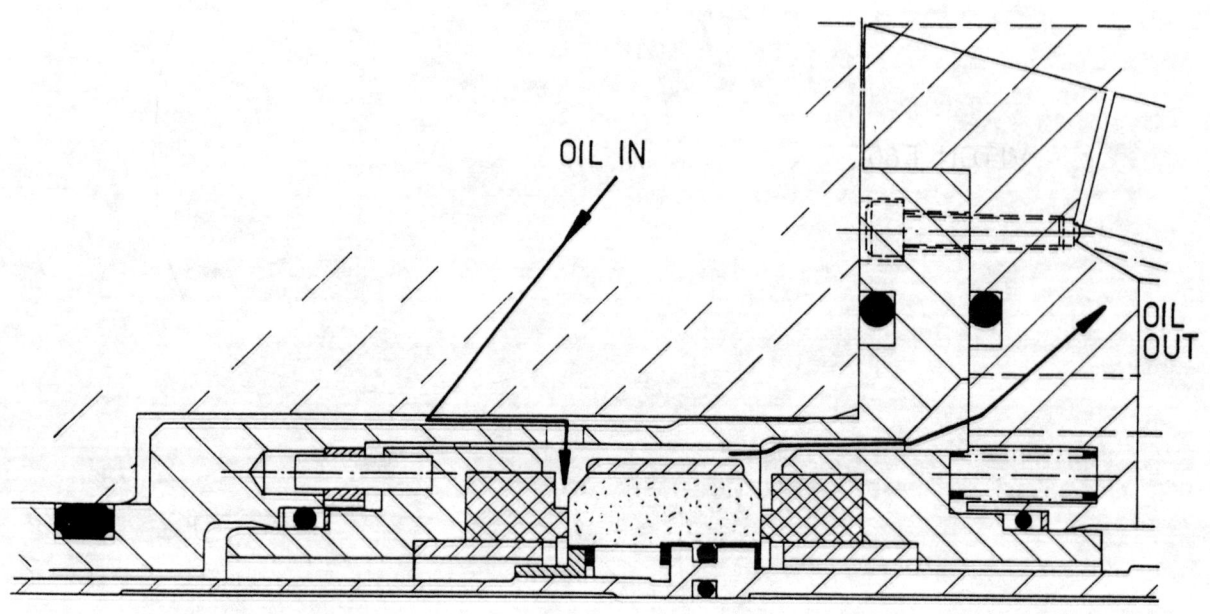

1. SEAL UNDER REVERSE PRESSURE

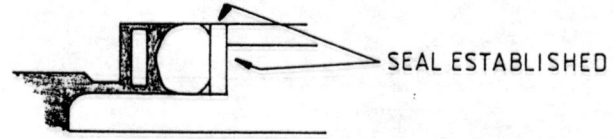

2. BARRIER PRESSURE APPLIED

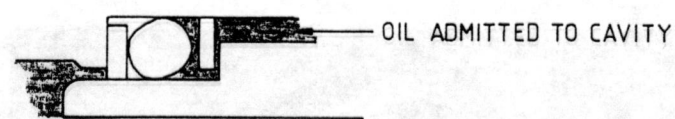

3. UNDER NORMAL RUNNING CONDITIONS

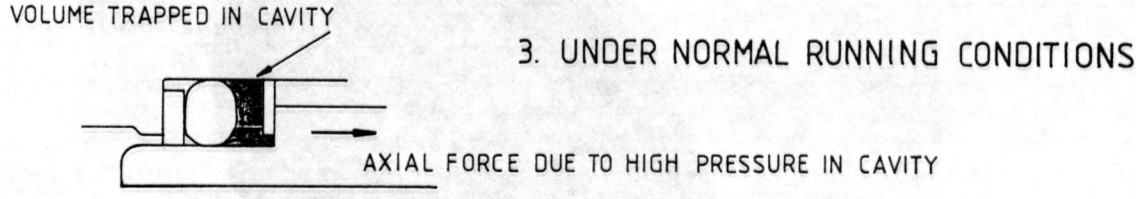

Figure 14a Pressure Trapping Mechanism

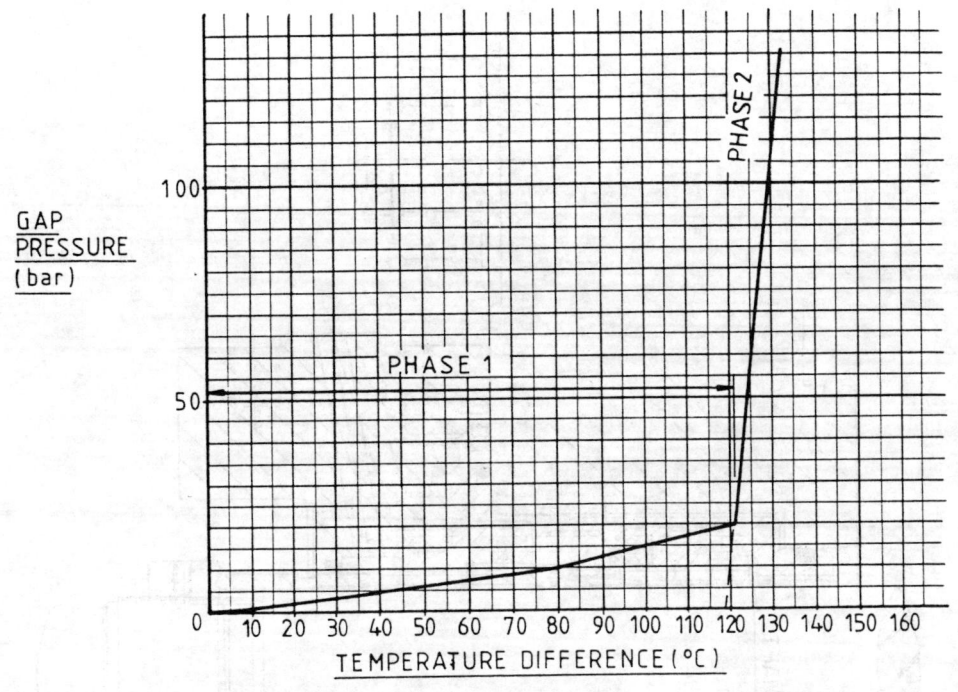

Figure 14b Theoretical Pressure 'build-up' in Gap

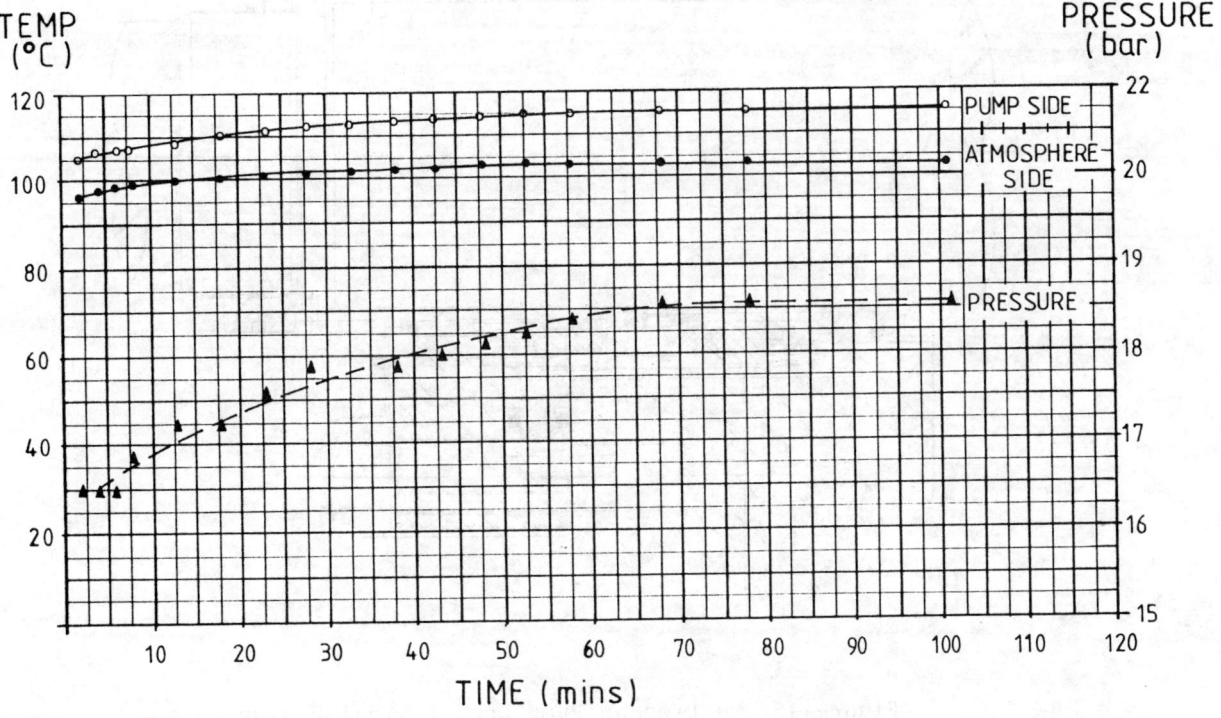

Figure 15 Temperature/time Relationship of Inboard and Outboard Seals. Also shown is pressure build up in gap with $P_b = 9.65$ bar.

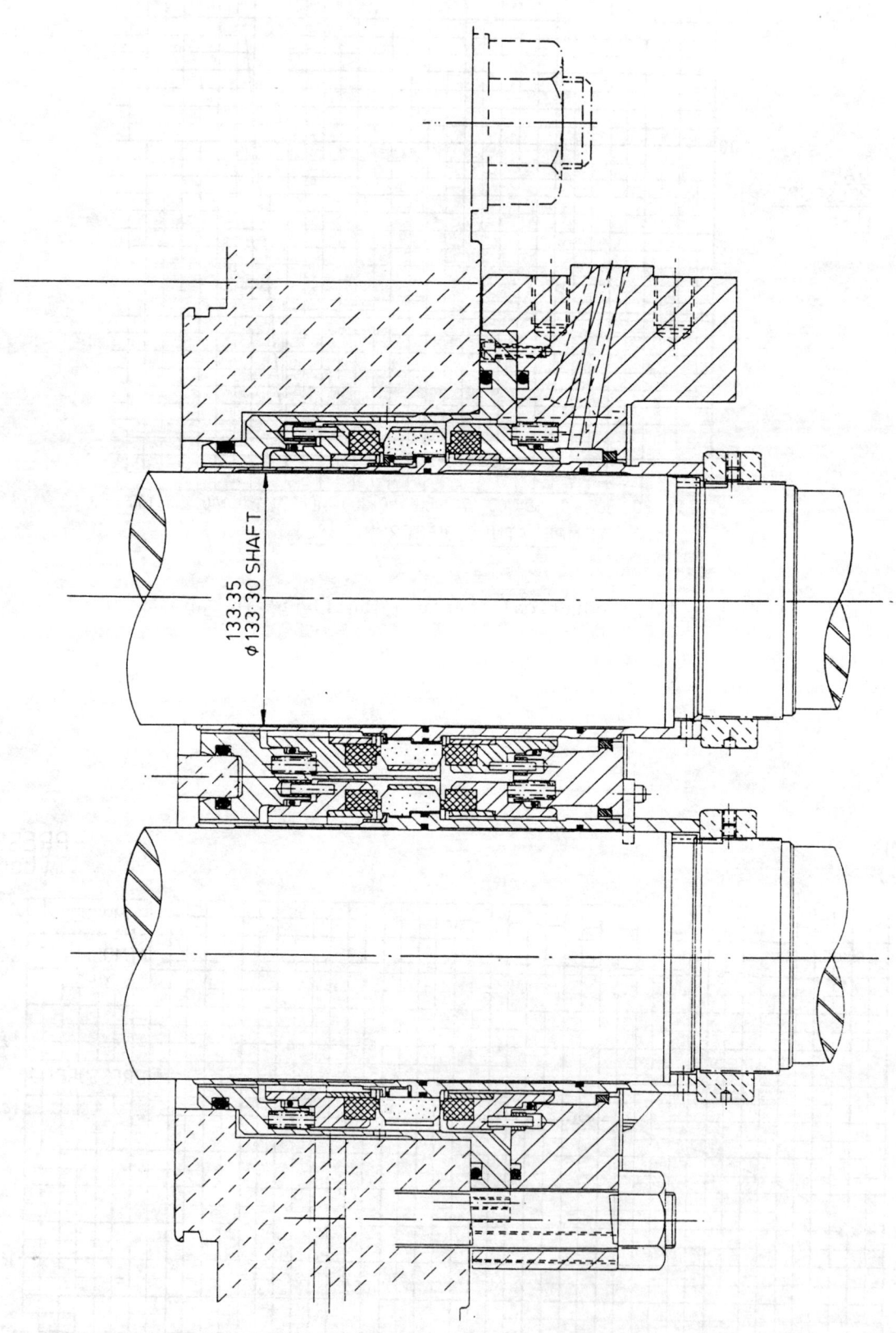

Figure 16 Multiphase Pump Seal - Final Design

11th International Conference of the
British Pump Manufacturers' Association

New Challenges – Where Next?

18-20 April, 1989
Churchill College, Cambridge

PAPER 15

THE DEVELOPMENT OF A SEAL FOR SUBSEA PUMPS
– A CHALLENGE FOR THE SEAL MANUFACTURER

W. Schmoeller
Dr. D. Zeus
G. Sivertsen

Feodor Burgmann Dichtungswerke GmbH & Co., D-8190 Wolfratshausen
FRAMO A/S, Nesttun/Norway

Summary

The mechanical seal HRS4/64 was developed for a down hole-pump.
The maximum operating limits of this seal are as follows:

- Speed   6000 rpm
- Differential pressure   15 MPa
- Buffer fluid temperature   130 °C

Because of the down hole operating condition an extremely high
reliability was required.

Deposits of crude-oil endanger proper seal function. Therefore
an internal pressurized design concept was chosen. The face
materials had to be corrosion resistant, therefore the only
material to be used was silicon carbide.

To meet the above requirements the design was optimized by means
of an FE-analysis and excessive test series on different test
rigs. The results of the analysis and of the tests will be
presented and discussed.

    1.   Introduction
    2.   Problem
    3.   Solution of the problem
    4.   Design
    5.   Finite-Element-Analysis
    6.   Test and test results
    6.1 Test procedure
    6.2 Test conducted by F. Burgmann Company
    6.3 Tests conducted by Framo A/S
    6.4 Final discussion of the results
    7.   Conclusion

Conference organised and sponsored by the British Pump Manufacturers' Association
in conjunction with NEL and BHRA, The Fluid Engineering Centre.
Co-sponsored by the Process Industries Division of the Institution of Mechanical Engineers.

1. Introduction

During the last four years a completely new electrically driven downhole pumping system was developed by Framo Engineering A/S. High reliability and flexibility in downhole pumping applications like

- oil production
- water lift for re-injection
- water supply
- geothermal reservoirs

were the main design criteria. Oil production from deep, high temperature, sour and often highly deviated wells makes extremely high demands on the pumping system. These extreme requirements define therefore the layout of the pump.

The basic system is shown in Fig.1 as a typical two-pump offshore installation. The pumps are installed inside a 7" OD production tubing and connected with each other and with an external supply unit by means of an electric/hydraulic pipe-stack. At the desired pump setting depth the lower pump will be landed on a seal nipple landing shoulder allowing for predetermined space-out for the operation of a deep-set downhole safety valve. The well output is transferred in the annulus between the ID of the production tubing and the OD of the pipestack.

The pump unit (Fig.2) is of an axial flow design with the following specifications:

- Capacity   90 m^3/h = 13600 BPD
- Head   310 mlc = 1017 ft
- Maximum speed   6000 rpm
- Power consumption   100 kW
- Number of stages   12
- Length   3100 mm
- Maximum OD   140 mm

The pump unit consists of two main sections, the electrical motor assembly and the impeller/diffusor assembly.

The motor assembly consists of a rotor/stator package, a mechanical seal, one thrust bearing and two journal bearings. The impeller/diffusor system is mounted on a rugged hollow shaft and supported by two oil-lubricated bearings. It also includes a second mechanical seal mounted to the lower pump end.

For motor cooling and bearing lubrication turbine oil is pumped through the pipestack to the pump-unit and back again.

This new downhole pump concept requires special high duty mechanical seals with high reliability. The design requirements are based on the following seal operating conditions:

- Differential pressure   15 MPa (dynamic, buffer fluid/crude oil)
                           5 MPa (static, crude oil/buffer fluid)

- Speed   < 6000 rpm
- Temperature   60 - 130 °C
- Medium   crude oil
- Buffer fluid Aeroshell Turbine Oil 750

2. Problem

Generally mechanical seals are applied at nearly constant operating conditions. This is ideal, because after a certain running-in period and dependent on design and face materials the face distortions are compensated. The sliding faces have adjusted themselves to the obvious operating conditions. Wear and leakage decrease. When conditions change the sliding faces have to readjust themselves. Wear and leakage will increase.

The operational data - speed, pressure, temperature - of the down hole pump can vary within the mentioned range during the running operation. Corresponding face distortions are the result. The running-in period will become the ordinary continous operation and will therefore define leakage and wear.

All this results in the demand that the influence of the operating parameters on the distortion should be minimized to establish good running behaviour over the large operating range.

The above mentioned correlation and the demand of an extremely high reliability are intensified by the medium crude oil to be sealed off. Due to solid content and disolved salts this medium is abrasive and chemically aggressive. In addition to this it is prone to build up deposits in the area of the dynamic O-ring and in the gap between shaft and seal face. Free axial movement of the seal face is hampered.

The high medium temperatures combined with the frictional heat in the seal gap can cause thermal decomposition of the buffer fluid oil. The oil film in the seal gap may disintegrate into cracking and polymerisation products. These may form a lacquer-like hard and uneven layer on the sliding faces. Both processes lead to excessive leakage.

3. Solution of the problem

The solution of the described problems is mainly influenced by

- the face materials
- the basic design concept of the mechanical seal
- the circulation of the buffer fluid
- the buffer fluid

Basically the resistance against abrasion and corrosion of the face materials should be very high. Additionally the face materials should be of high strength to minimize the load-dependent distortion of the sliding faces. Tungsten carbide is not suitable because the chemical quality of crude-oils varies. It is greatly influenced by a certain number of disolved salts, which bring the pH-value below 7. Selective corrosion of the binder will be the result. Consequently ceramics are the only materials to be used. Due to low thermo-shock resistance and high brittleness aluminium oxide as well as directly sintered silicon carbide are out of question.

Reaction sintered silicon carbide (SiC-Si) and silicon impregnated electrographite (SiC-C-Si) however meet all the demands. Therefore the following combinations of face materials are possible

- SiC-Si / SiC-C-Si
- SiC-Si / SiC-Si

The physical data of these mentioned face materials are listed in Table 1.

The reaction sintered silicon carbide has a bimodal grain structure (Fig.3). This structure has a positive influence on the sliding behaviour especially at high pressure and speed. In addition to this, brittleness is smaller than with fine grain material.

The silicon impregnated electrographite is a compound material. It consists of 60 % Sic, 35 % C and 5 % Si. The structure of this material can be understood as a SiC coarse screen cross-linked with a C coarse screen (Fig. 4).

To overcome the problem of a jammed seal face the seal must be internally pressurized by the cooling oil. The seal face has to be stationary. That means that the dynamic O-ring is arranged between the housing and the outside of the seal face. Thus deposits in the O-ring area cannot hamper axial movements of the seal face. Additionally the spring compartment is protected against sedimentation.

The circulation of the buffer fluid must be optimized to obtain good heat dissipation and uniform cooling.

As buffer fluid thermally stable turbine oils with metal-free additives must be used. Thus lacquer-like uneven layer on the sliding faces will be avoided.

4. Design

The original of the HRS1/78 seal (Fig. 5) is based on this concept. During the tests the design of seal face and counter face were modified. This led to a reduction of diameter.

The modified seal HRS3/64 (Fig. 5) is characterized by the following features:

- The HRS3/64 is a single-acting, pre-assembled, mechanical seal with a rotating counter face and a stationary seal face. It is pressurized internally.

- The rotating counter face is solid and made of reaction sintered silicon carbide. It is freely inserted into a housing ring. The ceramic ring is protected against fracture, which would cause sudden seal failure, by a metallic safety ring. The torque transmission is reached by an anti-rotation device between housing ring and metallic safety ring as well as by shrink fit between safety ring and seal face.

- The design concept of the stationary seal face corresponds to that of the counter face.

- Small seal width leads to low power consumption. This and the high heat conductivity of the face materials result in moderate temperatures in the seal gap.

- As secondary seal between face housing and seal housing an O-ring with a metallic back-up ring is installed.

- Multiple springs act on the housing ring via a thrust-collar. Two torque pins act as anti-rotation device.

- The distribution of the buffer fluid is optimized with regard to low turbulence losses and high heat dissipation. Smooth contours and surfaces reduce turbulence losses. The buffer fluid can be pumped via radial bores in the shaft sleeve from the hollow shaft to the seal gap or vice versa. Due to centrifugal forces no flow guide is required. All seal components, especially the face materials, are cooled uniformly. Local hot spots are avoided.

5. Finite-Element-Analysis

To reduce the number of tests as well as the test time a FEA was completed. For structure-, stress- and thermal analysis the FE program ANSYS was used.

Ceramic materials are very sensitive to tensile strain. In our case the situation is intensified by the internally pressurisation. Therefore we examined the cross sections of the counter face and the seal face as well as of the housing rings and of the safety rings as far as shrink-condition and load range were concerned. We could nearly eliminate the tensile stress in the ceramic face materials by modifying the cross sections of the rings and the shrink condition (Fig. 6).

The sliding face distortion is influenced by design, speed, differential pressure and temperature distribution. The gap shape of the original design (Fig. 5) was not stable over the whole load range. Pressure and temperature effects caused a change from A- to V-gap. Contrary to this the sliding faces of the redesigned HRS3/64 seal were always distorted to a V gap. We took this distortion into account when we specified the necessary radial taper of the sliding faces so that a seal gap with parallel faces at full load was reached.

Thermal analysis was necessary to obtain information about seal gap temperature and temperature distribution. 170°C in the outer part of the gap were found to be the maximum temperature achieved. The high thermal conductivity of the face materials resulted in a low temperature gradient (Fig. 7).

6. Test and test results

6.1 Test procedure

To find the operating limits of the seal and to optimize them due to the given performance specifications, the seal was tested on a test rig at the seal manufacturer's. After this first qualification step an acceptance test with a pump-like arrangement was carried out by the pump manufacturer. A final field test in an original pump should be the last qualification test. This test series carried out step by step together with the pump manufacturer was necessary because of the extreme performance specification and the high reliability asked for.

6.2 Tests conducted by F. Burgmann Company

The HRS design together with the combination of face materials were tested on a high-pressure test rig, which was equipped with a static frequency transformer for speed variation. To compensate the axial thrust caused by pressure, one seal was mounted on each end of the shaft (Fig. 8). The test conditions complied with the performance specifications. Speed, pressure and temperature were varied according to Fig. 9.

During the test series of the HRS1/78 the shapes of counter face and seal face were modified in order to minimize the influence of temperature, speed and pressure on the leakage. The result was not satisfying because seal face and counter face distortion changed between A- and V-gap depending on the running condition. Further modifications analysed by FEA signaled only gradual improvements. At this time it was decided to redesign the seal. HRS2/64 and HRS3/64 were born (Fig. 10).

Seal-arrangement, test set-up and procedure were identical to Fig. 8 and 9. Under absolutely identical conditions the leakage rate of the new design was 30 % of that of the original design. To guarantee a non-shifting shrink-fit between safety ring and face seal made of silicon impregnated electrographite, a support ring was necessary (Fig. 10, HRS2/64). In relation to the HRS3-design this solution caused an increase of leakage. Therefore the HRS3/64 was favoured. Fig. 11, 12 and 13 demonstrate the influence of differential pressure, speed and temperature on the leakage rate. This result is based on a test series of 6 seals. No traces of wear could be detected on the sliding faces.

Based on these results further tests were carried out at the test lab of Framo A/S.

6.3 Tests conducted by Framo A/S

The test set-up is shown in Fig. 14. The hollow shaft is driven by a hydraulic motor. The arrangement of the HRS3/64 seals on the hollow shaft corresponds to the later arrangement in the pump unit. The same applies to the guidance of the buffer fluid. The test parameters pressure, speed and temperature were varied according to Table 2.

During all tests the leakage rates at pressure (Fig. 15), speed and temperature below maximum load were higher than the rates measured at the Burgmann test-lab. Corresponding to the two anti-rotation pins two contact zones on the sliding faces were visible. This fact and the desire to further minimize the leakage brought about another modification of seal face and counter face (Fig. 10, HRS4/64). An excessive drop of leakage was the result (Fig. 15).

After a 100 hours test at full load only a faint trace of wear could be detected at ID. There were no visible local contact zones, that means there was no face waviness. Cracking of the turbine oil was avoided due to the high thermal conductivity of the face material and the effective cooling by the buffer fluid.

6.4 Final discussion of the results

When we compare the results of the HRS3/64 seals tested at Framo A/S and the F. Burgmann Company a difference in the measured leakage rates is given. This difference is due to

- different test rigs
- different test set-ups
- different flow rates of the buffer fluid (ratio: 1:10)

Temperature distribution and the temperature itself were influenced by these variations as well as the distortion of the sliding faces.

The seal face and the counter face of the HRS4/64 seal have metallic safety rings with symmetrical cross-sections and a radial torque transmission between safety rings and housing rings. Any moment about the shear center of the cross-section is avoided. These design features caused the reduction of loaddependent face distortion as well as the drop of leakage.

7. Conclusion

The design of the internally pressurized HRS4/64 seal and the combination of the face materials SiC-Si/SiC-Si were successfully tested up to 6000 rpm, 130°C and an internal pressure of 15 MPa.

The design of seal face and counter face - solid ceramic rings with metallic safety rings - guarantees a high reliability, which is necessary for off-shore application.

The selected face material - reaction sintered silicon carbide - is corrosion proof against any type of crude oil.

Disintegration of the turbine oil is avoided due to the high thermal conductivity of the reaction sintered silicon carbide and due to effective cooling of seal face and counter face.

This development shows that for a special pump application it is absolutely necessary for the seal and the pump manufacturer to work closely together right from the beginning in order to optimize the whole system. The seal is a main part of the pump.

Table 1: Physical data of different silicon carbides

Face Material / Parameter		Solid-material		Compound-material	
		SiC 98-100% SiC <2% C	SiC-Si 86-92% SiC 8-16% Si	SiC-C-Si 60% SiC 35% C 5% Si	C-SiC 80% C 12% SiC 8% resin
Density ρ	(g/cm^3)	3.05 - 3.2	3.05 - 3.1	2.65	1.95
Compressive strength σ_D	(N/mm^2)	2000 - 3900	1200 - 3000	600	82.7
Modulus of elasticity $*10^4$	(N/mm^2)	39 - 45	30 - 41	13.5	1.59
Hardness	(Skler)	2500 - 2600	1500 - 2900	125 (HR)	90 (HR)
Coefficient of thermal expansion α $*10^{-6}$	(1/K)	2.9 - 4.5	2.9 - 4.6	3.0	2.7 - 5.8
Thermal conductivity λ	(W/mK)	89 - 115	100 - 130	125	50 - 52

Table 2: Test parameter of the Framo acceptance test

Temperature t (°C)	60						100					
Speed n (rpm)	4500			6000			4500			6000		
Pressure p (MPa)	5	10	15	5	10	15	5	10	15	5	10	15

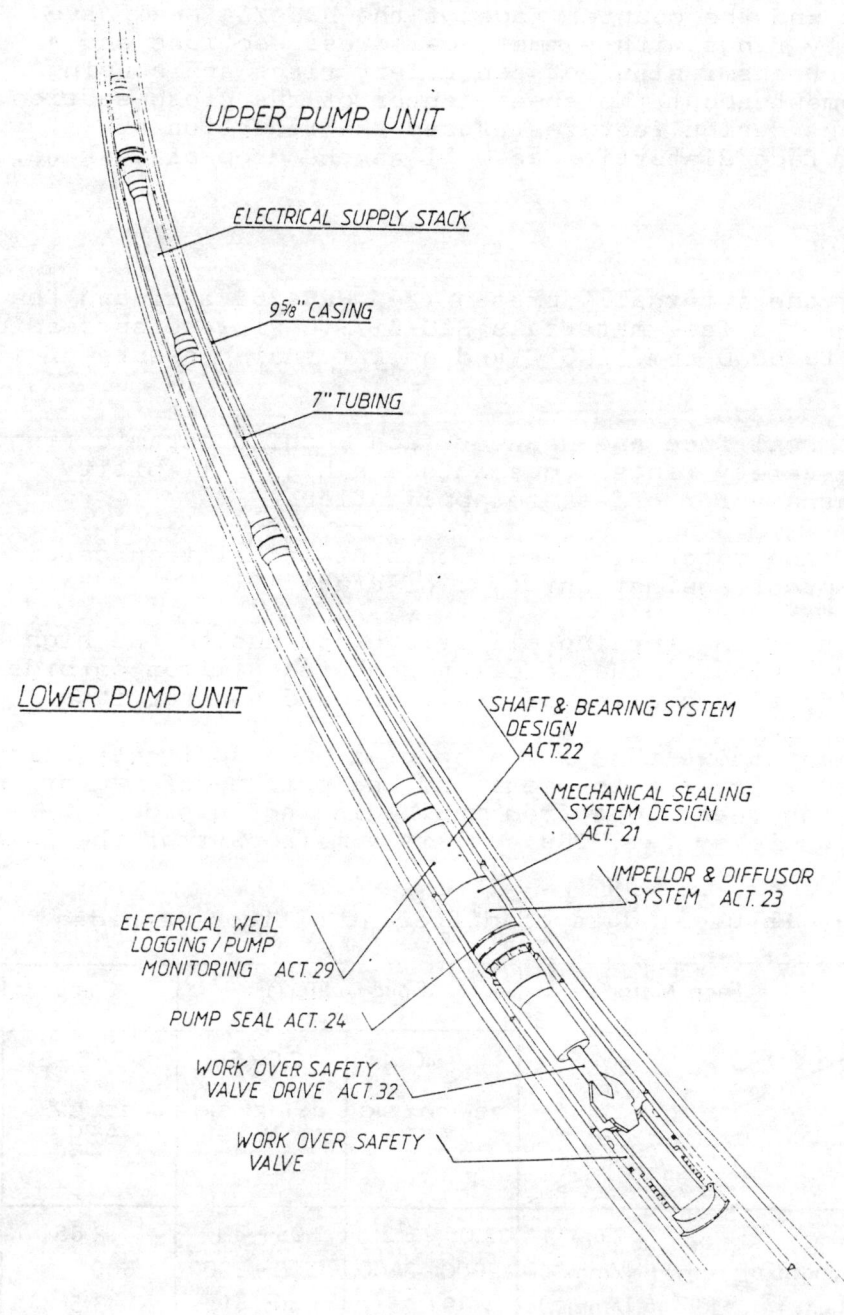

Figure 1: General System concept for off-shore installation of the down hole pump

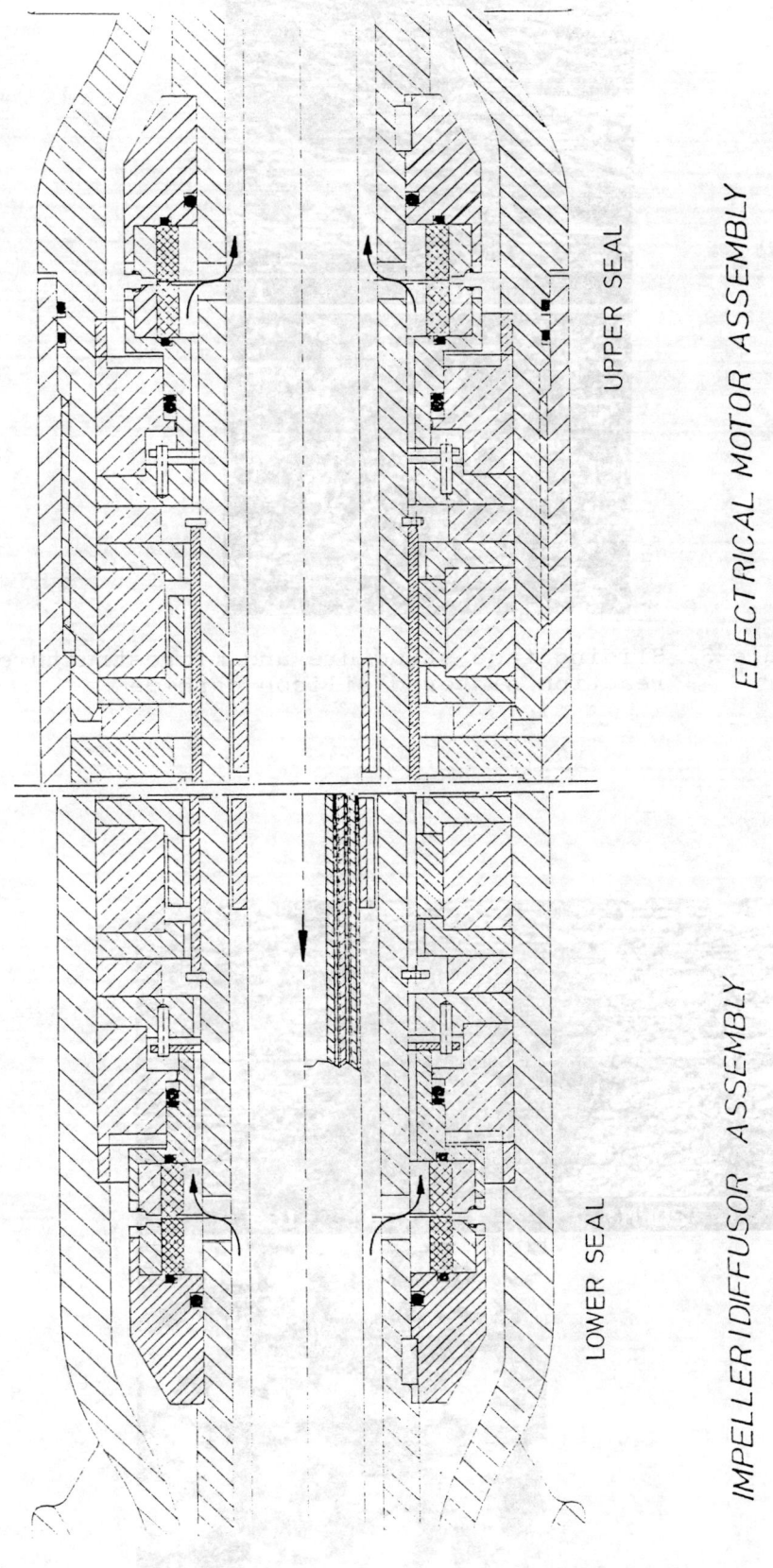

Figure 2: Cross-section of the down hole pump

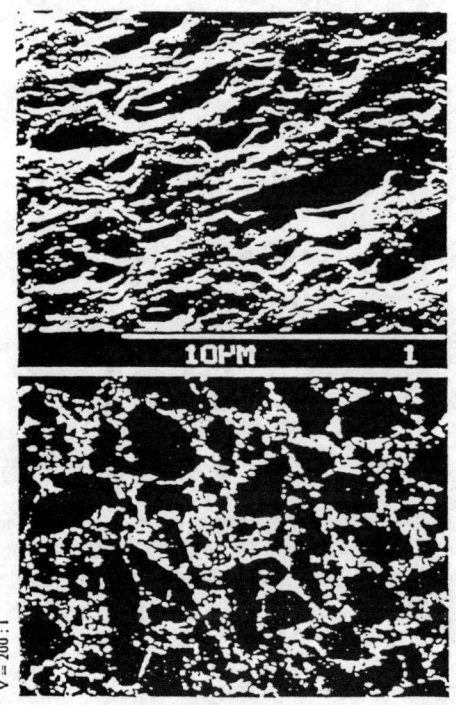

Figure 3: Sliding face structure and microstructure of reaction sintered silicon carbide

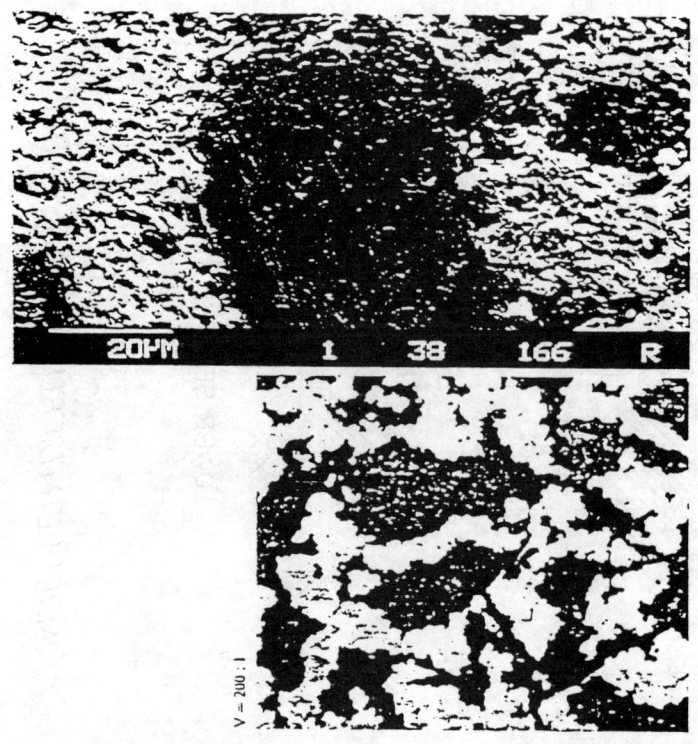

Figure 4: Sliding face structure and microstructure of silicon impregnated electrographite

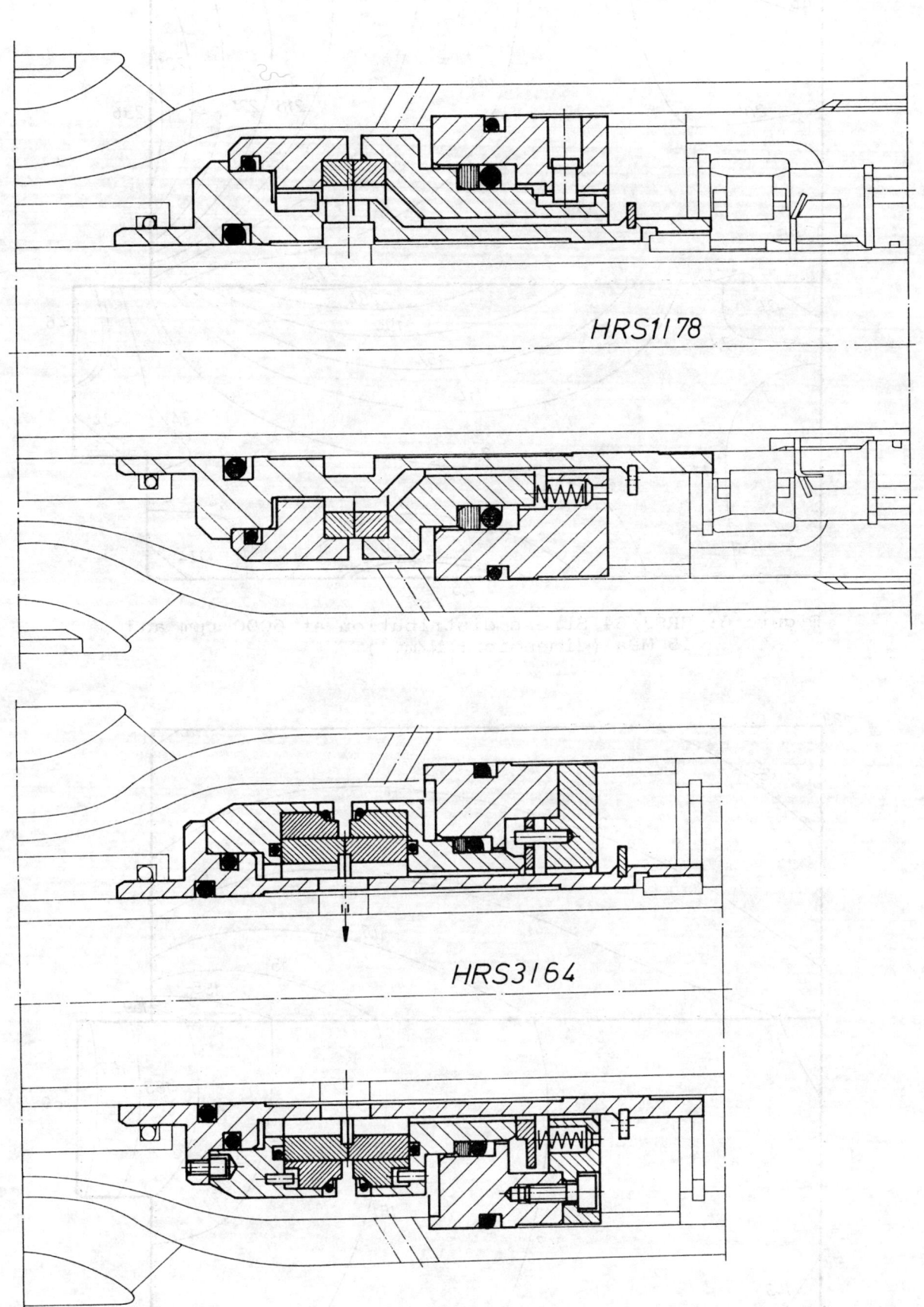

Figure 5: Comparison of the HRS1/78 and the HRS3/64 seal

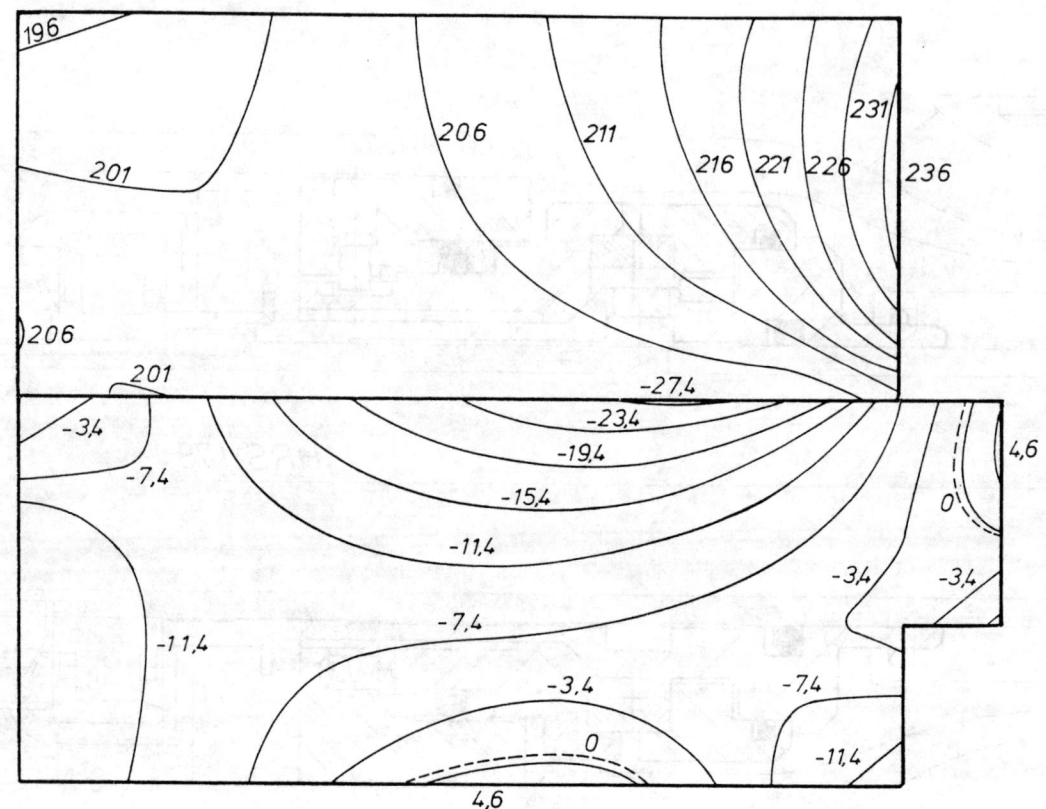

Figure 6: HRS3/64 Stress distribution at 6000 rpm and 15 MPa (dimension: N/mm²)

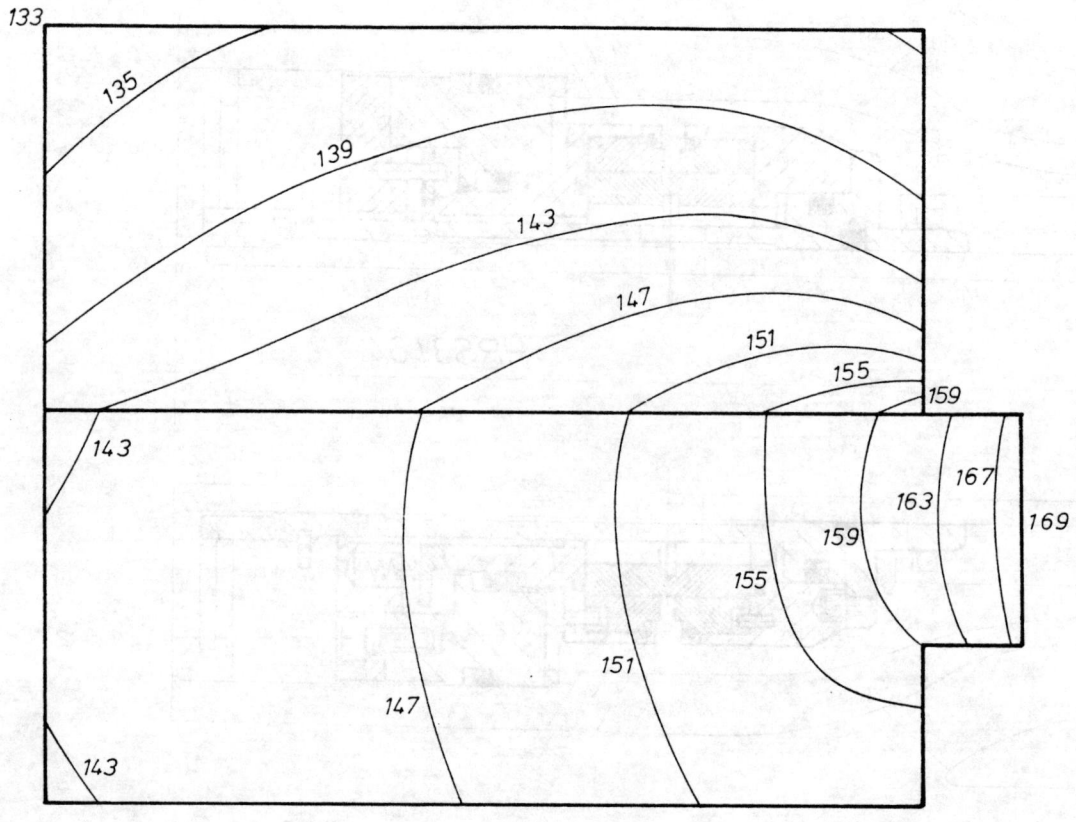

Figure 7: HRS3/64 Temperature distribution at 6000 rpm and 15 MPa (dimension: °C)

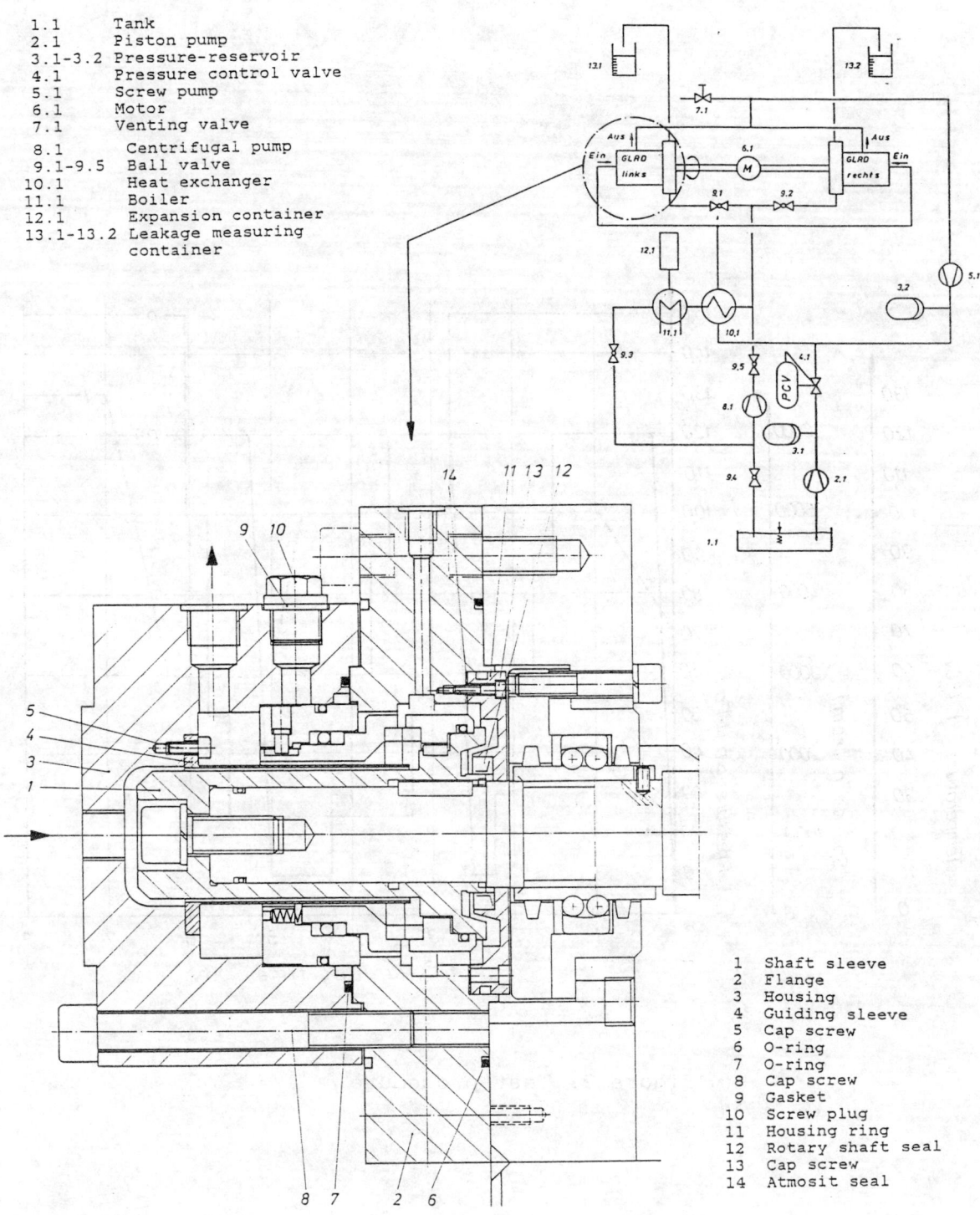

1.1	Tank
2.1	Piston pump
3.1-3.2	Pressure-reservoir
4.1	Pressure control valve
5.1	Screw pump
6.1	Motor
7.1	Venting valve
8.1	Centrifugal pump
9.1-9.5	Ball valve
10.1	Heat exchanger
11.1	Boiler
12.1	Expansion container
13.1-13.2	Leakage measuring container

1	Shaft sleeve
2	Flange
3	Housing
4	Guiding sleeve
5	Cap screw
6	O-ring
7	O-ring
8	Cap screw
9	Gasket
10	Screw plug
11	Housing ring
12	Rotary shaft seal
13	Cap screw
14	Atmosit seal

Figure 8: Seal arrangement

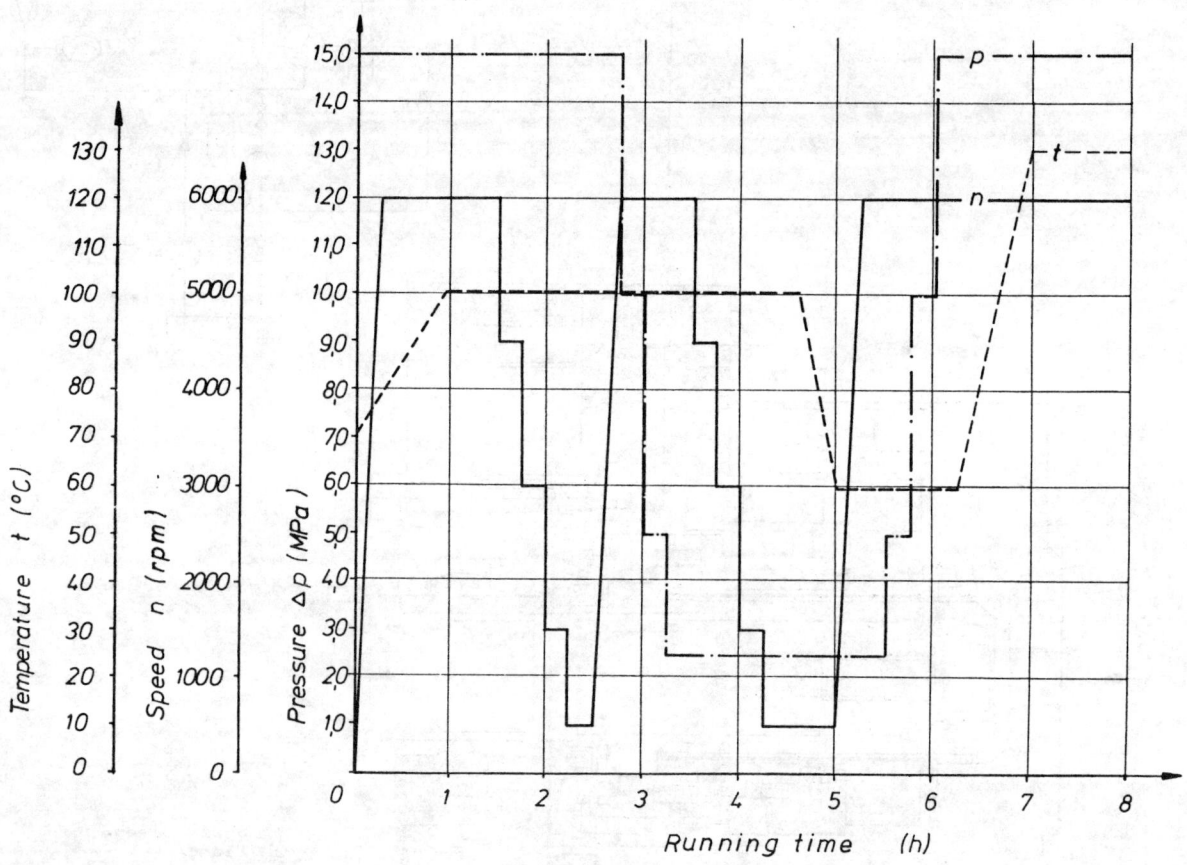

Figure 9: Test procedure

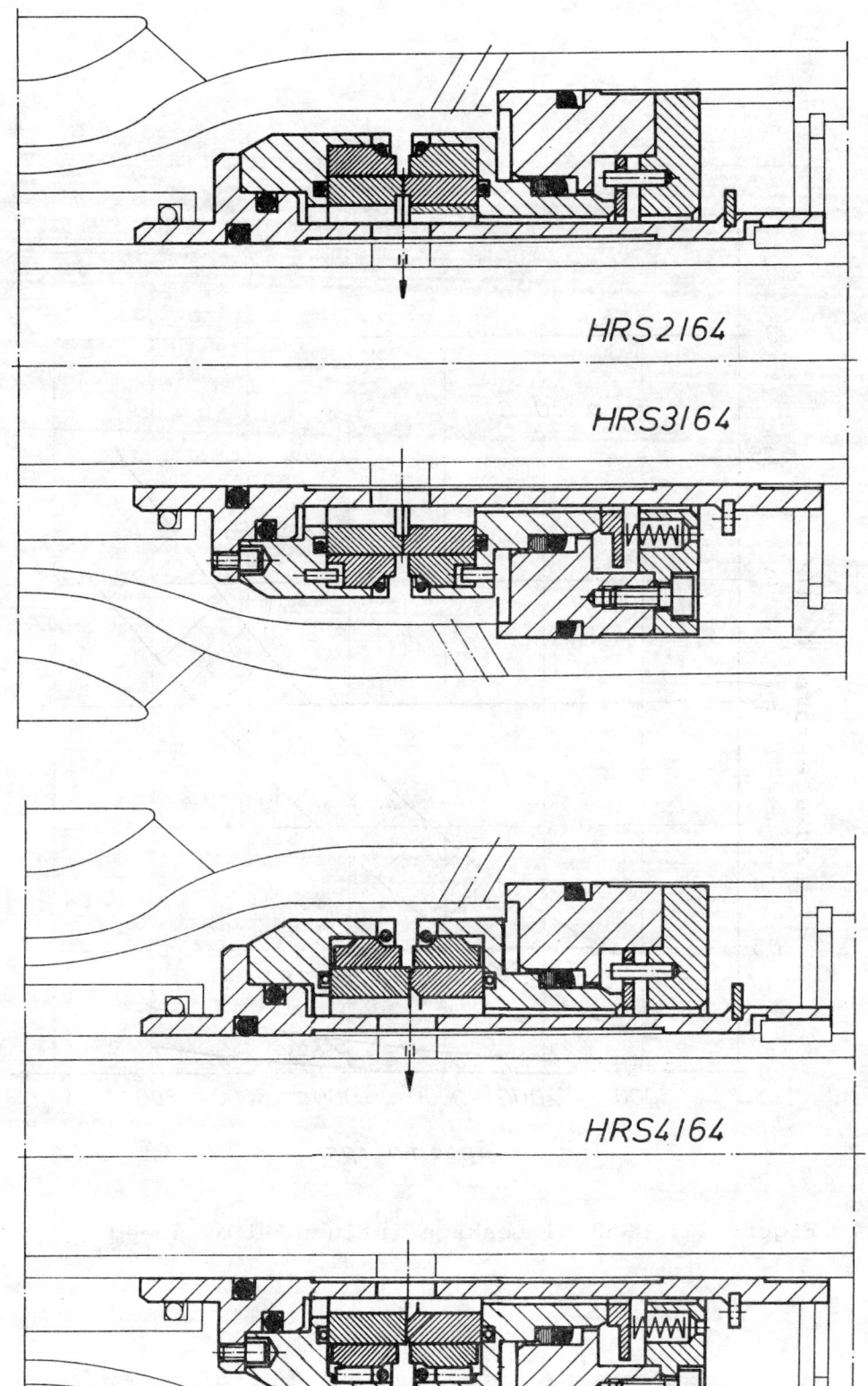

Figure 10: Comparision of the HRS2/64, the HRS3/64 and

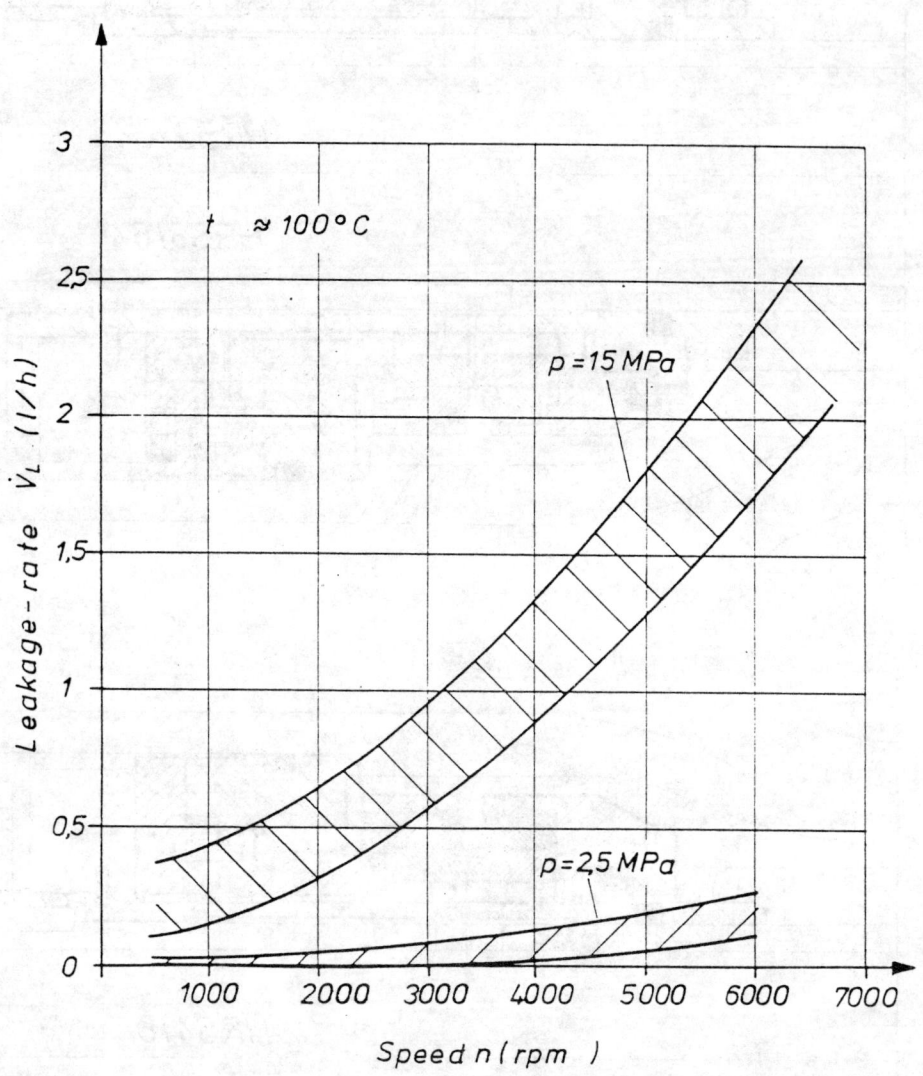

Figure 11: HRS3/64 Leakage influenced by speed

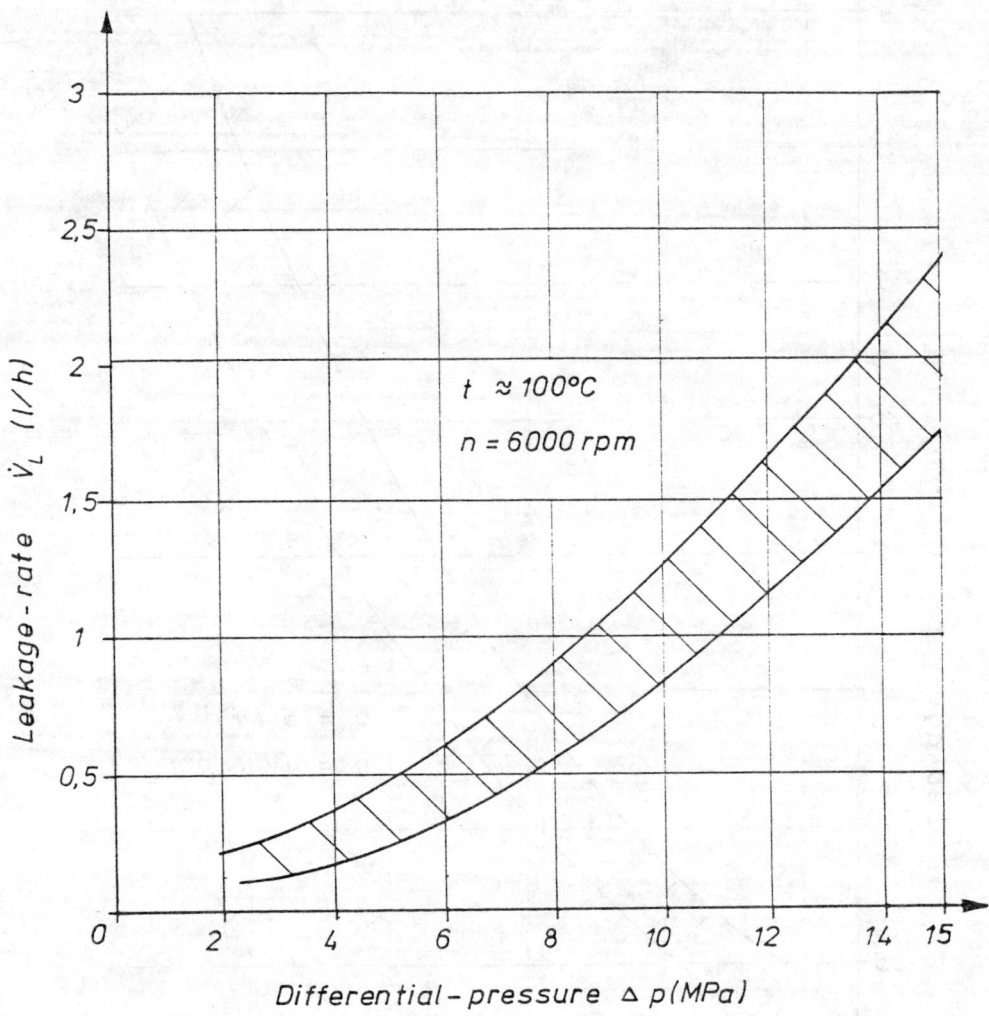

Figure 12: HRS3/64 Leakage influenced by pressure

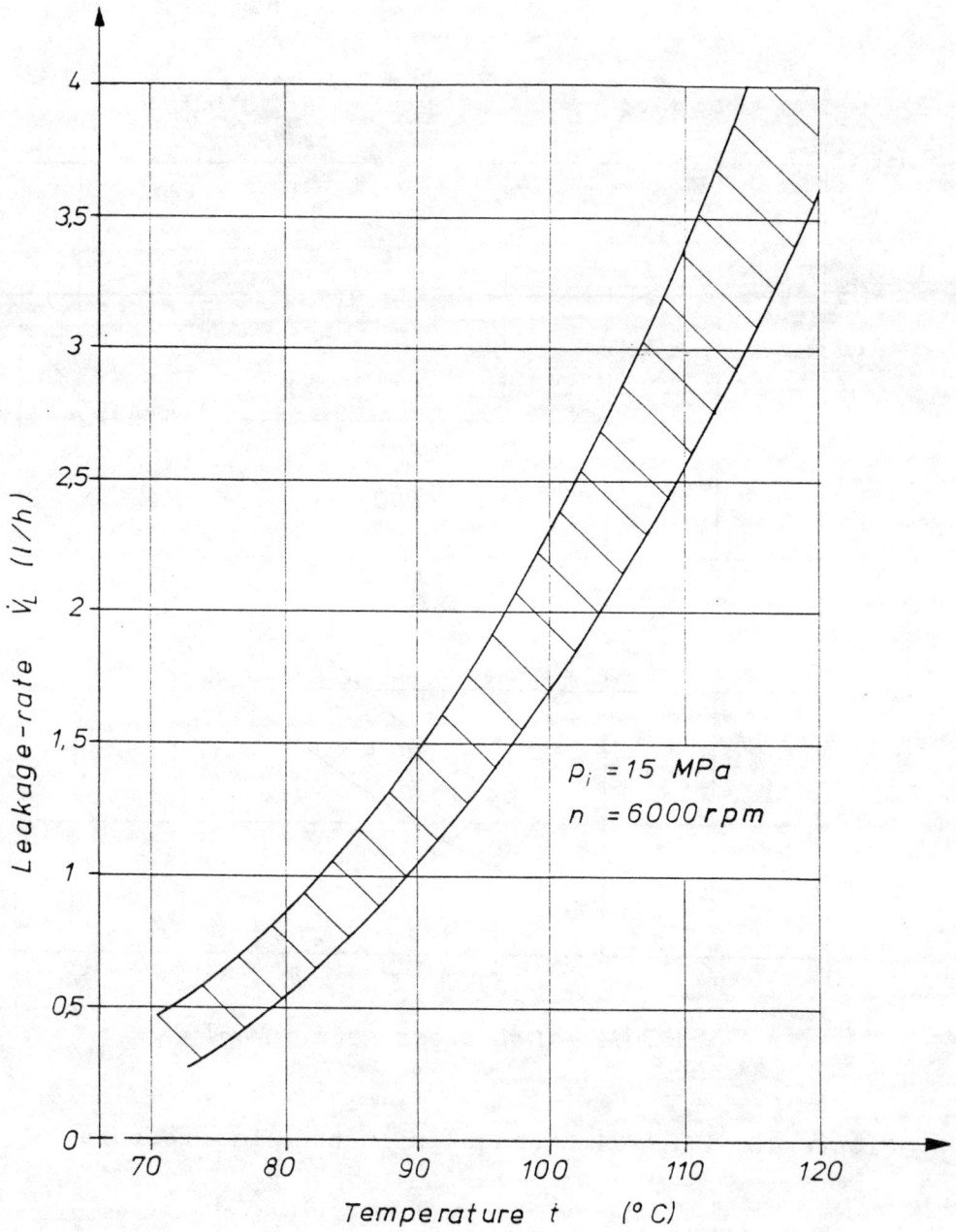

Figure 13: HRS3/64 Leakage influenced by temperature

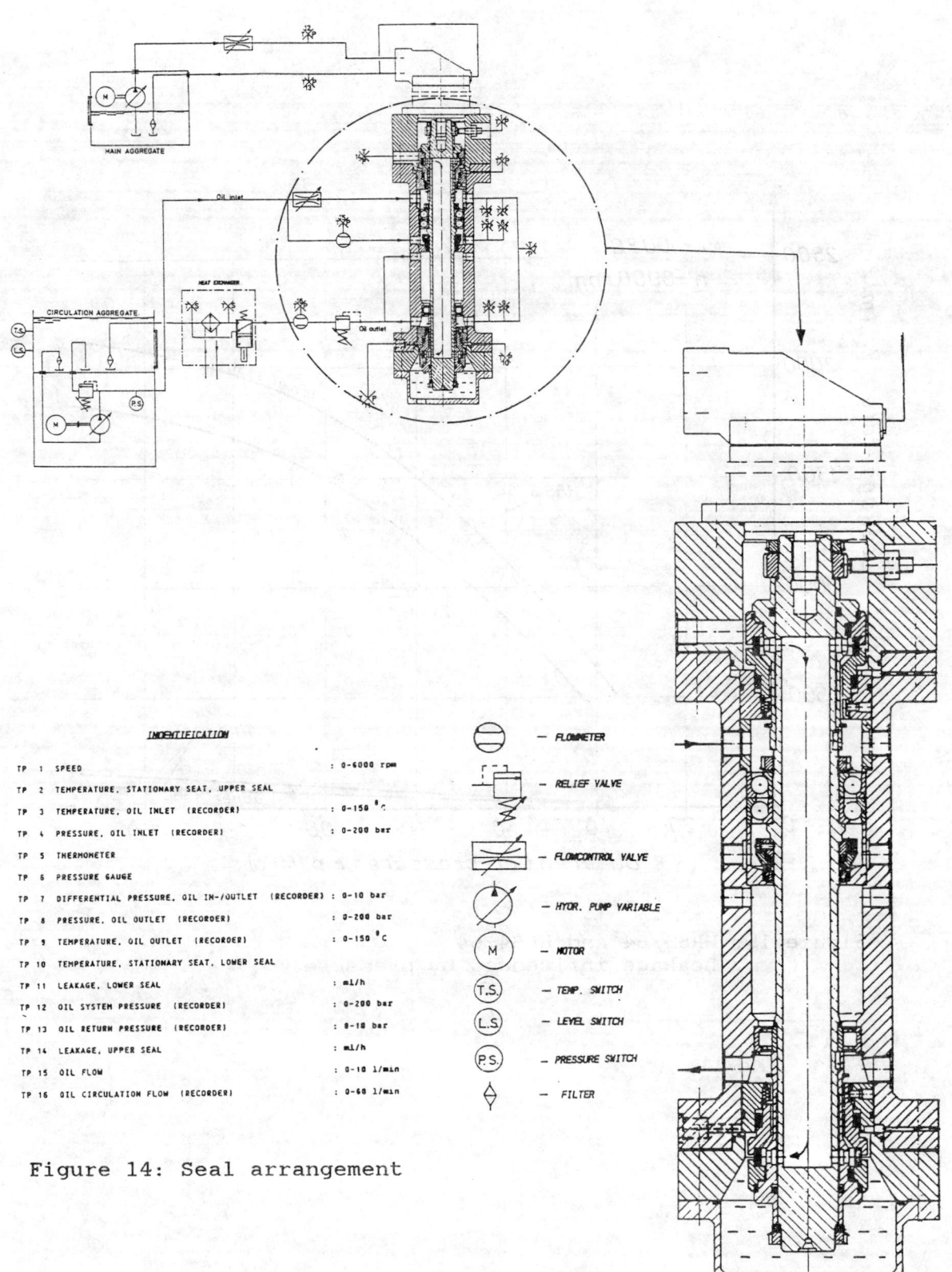

Figure 14: Seal arrangement

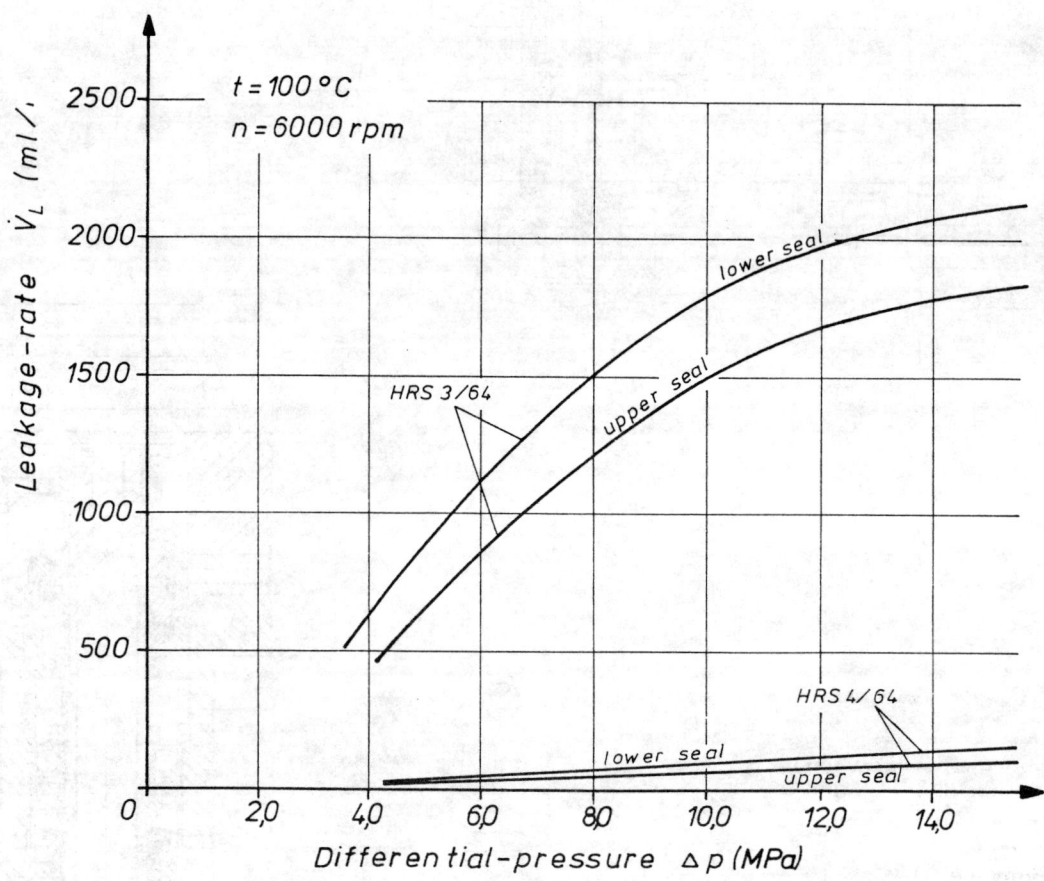

Figure 15: HRS3/64 and HRS4/64
Leakage influenced by pressure

11th International Conference of the
British Pump Manufacturers' Association

New Challenges – Where Next?

18-20 April, 1989
Churchill College, Cambridge

PAPER 16

DEVELOPMENT IN PUMP CONDITION MONITORING

B G Murray
G A Ratcliffe
S Palmer

Development Engineering International Ltd

1. INTRODUCTION

The objective of the following is to review some recent developments in machinery condition monitoring, with particular reference to pumping applications. Interest in machinery condition monitoring has grown in line with developments in the associated technology and in recognition of the substantial cost savings that it can achieve.

It is convenient to consider condition monitoring under the following three headings:

Vibration Monitoring

Lubrication Oil Monitoring

Machine Performance Monitoring

None of these techniques are new, but in each case, developments in instrumentation and computer equipment has transformed the scope and effectiveness of the techniques.

In the following, recent developments under each of these headings are summarised.

2. VIBRATION MONITORING

Vibration monitoring has a long and checquered history. At its simplest, vibration monitoring may be based on recording the overall vibration level at predetermined points on a machine (generally bearing pedestals) and noting the trend in the vibration level over an extended period. While many successes were recorded by this route, it is also clear that this simple approach was relatively insensitive and, in many situations, vibration monitoring lost credibility as a result.

Conference organised and sponsored by the British Pump Manufacturers' Association
in conjunction with NEL and BHRA, The Fluid Engineering Centre.
Co-sponsored by the Process Industries Division of the Institution of Mechanical Engineers.

With the development of a number of micro-processor based vibration data collectors, it is now both possible and practical to routinely capture vibration spectra. By this route, the sensitivity and reliability of vibration monitoring is greatly enhanced. Figure 1 illustrates such a system. Vibration spectra are collected and stored during a vibration survey by the data collector. This data is then then transmitted, either via modem or on return to base, to a desk top computer for archiving, exception reporting and graphical display.

Figure 2 shows vibration measurements taken from a 300 kilowatt seawater injection booster pump on an offshore oil production platform which illustrates this point. The data is presented in the form of a "waterfall" plot showing vibration spectra obtained over a six month period. The following peaks appear in the spectra:

(i) The running speed of the machine (30 Hz) can clearly be seen, varying little over the period,

(ii) A peak is evident at twice running speed (60 Hz). While this could correspond to a degree of misalignment between the motor and pump, it is more probable that it is associated with mains frequency.

(iii) The impeller blade passing frequency (240 Hz) is clearly in evidence. It can been seen that this fluctuates considerably from month to month, without any significant change in the machine condition. It can also be seen that this component is substantially higher than the running speed vibration.

(iv) The dominant peak in the vibration spectrum is at 110 Hz. This frequency is the running speed of a nearby 4 MW water injection pump. This vibration is being transmitted via the machine foundations by the pump in question.

The point about this example is that overall vibration level monitoring would have been totally misleading, since this would be dominated by the foundation transmitted vibration and the blade passing component. Substantial changes could occur at and around machine running speed without a noticeable change in the overall level. Conversely, overall levels would be highly unstable from month to month.

The hardware required to implement such a systems is as follows:

(i) A robust portable data collector, capable of obtaining vibration spectra, storing it during survey and uploading the information to a desktop computer. A variety of such devices are now commercially available.

(ii) A software system giving convenient access to the vibration data, allowing trends to be displayed graphically, and allowing exception reporting functions.

(iii) Means of attaching a vibration transducer to each position in turn during the survey. While the first two objectives are well served by a variety of commercial system, it is remarkable how carelessly the question of transducer attachment is often addressed. It is well established both from practical experience and published literature, that hand held vibration probes are extremely unreliable. while magnetic clamps can offer some improvement if used carefully (on Ferro-magnetic surfaces!) the optimum solution is a quick-lock bolted down transducer attachment.

3. LUBRICATION OIL MONITORING

Spectrographic oil and particle analysis is a well established technique, entirely complementary to vibration monitoring. It can be carried out at two levels.

At its simplest, an oil analysis laboratory will receive samples from the field at regular intervals and report, usually by telex, on the level of additives and contaminants in the oil. Based on preset high and low alarm levels for each contaminant/additive, the oil is "passed" or oil change is recommended.

However, the lub oil analysis data can also be used as a **machine** condition monitoring tool if trends in contaminants are monitored at levels below the laboratory's pass/fail level. An obvious example is changes in the white metal content as an indicator of bearing wear. Another example is shown in Figure 3 where a leak in a seawater/oil heat exchanger shows up as a rising sodium content. Changing the oil brought the sodium level back within specification but equally important, is to trigger the required maintenance activity.

Although spectrographic oil and partical analysis has been available for some time, a new development in this area is software to carry out trending and exception reporting to make oil based condition monitoring a practical proposition. This work probably requires to be carried out by the engineer or condition monitoring specialist responsible, since proper interpretation of the data requires an intimate knowledge of the machine, its operating history etc.

The software package which generated Figure 3 allows direct transmission of lube oil analysis data to the condition monitoring engineers computer. Automatic exception reporting then alerts the engineer to a small subset of the total data which warrants further investigation. Obviously, the alarm levels which would initiate an exception report are completely under control of the condition monitoring engineer.

4. PERFORMANCE MONITORING

4.1 Need For Performance Monitoring

Perhaps the most fundamental condition monitoring data is the daily equipment log sheet. Too often, perhaps, this is neglected until a fault appears. As with lube oil analysis there is considerable scope for automatic computer based trending and alarm generation. Obviously, what is required is to compare the actual head and flow achieved by the pump with its "as built" characteristic. Monitoring trends in this way leads to a quantitative basis for determining maintenance intervals. Performance data of this sort is entirely complementary to vibration measurements. On the one hand, vibration data gives a good indication of the mechanical condition of the machine, particularly bearings, balance, and alignment. On the other hand, performance monitoring indicates the condition of wear rings, interstage seals and other hydraulic aspects. A further advantage of performance monitoring is that undesirable operating conditions can be readily identified, for example if the machine is operated at or near to cavitation, this may be identified. Taken together, vibration and performance monitoring give a complete picture of the operating condition of the machine.

The "alarm level" for machine performance can be condition based once a correlation is established between measureable performance parameters and the acceptable level of internal wear. Alternatively, there may be clear economic criteria for carrying out maintenance if the increase in pumping costs as a result of wear within the machine becomes large compared with overhaul costs. In a third senario, the loss of product value resulting from inefficiencies in the pumping plant requires to be compared with overhaul costs. For example in the Oil Industry, it has been estimated that a 2% loss in efficiency of a large water injection pump would justify the cost of overhaul, when assessed in terms of reduced oil production.

4.2 Practical Implementation

In some situations, pump performance monitoring is as simple as described above. That is, the head rise across the machine can be measured by the means of pressure guages and flow metering is available to allow the operating point of the machine to be plotted on a head-flow characteristic. If motor power is also recorded, then a complete picture of the condition of the machine is obtained. However, in perhaps the majority of pump applications, flow measurement on individual machines is not readily available and pump efficiency measurement becomes of interest. Even if flow measurement is possible, there can still be advantages in recording the pump efficiency.

The thermometric method of pump efficiency determination has a long history extending over thirty years or more. Its use for maintenance scheduling was first described in detail in 1972 (Reference 1). However, it has not been widely implemented, possibly because of limitations in available instrumentation. The essence of the method is as follows:

(i) The head rise across the machine is recorded either by means of conventional instrumentation or by a differential pressure transducer.

(ii) The temperature rise across the machine is recorded by means of specialised instrumentation attached for the purpose. The temperature rise involved is relatively small typically, less 0.1°C on a 10 bar head rise pump.

(iii) The pump effiency is computed based on the temperature rise (which measures the losses in the machine) and the head rise (which measures the useful pumping power delivered by the machine).

Figure 4 shows the relationship between the measured parameters and the pump effiency for water as the working fluid. Similar graphs have been drawn for seawater, crude oil, hydraulic oil etc, based on a knowledge of the appoprapriate thermodynamic properties.

A limitation of instrumentation systems available until recently was that they require the use of temperature sensors inserted into the pipeline at the suction and discharge of the machine. Frequently, this is inconvenient, particularly when retro-fitting equipment. in practice, this difficulty can be avoided by careful use of surface contact temperature sensors with suitable thermal insulation. Figure 5 illustrates this based on finite element calculations of the temperature distribution from inside the pipeline to the surrounding air. When a carefully designed insulation system, the measured temperature on the pipe surface is practically identical to the true bulk fluid temperature. using this technique, temperature sensors can be fitted to the suction and discharge of the machine while it is running and an indication of its efficiency and hence operating condition obtained.

One of the major advantages of the thermometric method is that efficiency can be obtained economically without the need for flow or torque measurements. A useful enhancement of the technique is that it allows the flowrate through the machine to be **calculated** based on measurements of the head, efficiency and motor power. In this way, the operating point of the machine (head/flow) and efficiency/flow) can be compared with the as built condition of the machine.

4.3 Some Case Studies

4.3.1 Multi-Pump Installation

Figure 6 shows results obtained from a plant on which 4 large coolant pumps ran in parallel. each pump had a nominal delivery of around 6500 gal/min and a head rise of approximately, 150 psi. The pumps operated between common suction and discharge headers. Accordingly, conventional instrumentation gave little indication of the condition of individual machines since the flow to individual pumps was not metered.

After arriving on site, sensor pads were attached to the suction and discharge of each machine and the pump efficiency derived by the thermometric method. having calculated the efficiency of flow through the machines could be estimated since

$$\text{flow} = \frac{\text{Hydraulic Power}}{\text{Head Rise}}$$

$$= \frac{\text{Electrical Power Consumption}}{\text{Motor Efficiency} \times \text{Pump Efficiency} \times \text{Head Rise}}$$

Figure 6 compares the data derived in this way with the manufacturer's characteristic. The test results showed clearly that pump No 4 was performing well below the other units and the manufacturer's characteristic. This result was verified by maintainance records.

4.3.2 Offshore Water Transfer Pump Test

Figure 7 shows results obtained during a trial on a water transfer pump situated on an offshore oil production platform. The purpose of the test was to determine the acceptable operating conditions for the machine. When the test engineers arrived on site, the machine was running. Without the need for shut down, the necessary instrumentation was attached to the pump. The flowrate through the machine was then varied by throttling the discharge. As can be seen, both the head and efficiency measurements are in close agreement with the manufacturers test curve up to a critical flowrate beyond which cavitation occurred as a result of certain aspects of that particular installation.

The importance of this example is that it illustrates the quality of information which can be obtained rapidly, on-site, without the need for machine shutdown.

4.3.3 Crude Oil Pump

Figure 8 shows the head and efficiency characteristic obtained from measurements on a 14 bar head rise 3 MW gas turbine driven crude oil pump. It can be seen that a good correlation was obtained between the manufacturers characteristic and site measurements using the thermometric method. This data was in fact a baseline measurement for a condition monitoring programme.

The relevance of these measurements is that they were obtained using crude oil as the working fluid, where the thermometric method has not, to date, been widely applied. In practice, the technque can be applied to any fluid if the following data is available:

(i) Density

(ii) Specific heat

(iii) Coefficient of cubical expansion

In situations where fluid properties are uncertain, more sophisticated instrumentation techniques can be used to overcome the problem.

5. CONCLUSIONS

The fundamental techniques underpinning condition monitoring of pumps are not new. Vibration monitoring, lubrication oil monitoring and machine performance monitoring are all many years old. However, recent developments in instrumentation and computer hardware have taken these techniques out of the test laboratory and into the field. Using modern equipment it is possible to obtain:

(i) Reliable vibration spectra to give sensitive and precise mechanical condition information.

(ii) Software and data transmission systems are now available to allow detailed lubrication oil monitoring to be carried out by machinery engineers.

(iii) Recent developments in pump performance monitoring have made it inexpensive, convenient and accurate.

REFERENCES

(i) Whillier A   Site testing of high lift pumps in the South African Gold Mining Industry I Mech E Conference (1972)

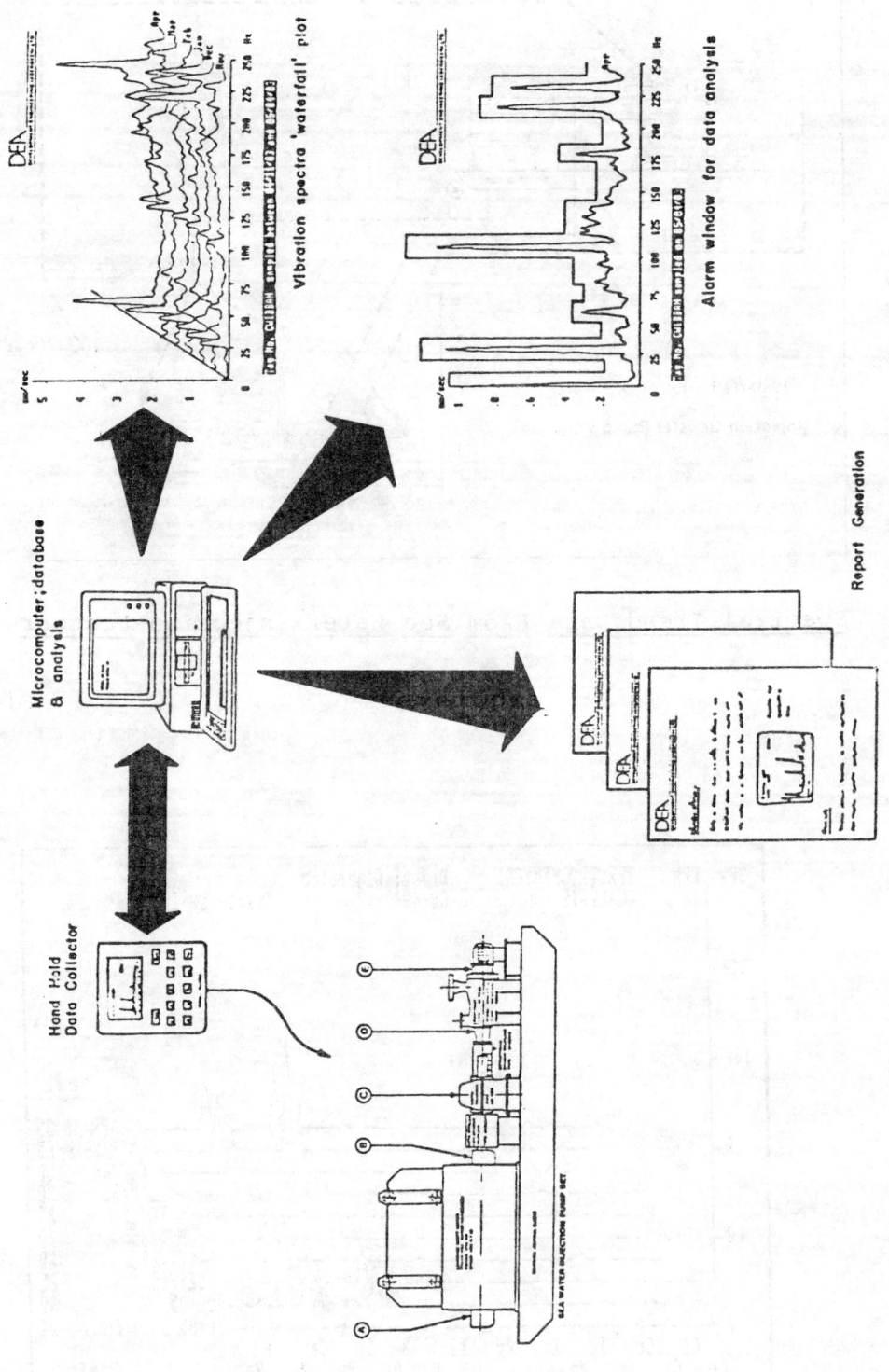

Schematic of Vibration Data Collection System

Figure 1

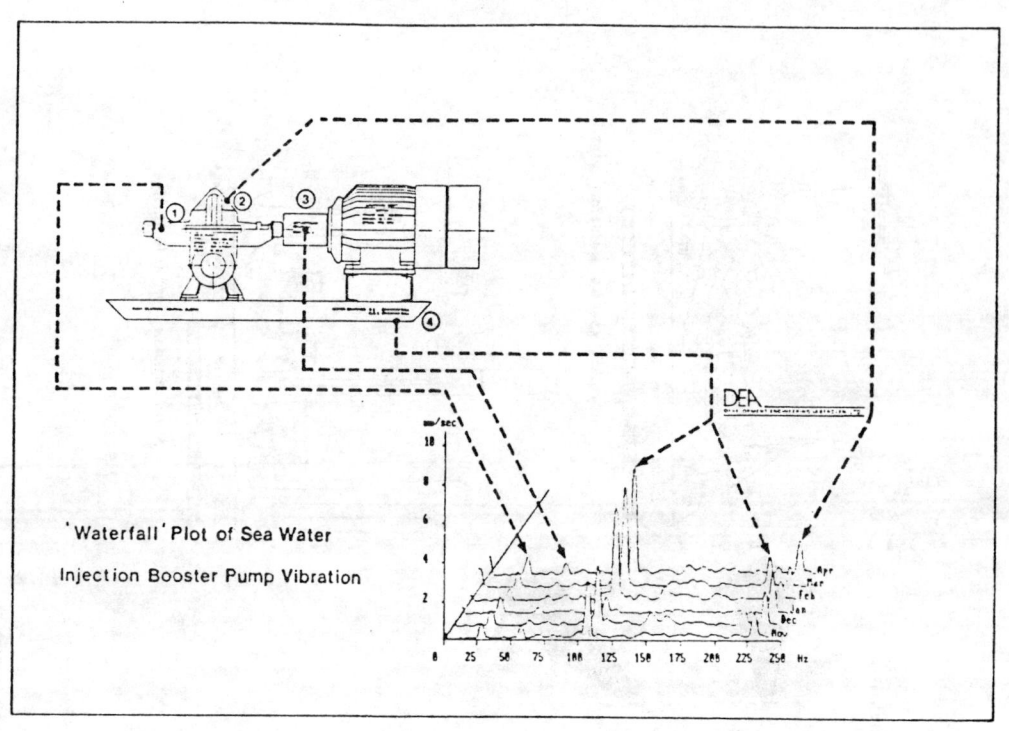

Spectral Trend Data From Sea Water Injection Booster Pump

Figure 2

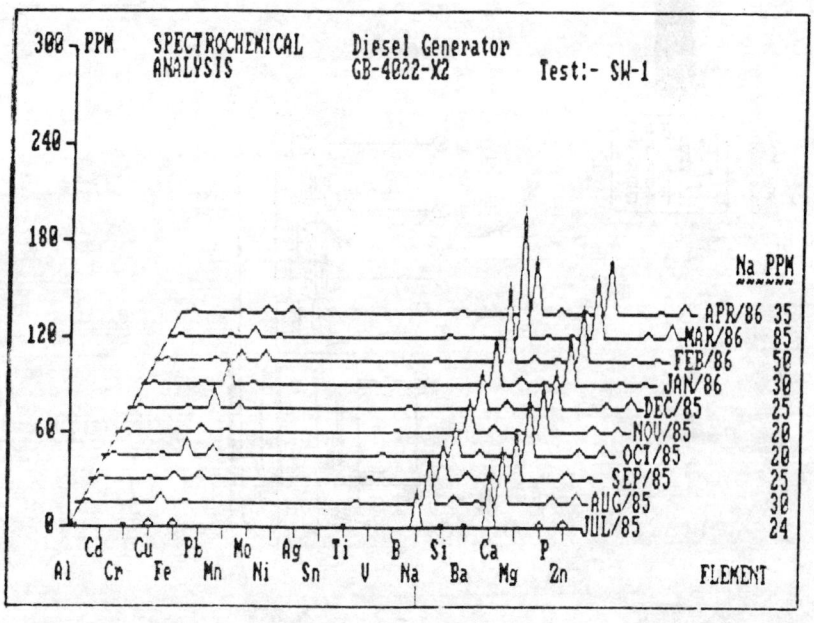

Spectrographic Oil Analysis Waterfall

Figure 3

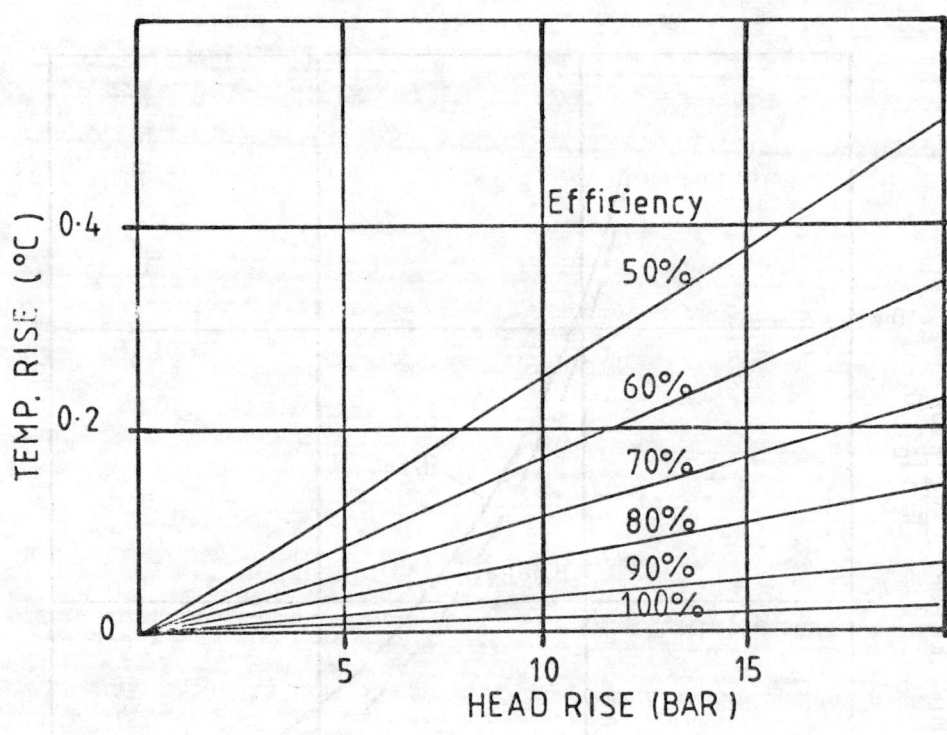

Temperature Pressure Efficiency Characteristic for Water (20°C)

Figure 4

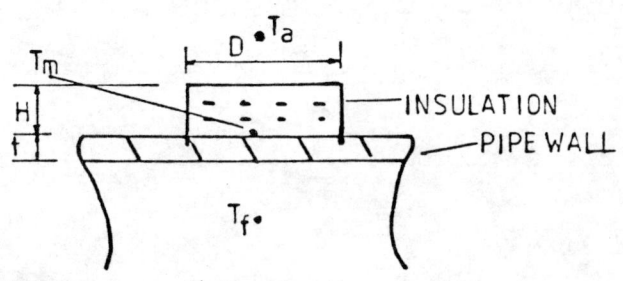
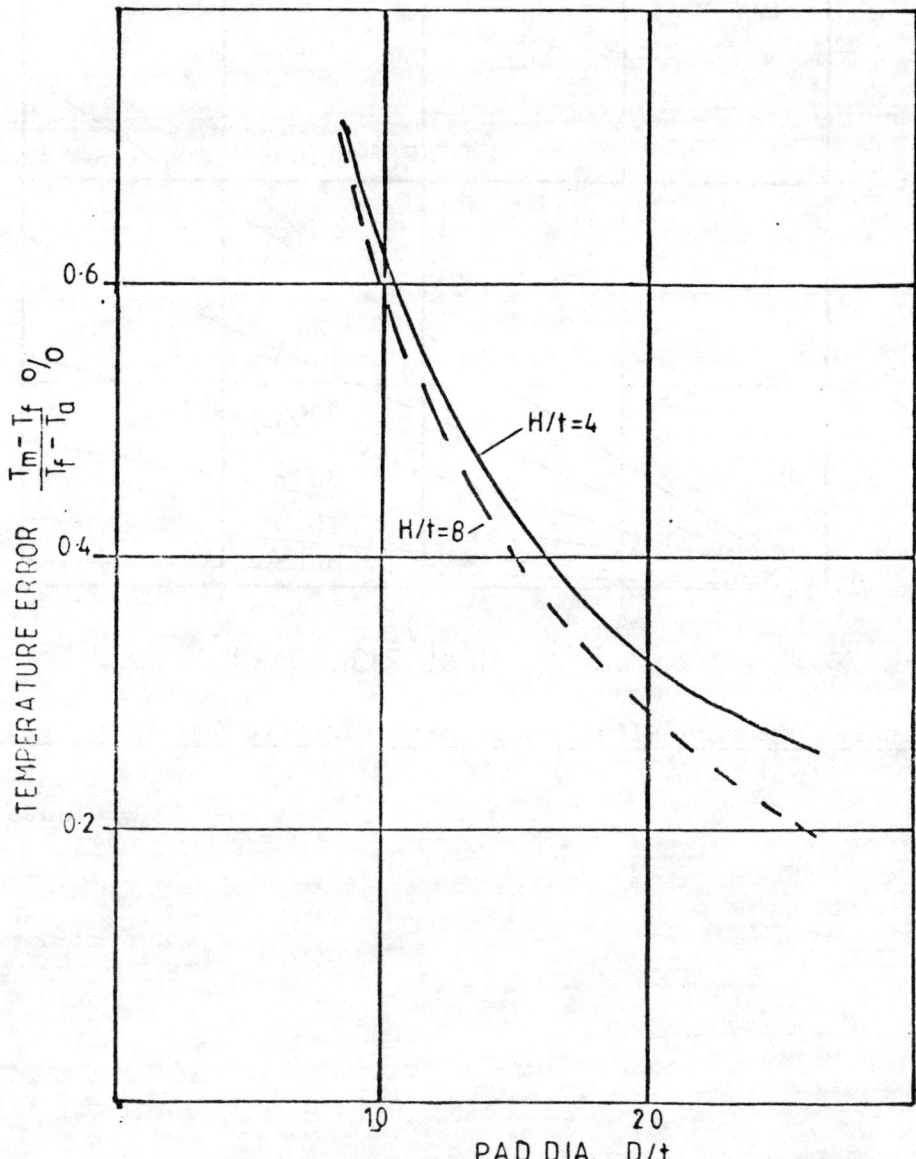

Error in Surface Contact Temperature Measurements
As a Function of Insulating Pad Size

(From Finite Element Simulation)

Figure 5

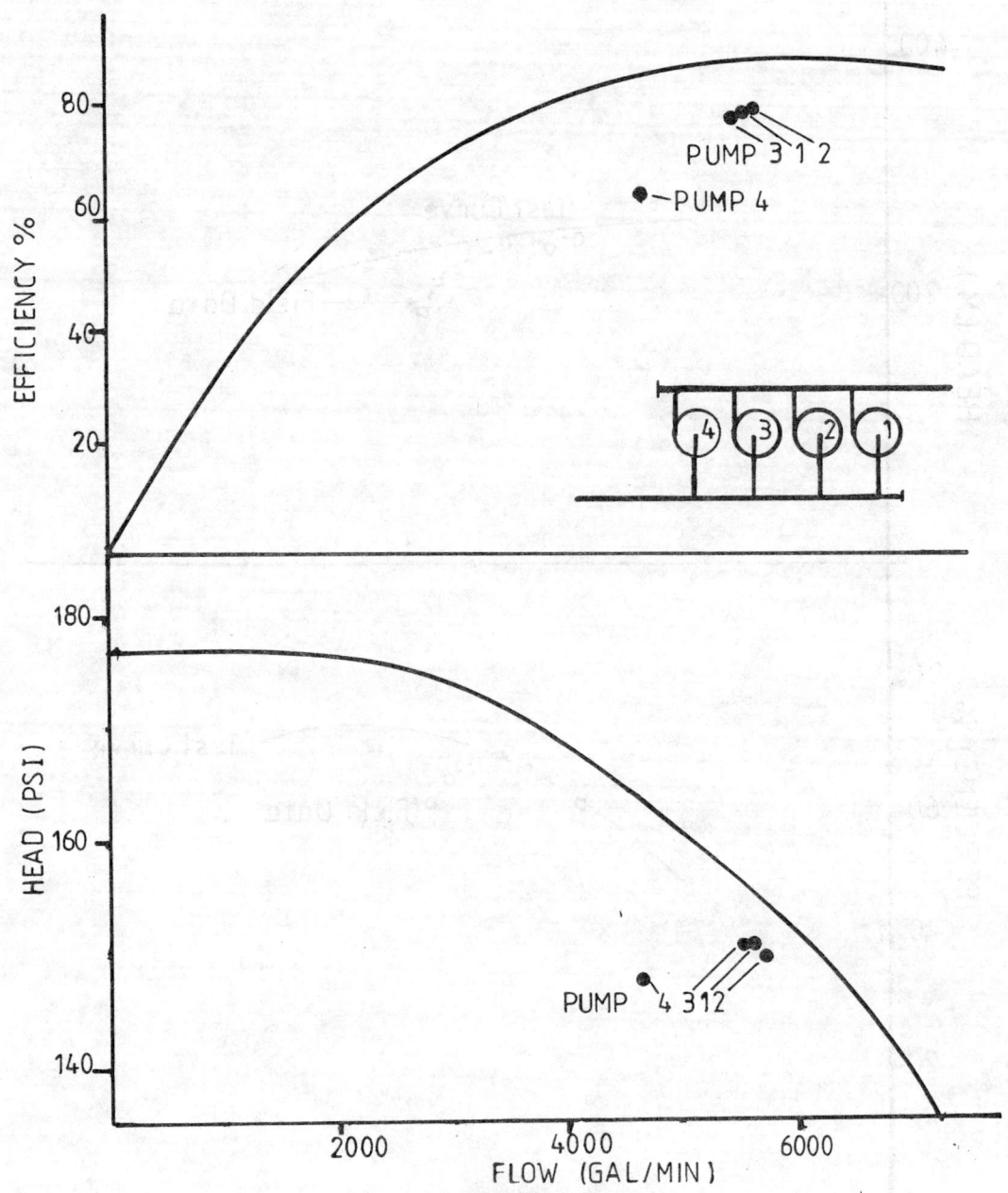

Performance Characteristic - 4 Pump Installation

Figure 6

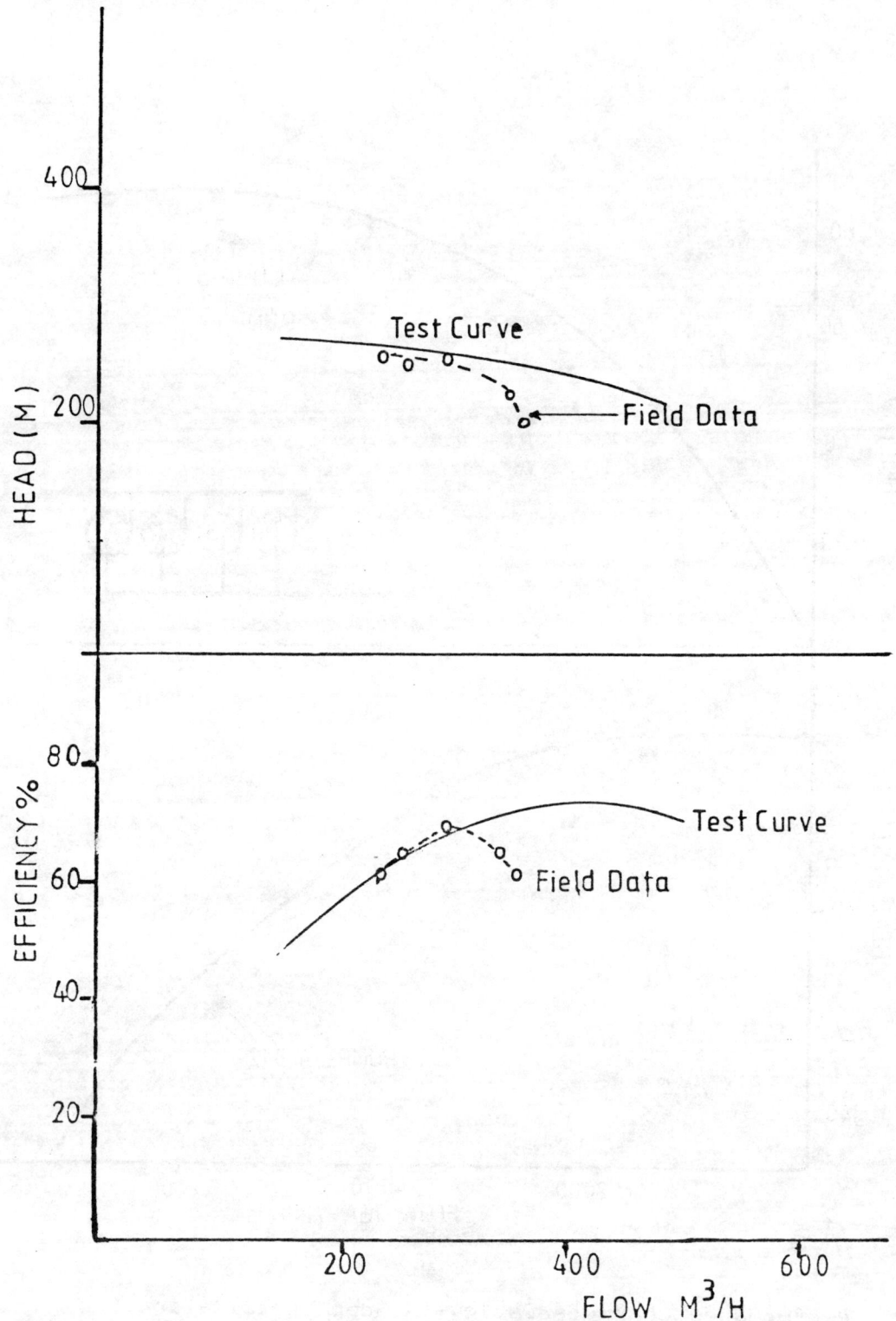

Field Data/Test Shop Data Comparison

Figure 7

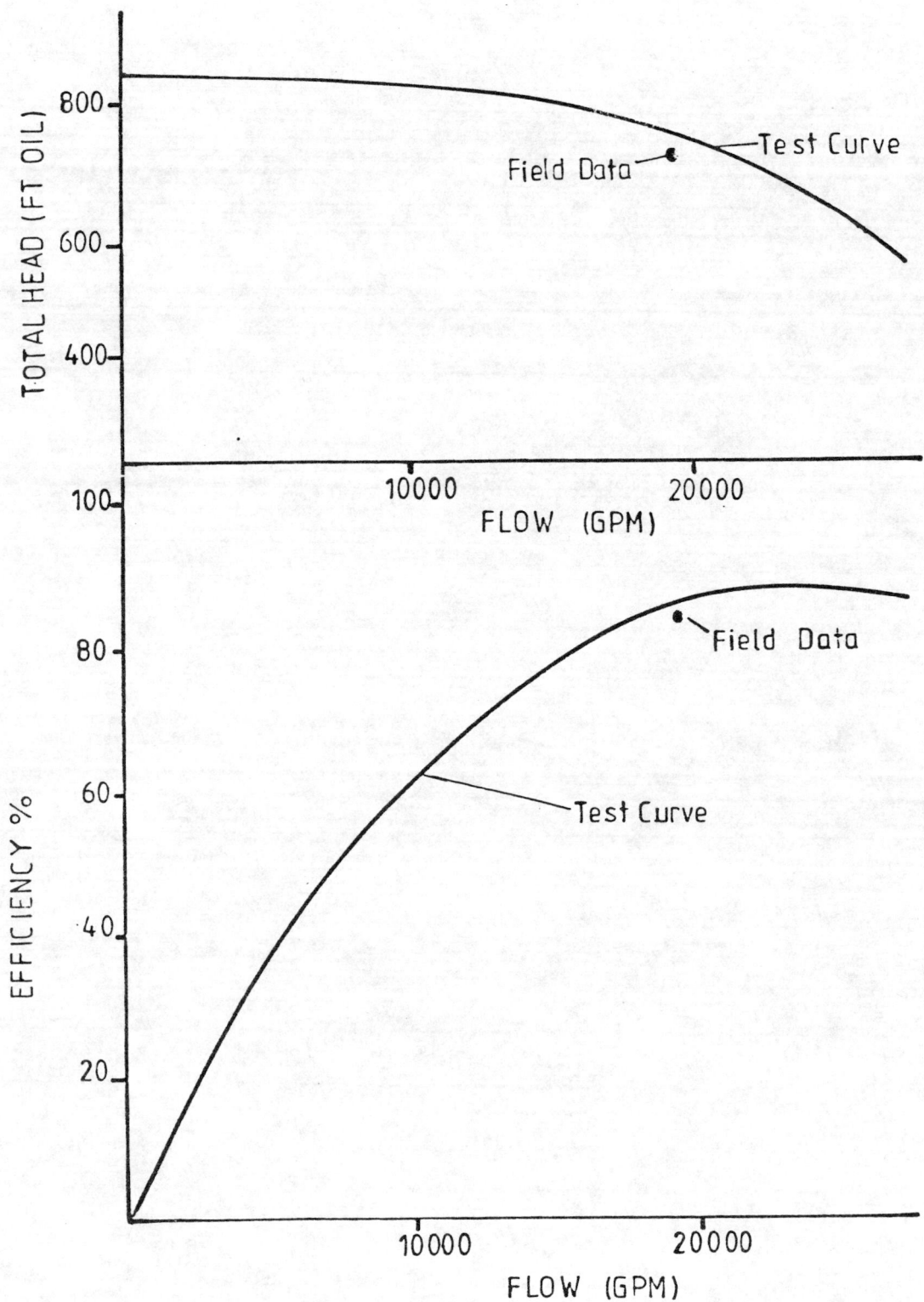

Field Data/Test Shop Data Comparison for Crude Oil Pump

Figure 8

11th International Conference of the British Pump Manufacturers' Association

New Challenges – Where Next?

18-20 April, 1989
Churchill College, Cambridge

PAPER 17

Thermodynamically Based Pump Performance Monitoring

Maurice A Yates
Director
Advanced Energy Monitoring Systems Ltd
Ottery St Mary
Devon
EX11 1NR

Summary

The assessment of the performance of pumping plant on site has long been a problem for the purchaser. However, with the advent of real time monitoring of all the pumping parameters he can now assess the plant's performance and the optimum time for refurbishment.

List of Symbols

Symbol	Definition	Unit
a	Isothermal Factor	m^3/Kg
Cp	Specific Heat	$J/Kg/K$
DT	Differential Temperature	K
Eh	Hydraulic Energy/Unit Mass	J/Kg
Em	Mechanical Energy/Unit Mass	J/Kg
Ex	Contractural Losses/Unit Mass	J/Kg
P	Pressure	Pa
Q	Flow	m^3/sec
U	Velocity	m/sec
Z	Height	m

Subscript

1	Suction
2	Delivery

Conference organised and sponsored by the British Pump Manufacturers' Association in conjunction with NEL and BHRA, The Fluid Engineering Centre.
Co-sponsored by the Process Industries Division of the Institution of Mechanical Engineers.

History

The thermodynamic method of hydraulic machine testing was first conceived in 1914 by a Frenchman, Poirson. He found by applying the first law of thermodynamics to the efficiency problem that the measurement of loss is a more accurate way of evaluating a machines performance than that of measuring total performance.

His work was continued in France and various reports of research work, particularly in the 1920's were produced.

The method however took a great step in 1954 when an account by Willm and Campmas of investigations made by Electrocite de France stimulated additional work in Europe. This was particularly true of the City of Glasgow where the close proximity of a major University and the U.K. National Engineering Laboratory led to some very rewarding research work into the method.

This association produced improvements both in the theoretical awareness of thermodynamic properties and in the design of construction of high accuracy a.c. resistance bridges, which were consistently reading in the mK region.

The method stayed in the theoretical/very highly skilled field for over twenty years with the majority of the practical work being carried out on a few large hydro electric schemes using manual sampling techniques.

The 1980's and the computer revolution, coupled with a growing awareness of the controllable aspect of energy consumption, led to a market potential for a low cost pump efficiency measurement service/product facility.

The thermodynamic method of measuring hydraulic efficiency is based on the conservation of energy in that the mechanical shaft power applied to a hydraulic machine is converted into hydraulic energy (useful energy) and thermal energy (loss).

The efficiency of a machine is defined as

$$= \frac{\text{Useful Workout}}{\text{Total Work Input}}$$

$$= \frac{\text{increase in hydraulic energy}}{\text{input mechanical energy}} = \frac{mE_h}{mE_m} = \frac{E_h}{E_m}$$

Therefore we can now talk in energy/unit mass terms.

Considering firstly the increase in hydraulic energy/unit mass (E_h).

From Bernoulli hydraulic energy of a fluid has three parts:

1. Pressure energy
2. Kinetic energy
3. Potential energy.

The increase/unit mass of each of these terms may be expressed as:

$$Vm(P_2-P_1) + \frac{U_2^2-U_1^2}{2} + g(Z_2-Z_1) = Eh$$

To obtain the value of the input mechanical energy we should consider the perfect operating condition ie 100% efficiency. In this case all of the mechanical energy is converted directly into fluid energy and the increase in fluid energy/unit mass may be defined as:

$$a(P_2-P_1) + Cp(Q_2-Q_1) + \frac{U_2^2-U_1^2}{2} + g(Z_2-Z_1) = Em$$

However, some of the losses will not appear in the fluid and here we should specifically consider radiation, either heat gain or heat loss and external bearing losses.

For those losses we can use a value Ex defined in ISO 5198 as contractural losses.

Therefore Efficiency = $\dfrac{Eh}{Em + Ex}$

giving

$$\frac{Vm(P_2-P_1) + \dfrac{U_2^2-U_1^2}{2} + g(Z_2-Z_1)}{a(P_2-P_1) + Cp(Q_2-Q_1) + \dfrac{U_2-U_1^2}{2} + g(Z_2-Z_1) + Ex}$$

To simplify the above we can consider the following approach

Efficiency = $\dfrac{\text{Work Out}}{\text{Work In}} = \dfrac{Wo}{WI}$

WI = Wo + Losses

Efficiency = $\dfrac{Wo}{Wo + Losses} = \dfrac{1}{1 + \dfrac{Losses}{Wo}}$

Losses = heat gained by the fluid pumped = $f(QxDT)$

Wo = $f(QxH)$

$$\frac{\text{Losses}}{\text{Wo}} = \frac{f(Q \times DT)}{f(Q \times H)} = f(DT/H)$$

Therefore Efficiency $= \dfrac{1}{1 + f(DT/H)}$

This simpler technique, whilst explaining the workings of the thermodynamic principles, has the problem of the Joule-Thompson effect, in that all the temperature rise is not due to the internal losses some of the temperature rise is from the isentropic compression of the fluid.

Methods Adopted

All the earlier work relied on the use of either ac or dc resistance thermometer bridges for measuring temperature which were manually balanced.

The pressure measurements were made using bourdon tube gauges or a standard dead weight pressure tester.

Whilst time consuming this method was successful when applied to large turbines which were running stably under well defined conditions.

When this approach is applied to pumps which may be relatively small in power terms and subject to instantaneous change due to system influences, the reading time is too long for accurate results.

In view of this in the early 1980's I commenced work on an instrument which would read all the necessary parameters in real time and also carry out data aquisition and computation of results. These results were then displayed on an alpha numeric LCD screen and printed out each minute on a dot matrix printer.

The speed of response of such a system has led to an increased knowledge of pumping plant operation.

Experience

My experience of testing pumps in the U.K. has covered all shapes and sizes and almost every pump tested has its own particular story. There is, however, a common thread running through all the stories in that in the main the plant exhibits the following:-

1. Incorrectly sized.
2. Running away from B.E.P.
3. Difficulties in parallel operation.
4. Poorly maintained plant.

5   Poor station layouts.
6   Recirculation in stations.
7   Poorly maintained instrumentation.

Incorrect Sizing

In the U.K. Water Industry this is historic. Pumps and pumping stations are designed for 20-30 years predicted growth.

Growth prediction in the U.K. is notorious. One only has to look at the electricity supply industry where our 1988 demand is barely above the demand predicted in the early 60's for the late 60's. Whilst it is very fair to design the actual pumping station and the pipework for these futuristic predictions, the pumps themselves should be looked on in much narrower terms, perhaps no more than five years hence.

Running Away From B.E.P.

This is a continuation of (1). In this case the designer will predict a 30 year head and then add a 10% margin. This will go on to the pump manufacturer who, to be certain to achieve duty, will allow a tolerance on the pumps head performance. The result of this will be that the pump will run far out to the right.

Difficulties in Parallel Operation

This is often due to the tendancy to regard a pump as a constant flow device i.e. if you have 3 - 5ML/Day pumps in a station each will pump 5ML/Day operating singularly and a total of 15ML/Day when all three are operating.

In specifying duties it is imperative that the generated head is also given.

Poorly Maintained Plant

In this case the pump is its own worst enemy because in the majority of cases the pump is very tolerant and will run in all sorts of running conditions for many years before failing. In fact very few pumps in the Water Industry actually fail in service.

This is one area, however, where rapid strides are being made and operators are far more aware of the cost of running pumps under poor operation conditions.

Poor Station Layout

This again is historic and may be simply defined as putting a quart into a pint pot.

Stations are designed without reference to the potential head loss in either the suction or the discharge pipework. The valve gear is often put right up against the pump and sized on pump discharge branch sizes, apparently to save money.

Instead the losses which are approximately proportioned to velocity squared are greatly increased.

In this area we are seeing an increasing use of butterfly valves because of their lower space requirements, however, the head loss through this is in the 1-2m range for normal velocities and at higher velocities this is greatly increased.

An average flow rate of 100l/s through a 1.5m head loss will cost £736/year or say £5/l/s/m/year.

Recirculation in Station

This is one of the latest problems to come to light here when we have parallel pumping plant the inability of the non-return valves to maintain drop tight over many years leads to recirculation back to the suction via a non-running pump.

The effect has been determined when individual pump performance tests have been carried out using the Yatesmeter and flow comparisons made with recently calibrated station instrumentation.

Instrumentation

Good quality instrumentation has often been installed and not checked or recalibrated throughout the life of the station.

For many reasons it is essential that all instrumentation is checked and calibrated on a regular basis.

Taking power stations as a basis here each and every instrument is checked at least once per year and some, dependant upon their working environment, are checked many times a year.

There are already signs that this is an area that the U.K. Water Industry is addressing itself. But we are a long way away from duplicate instrumentation and automatic checking.

Experience with Thermodynamic Testing

Throughout the last five years we have a steady improvement with the techniques adopted. These improvements have been in both equipment and operator skills.

It is now possible under very good conditions to test pumps in the region of 10m head and still obtain accuracies well within the current test standards.

Measurement of high head pumps is now possible with very little operator knowledge and future development will see this approach being applied to the lower head pumps.

It is becoming increasingly clear that the development of the instrument was slower in the early years due to the belief that variations in the minute temperature changes were entirely due to the development equipment, however, this is not so.

For instance a significant proportion of pumps in the field are running under unstable conditions, not necessarily due to an unstable pump characteristic but a combination of several factors such as pump characteristic, system characteristic, pump condition and pump speed.

Conventional instrumentation is slow and does not record what are often very frequent oscilation.

There was one notable example when a pump was running under test and another pump started on the system and, although the apparent surge was dissipated in a few minutes, it was twenty minutes before the pump under test returned to normal operation.

This type of unstability can lead to premature failure of the pump.

The increasing use of frequency controllers to control the pump speed and hence its head or flow increases this problem and may induce premature pump failure.

I and my associates have now carried out over 500 pump tests using the Yatesmeter and other users of the meter have carried out many times that number again.

Experience has shown that the method is quick, responsive to change and inherently accurate.

It has highlighted many pumping/system problems, it has also shown that age is no criteria for judging a pump's condition.

For instance a pump which had run almost continuously for 55 years was almost as good as the day it was built showing only a 8% fall off in efficiency.

Conversely a pump which had run for only six months had shown an efficiency loss of 20% and an additional annual running cost of £50,000.

Pump Performance Monitoring Equipment

So far I have discussed just the one off tests mainly for acceptance or refurbishment decision.

This is probably only the tip of the iceberg because pumps and system are interactive in their operation and a snapshot of one particular circumstance does not tell the whole story.

The thermodynamic technique can be very effectively used as a performance based monitoring by measuring the change in enthalpy of the pump fluid as it passes through the pump.

Systems of this type are already in operation in the U.K. and are identifying areas of improvement in pump selection and operation.

These systems measure the following parameters to a high degree of accuracy in real time:

- Relative enthalpy change
- Suction pressure
- Discharge pressure
- Fluid temperature
- Input power to motor.

The onboard computer then computes the thermodynamic measurements to give pump efficiency in real time.

It carries a programme to determine the actual motor power from the station motor efficiency, its performance characteristics and its input power.

Having assessed pump efficiency, motor efficiency and measured input power it then computes the fluid flow.

The Future

To see into the future we must consider the past. In 1981 I presented a paper to this body outlining the thermodynamic techniques. At the 1983 conference I expanded the cost benefits of good procurement of plant. Today I have tried to bring the two papers together coupled with very active field experiences over the last six years.

These six years have seen a progression from direct consultancy in pump efficiency to portable efficiency monitoring equipment to permanently installed monitoring equipment. These three arms will undoubtedly continue but the fourth arm will be real time control of a pumping system at its optimum running point. The testing of pumps I see improving to enable manufacturers/purchasers to work closer together in obtaining better quality pumping plant with the major benefit being in the reduction of electricity consumed throughout the pumping world.

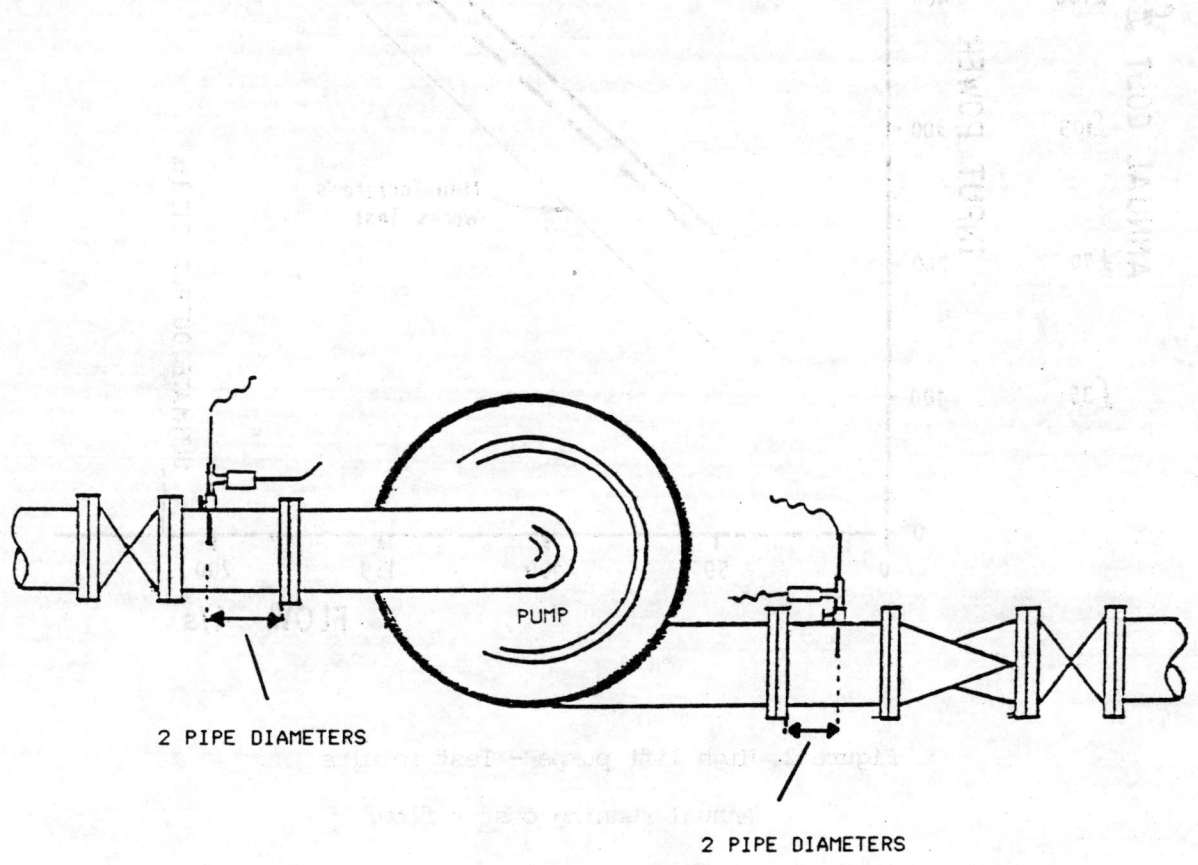

Figure 1. Where to measure

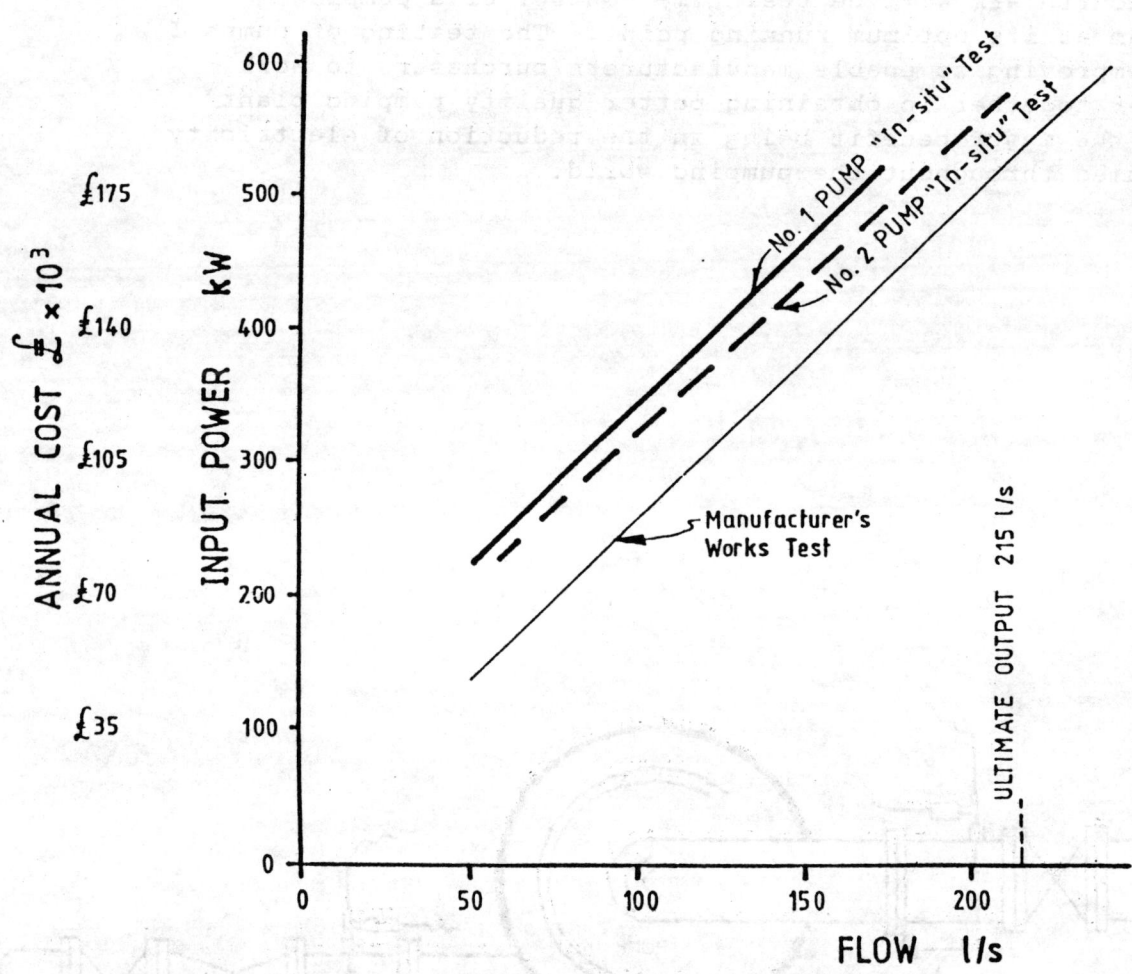

Figure 2. High lift pumps - Test results
Annual running cost v flow

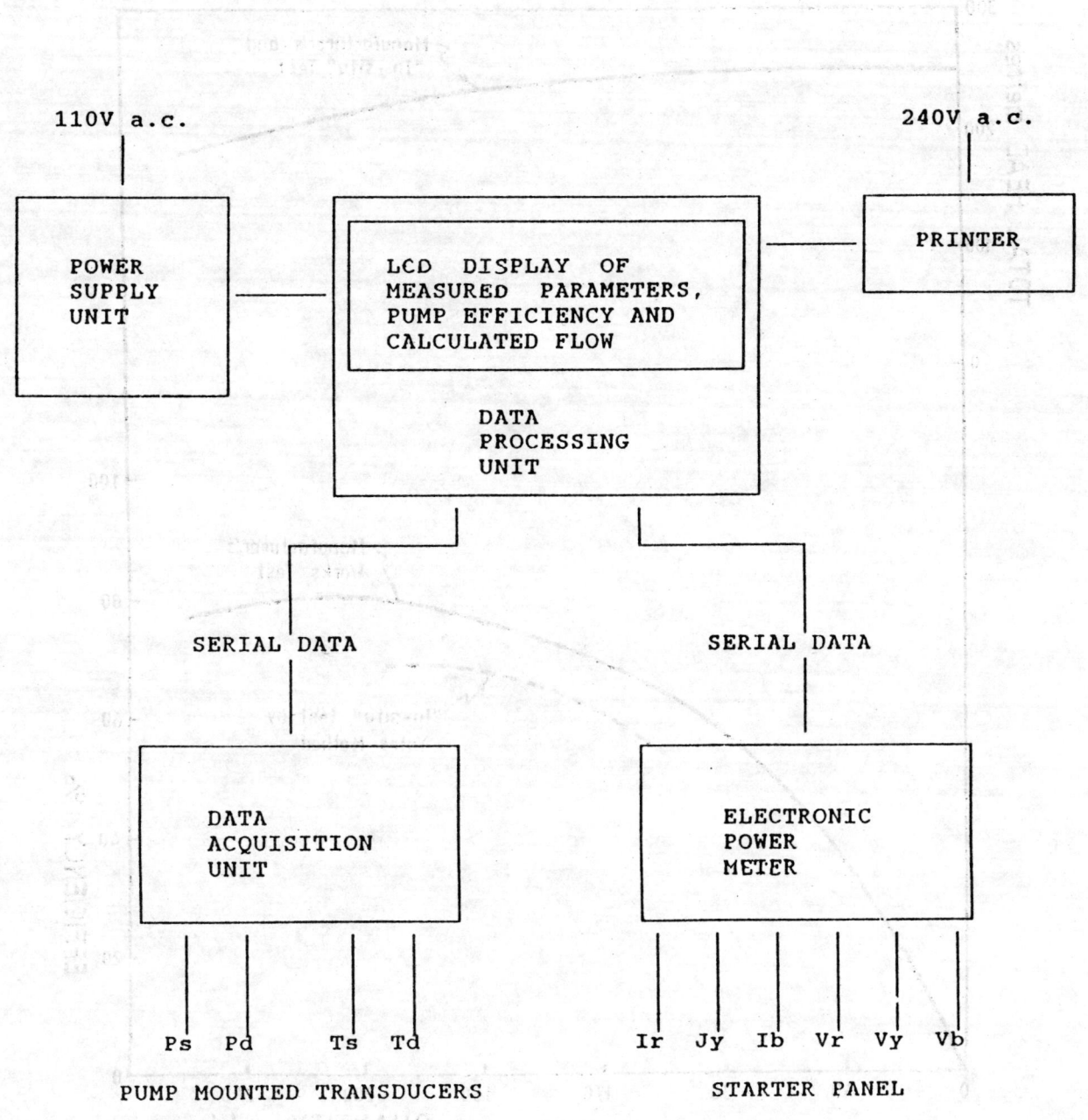

Figure 3. The Yatesmeter - pump efficiency monitor

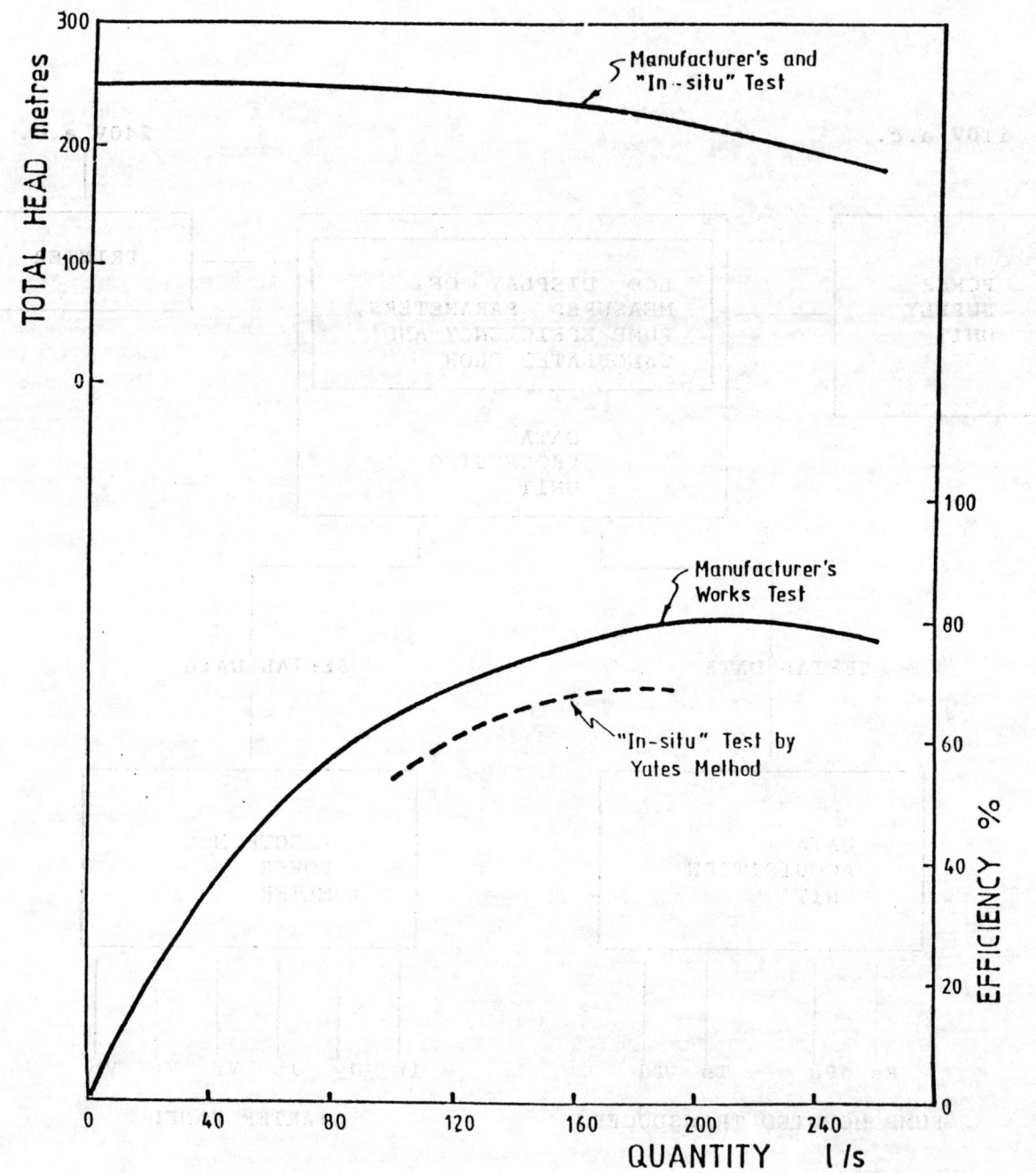

Figure 4. High lift test pump - 6 months old

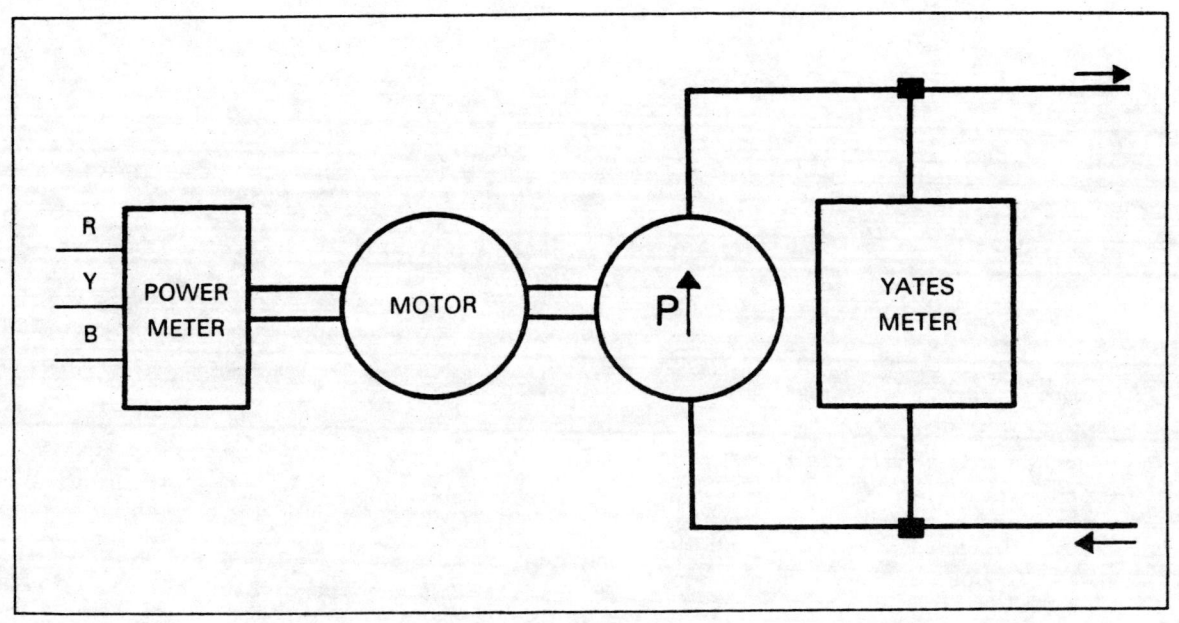

Figure 5. Pump condition monitoring

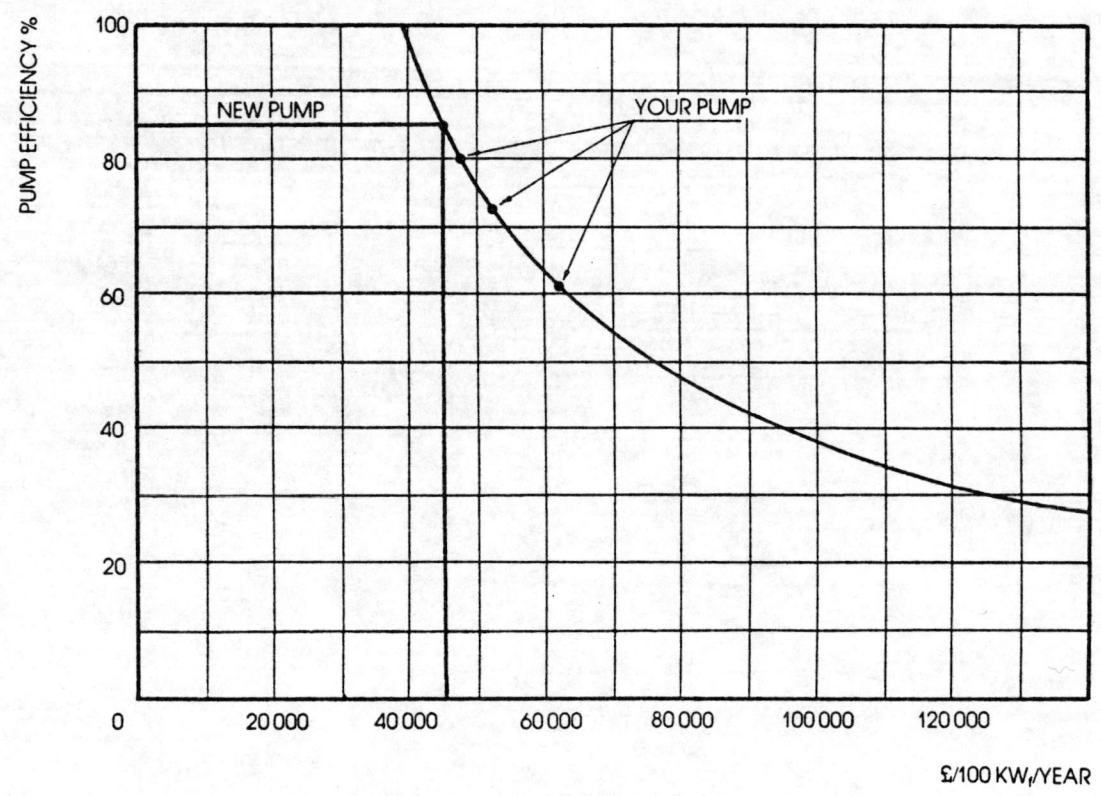

Figure 6. Pumping costs

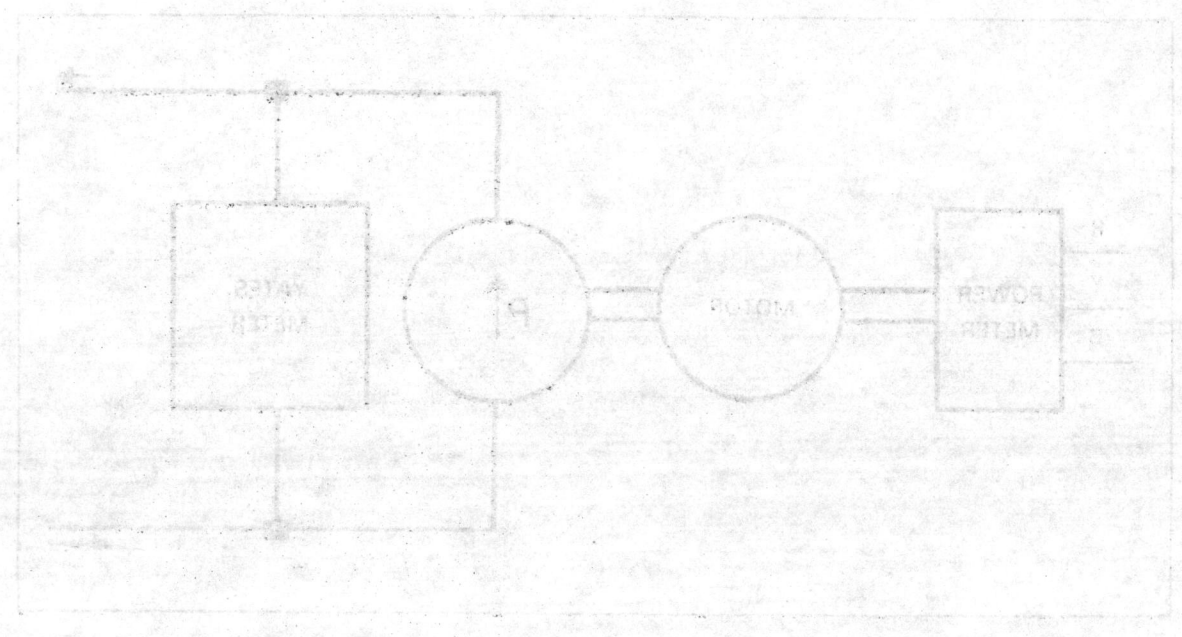

11th International Conference of the
British Pump Manufacturers' Association

New Challenges – Where Next?

18-20 April, 1989
Churchill College, Cambridge

PAPER 18

THE DEVELOPMENT OF PROFILED MECHANICAL SEAL FACES
FOR POSITIVE LIQUID TRANSFER
BY
P R ROGERS
JOHN CRANE UK LTD

ABSTRACT

Conventionally installed mechanical seals are arranged to prevent liquid surrounding the seal from escaping excessively by inward radial movement through the interface film. With compound seal installations often characterised as "double back-to-back" and "tandem" the effective sealed liquid is contained in the same way.

The basic idea behind most variations of the conventional single seal installation is to improve the condition of the interface film. One way to do this directly in clean but poor lubricating conditions is to provide the usual radially flat faces with relatively deep face cutouts or grooves which, while increasing liquid access to the faces to improve their lubrication hence seal performance, can also increase seal emissions.

In this paper we describe a seal face profile having some of the characteristics of a positive displacement pump which can be used to move liquid either with or against the pressure gradient. Used in the latter way it is an "upstream pumping" device which will carry liquid from a low pressure source at a controlled rate across the sealing faces and into the sealed liquid.

Mechanical seals profiled in this way can therefore be used to reduce unacceptable emissions and, because the seal interface film is provided with a very effective integral clean low volume flush, the effects of all forms of abrasive damage to the seal faces is very much reduced.

This development is therefore capable of meeting the particular requirements of users where safety, reliability, predictability and long life are key seal selection criteria.

The paper concludes with operating details of working installations.

INTRODUCTION

There are many different types of mechanical seals which, regardless of the differences in their appearance, fit into a fairly restricted number of categories. Classification is usually made first by design characteristics and then by positional arrangement (Fig. 1).

Conference organised and sponsored by the British Pump Manufacturers' Association
in conjunction with NEL and BHRA, The Fluid Engineering Centre.
Co-sponsored by the Process Industries Division of the Institution of Mechanical Engineers.

Wedge or "O" ring, balanced or unbalanced; these are design characteristics. The classification of seals by orientation such as rotating or stationary mounted, single, double or tandem, addresses positional arrangement regardless of their design.

The use of various seal arrangements was accelerated at a time before the advent of modern high performance very hard seal face materials which have come into common use because they offer to overcome the inadequate resistance to abrasion and poor lubrication of the materials then available.

These sealing arrangements are a totally accepted feature of sealing standards and industry, although there is clear evidence that seal users find that the reliability and initial cost of multiple seal installations is often less than totally satisfactory or acceptable.

A double seal arrangement is the basic response to "difficult" services which involve toxic, non-lubricating or abrasive liquids. Gas can be considered a particular example of a non-lubricating fluid. Two seals, mounted back-to-back (Fig. 2), operate in an artificial clean liquid environment created by an externally provided barrier liquid. In principle the life of the installation was then independent of the "difficult" liquid being pumped but there are some practical reservations.

The first is that the barrier liquid has to be maintained at a higher pressure than the pressure of the pumped liquid immediately adjacent to the seal. This pressure is difficult to define precisely under all operating conditions and in practice the pressure of the barrier liquid has to be considerably higher to ensure that the primary liquid between the seal faces is essentially that of the barrier. Some sort of instrumented support system is generally necessary the cost of which depends as much on the standards to which it is constructed as the duty requirements. In API 610 the double back-to-back system is Plan 53.

Some seal designs, notably metal bellows and "O" ring types can withstand, or be designed to withstand, some degree of pressure reversal i.e. sealed liquid pressure higher than barrier pressure. This is the wrong way for pressure to be applied to most seals because reversing the pressure tends to unload and possibly even open the faces leading to leakage into the barrier system. If this happens it calls into question the logic of the pressurised barrier double seal arrangement. Wherever possible therefore, seals designed to withstand pressure reversal should be used.

While the higher barrier pressure should prevent escape of sealed liquid into the barrier, cross migration up the pressure gradient undoubtedly occurs in practice. This means therefore however that contamination of the barrier liquid will occur and that controlled monitoring of it is necessary to establish the appropriate maintenance programme.

Pressure reversal can occur for a number of reasons:-

1) Pumped liquid pressure surging.

2) Unsatisfactory system maintenance.

3) Loss of pressure source.

4) Loss of barrier liquid.

To a greater or lesser degree, all these reasons are related to standards of maintenance and equipment operation so that a double seal system less prone to these upsets is desirable.

Such is the tandem arrangement which consists of two generally identical seals mounted one behind the other in the same direction (Fig. 3). The barrier liquid is, as before, in between the two seals but in this case it is at a lower pressure than that being sealed. The API code for this arrangement is Plan 54.

Tandem seals are used for three main reasons:-

1) Redundancy. In a tandem arrangement the inner seal effectively does all the work of sealing. In the event of primary seal failure the second seal will contain leakage and spillage into the local environment and may also allow extended operation if maintenance is inconvenient.

2) To contain a quench fluid used to assist the function of the inner seal. In this mode the outer seal can be considered to be an improved and effective quench bush for, typically, crystallising duties such as caustic or sugar.

3) Direct fugitive emissions. Normal emissions will occur from the inner seal which will enter the barrier fluid system. Arrangements are therefore needed either to vent the system appropriately where these emissions vaporise, or to monitor the condition of the system liquid and change it when necessary.

The inherently simpler system associated with the tandem seal arrangement has increased its popularity particularly for controlling fugitive emissions. This has been particularly so of light hydrocarbon duties both in refining and production. One basis of this popularity has been the idea that if the arrangement is quite simple in both design and operation then it is more likely to be effective and reliable. This can only be so if the whole sealing arrangement is properly engineered and the primary seal is fully up to its duty. Unlike the back-to-back double, the barrier liquid system for a tandem arrangement should be regarded as providing emission containment rather than essential operating support to the inner seal.

UPSTREAM PUMPING PRINCIPLES

It is not within the scope of this paper to cover the basic theory of spiral groove technology much of which has been published and summarised since its development began. The author's company has published a number of papers on the application to gas sealing. However it is appropriate to describe its application in a new way which offers significant improvements to rotating shaft mechanical seal performance.

Fig. 4 illustrates a spirally grooved flat lapped disc the radial width of which can be varied to suit the application. Grooving is only part way across the face from the inside diameter and this is the normal configuration for its application to liquid sealing. The harder material of the seal face pair carries the grooves and is normally the rotating element. This disc is then placed against a similar but ungrooved stationary flat face.

A liquid is placed in the bore of the disc will enter the grooves and, with the rotation shown, will be driven outwards towards the closed ends of the grooves mainly by viscous shear. In effect it is a pump operating at closed valve and considerable pressure can be generated.

These two flat faces are those of a conventional mechanical seal so that one of them is free to move axially to accommodate relative movements but is restrained by the closing forces derived from the sealed pressure outside the seal and the spring load. The pressure generated by the grooving therefore tends to open the faces by opposing the total closing load and this has two effects.

    1) A flow of the liquid from the bore to the outside is promoted.

    2) The pressure generated by the grooving reduces in relation to the face separation.

The result is that an equilibrium condition is reached which is inherently stable.

In a conventional mechanical seal the pressure drop profile across the faces has, in general, to be assumed but it will vary with the liquid and throughout the life of the seal as wear of the flat faces occurs. This is illustrated in Fig. 5.

Seal designers try to ensure that under the worst operating conditions the pressure penetration of the faces will not overcome the closing forces to the extent that gross leakage occurs. In some cases this can be in conflict with the force balance under normal operating conditions so that a compromise is made. It is therefore of considerable interest to the seal designer to be able to control the contour of the pressure variation across the faces. The term "variation" rather than "drop" is used because, if effect 1) is promoted, it must be because the pressure at the closed end of the groove is higher than the sealed pressure. This is shown in principle in Fig. 6.

The clear area of face running around the grooves provides the stop, or dam, which creates this higher pressure but it also has another function. This is to provide an effective seal on shutdown when, of course, the dynamic effect is lost.

SEAL PERFORMANCE AND CALCULATIONS

The key difference between an upstream pumping seal and a conventional type is the absence of contact between the two radial seal faces. There is therefore a complete axial force balance in the former case whereas with the normal seal the net axial closing force is made mechanically between the two faces.

With reference to Fig. ?, the closing forces in both seal forms are the sum of the net axial hydraulic force plus spring load.
This is represented by the following equation.

$$F_c = P_o(r_o^2 - r_b^2) + P_i(r_b^2 - r_i^2) + F_{sp} \quad (1)$$

The face opening force is simply the integration of the hydraulic penetration pressure profile radially across the face. The pressure profile is usually abbreviated to an average factor "k" of the hydraulic pressure. The opening force is represented by the following equation.

$$F_o = k(P_o - P_i)(r_o^2 - r_i^2) + P_i(r_o^2 - r_i^2) \quad (2)$$

It is generally assumed that the value of "k" for conventional seals can have a value from 0 to 1. Zero represents the full pressure drop at entry to the seal faces and 1 is full pressure penetration across the whole width of the faces. In the absence of definitive data "k" is often assumed to be 0.5 for parallel faces.

However with the upstream pumping concept the "leakage" direction is reversed and the pressure Pd achieved must be greater than the hydraulic pressure Po. The opening force is given by the following.

$$F_o = k_{di}(P_d-P_i)(r_d^2-r_i^2) + k_{do}(P_d-P_o)(r_o^2-r_d^2) \quad (3)$$
$$+ P_i(r_d^2-r_i^2) + P_o(r_o^2-r_d^2)$$

For there to be no mechanical contact, the opening and closing forces must be in stable equilibrium so that Fc/Fo = 1. This is unlike the conventional seal in which Fc/Fo usually lies between 1.2 and 3.0.

POWER CONSUMPTION

This is the sum of rotational turbulence, interface viscous shear and interface mechanical friction. The first factor can be controlled to a degree by design, particularly of the installation but is not generally considered to be very significant up to 3000rpm. Interface viscous shear is considerably reduced in the upstream pumping seal because of the slightly thicker interface film used and mechanical friction is eliminated. One benefit of this is that film temperatures are reduced so that simple dead ended quench arrangements can be employed up to quite a high seal performance requirement.

PUMPING RATE

Although a shear effect device, in practice it is relatively independent of viscosity and specific gravity. Pumping rate is influenced primarily by face velocity groove design parameters and the pressure gradient required. The calculation is beyond the scope of this paper but typical pumping rates lie between 1.0 and 15.0 ml/min.

THE EFFECTS OF PRESSURE AND HEAT GENERATION

In order to maintain control over the upstream pumping effect it is necessary to control (a) the key face component distortions produced by the hydraulic pressures surrounding them and (b) thermal movement created by temperature gradients through these same components away from the highest temperature region at the interface. If the two faces move significantly out of parallel the profile of the generated pressure can disrupt seal performance either by starvation or insufficient pumping effect caused by too large a face separation.(Fig. 8)

It is therefore necessary to adopt the appropriate analytical techniques to assess and optimise the effects of both pressure and temperature. One of the significant effects of upstream pumping is, however, a considerable reduction in power consumed and hence reduced face temperatures, compared with conventional seals.

SEAL ARRANGEMENT

Referring back to the descriptions of the back-to-back and tandem seals, it can be seen that this spirally grooved device produces the effect required of the back-to-back but with a liquid pressure in the bore which is less than that sealed, i.e. as for a tandem arrangement. All that is therefore necessary is to provide a low pressure containment which can be any of the standard forms such as gland packing or a lip seal. However, bearing in mind the potential requirement for reliability, a low pressure mechanical seal is preferred. It may also be appropriate for it to be in the form of a cartridge for practical reasons.

Fig. 9 illustrates such an arrangement in which the spirally grooved seat (1) is on a sleeve (2) and rotates with the shaft. The spring loaded counterface is stationary and is the configuration preferred when careful control of the seal function is required. This is most often seen with conventional seals when running at high speed. There are a number of incidental design features of this seal arrangement but the essential points are an internally pressurised outer conventional ungrooved seal (3) retaining the low pressure barrier liquid entering the interspace at (4).

In its simplest form, the barrier liquid is supplied from a header tank which is manually replenished. The next stage might be to provide some form of automatic filling but essentially the system remains a static one unless the operating temperature is high in relation to the atmospheric boiling point of the barrier liquid or the seal generated heat reaches the guidelines given later.

BARRIER LIQUID TRANSFER

The barrier liquid selected for a particular application will necessarily be compatible with the liquid being sealed as it will be pumped by the sealing mechanism into the process. The preferred viscosity range is between 0.4 and 10.0 cP with a vapour pressure less than 0.3 bar at pumping temperature.

Experience suggests that upstream pumping seals will operate satisfactorily at pumping rates frrom 0.1 to more than 15 ml/min. The particular rate chosen at the design stage will depend upon a number of factors. For process and maintenance reasons a pumping rate at the lower end of the range may be required but to ensure seal integrity a higher rate may be advised.

Typical barrier liquids are water, alcohol, hydraulic oil and kerosene but many others are possible provided they are clean; that is they contain no abrasives or dissolved materials which might settle out on the faces to inhibit the pumping performance.

SEAL POWER CONSUMPTION

The power consumption of the upstream pumping seal is considerably less than that for a conventional face contacting seal and would be calculated at the time of seal selection. That for the low pressure outboard seal is calculated in the normal way and depends upon size, speed and sealed pressure. It would be comparable to that of the same seal in a standard tandem arrangement. However as a rule of thumb, the total upstream pumping package consumption would be less than that for a single seal on the full duty.

It was pointed out earlier that barrier liquid supply arrangements were matched to the seal power consumption. These are indicated below but some variation may be expected arising from the choice of barrier liquid and the operating temperature.

kW	Piping Layout
<0.15	One line from the barrier reservoir to the seal.(Dead Ended) Fig. ?.
0.15-0.3	Flow and return lines to the barrier reservoir (Inlet and Outlet) to enhance thermosyphon circulation.
>0.3	Flow and return lines to the barrier reservoir (Inlet and Outlet) with positive circulation.

OPERATIONAL FEATURES

A quick calculation based upon a face separation of 3 microns and a pumping rate of 2 ml/min. will confirm that the interface film of a 50mm seal with a face width of 4mm is changed about once per second at 3000rpm. The basic feature of this new seal is therefore its ability to apply a clean flush at a predetermined rate precisely to the parts which benefit most from it. Previously this was only possible in a general way by providing a general relatively very much greater level of flush to the seal cavity with the consequent potential for unacceptable contamination of the pumped liquid.

This suggests that the upstream pumping principle has application to liquids which require highly controlled emissions and to liquids which are abrasive either through dissolved or undissolved solids.

The absence of mechanical contact between the faces in operation coupled with the cleaning action of the barrier liquid virtually eliminates wear and maintains the predictable characteristics of the device. This suggests that reliability, repeatability and service life should be improved over that of the equivalent conventional mechanical seal.

Another consequence is a reduction in total power consumption, not so much from the device itself as from the elimination of the loss of pumping efficiency arising from recirculation to a single seal or barrier liquid pumping power with conventional double seal arrangements.

The simplification of the barrier liquid system leads to improved operation and the control of the rate of contamination or dilution of the pumped liquid gives improved process control.

The essence of upstream pumping is that it provides the centrifugal pump with a simplified double seal arrangement which improves virtually all aspects which previous double seal arrangements have been criticised for.

OPERATING EXPERIENCE

From a number of installations already in operation, two will be used to illustrate extremes of application

	1.	2.
Application	Sour Water	11% Carbamate
Pump	ANSI Single Stage	Two Stage O/Hung
S.G.	1.0	1.15
Temperature	38 deg.C	80 deg.C
Viscosity	0.7 cP	1.0 cP
Speed	1760 rpm	15600 rpm
Suction Pressure	1.03 bar	--NA--
Discharge Pressure	5.86 bar	--NA--
Sealed Pressure	1.5 bar	83 bar
Barrier Liquid	Water @ 26 deg.C	Water @ 28 deg.C
Barrier Pressure	1 to 1.35 bar	7 bar
Barrier System	Dead Ended	Dead Ended
Seal Size	38 mm	50 mm
Stationary Seal Ring	Carbon	Carbon
Rotating Seal Ring	Tungsten Carbide	Tungsten Carbide
Effective Face Width	3.55 mm	8.9 mm
Pumping Rate	0.4 ml/min	16 ml/min

Installation 1. was required to reduce high flush water consumption.

Installation 2. is an abrasive service and the primary requirement was
to improve installed seal life.

Performance.   Both seals have been running between 1 and 2 years.

SUMMARY

The face profiling treatment described offers to users of conventional
centrifugal pumps a mechanical seal associated with a low pressure
sealant system with significant all round benefits.

(a) Improved pump efficiency.
(b) Extended seal life.
(c) Emission control.
(d) No barrier fluid contamination.
(e) Minimum process contamination.
(f) Effective seal performance on difficult liquids.

These translate into tangible potential reduction of first installation
and upgrading costs. It is foreseen that the advance in seal life
expectancy and reliability could bring similar improvements to
maintenance planning.

REFERENCES

Use has been made in this article of material from the following
references

1. Schoenherr, Karl "Fundamentals of Mechanical Seals" reprinted from
    Iron and Steel Engineer, reprinted in Engineered Fluid Sealing,
    John Crane UK 1976.

2. Sedy, Josef, "Improved Performance of Film-Riding Gas Seals Through
    Enhancement of Hydrodynamic Effects", ASLE Transactions Vol 23, 1
    35-33, 1978.

3. US Patent No 4,290,611 "High Pressure Upstream Pumping Seal
    Combination", Josef Sedy. Issued September 22, 1981

4. Buck, G S Selection and Design of End Face Mechanical Seals for
    Common Refinery Services, Louisiana State University, 1978.

ACKNOWLEDGEMENTS

The author would also like to thank Mr A Ali of John Crane Inc for his
support in the preparation of this paper.

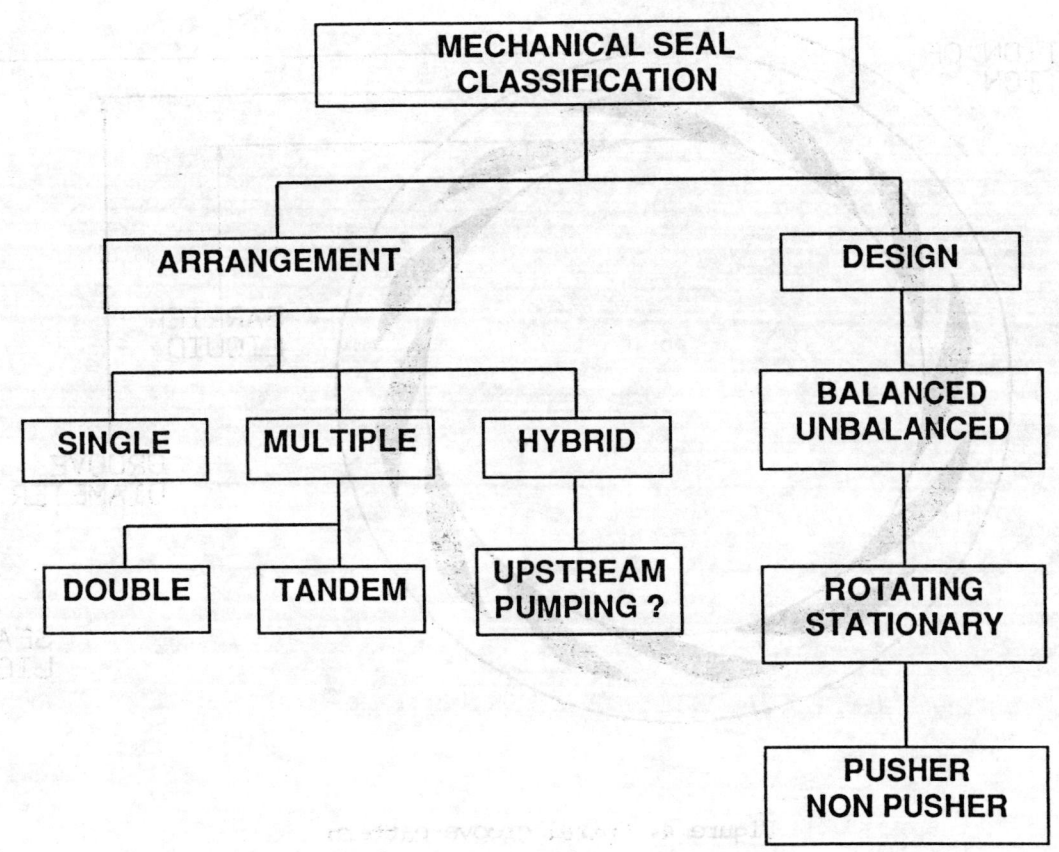

Figure 1. Mechanical seal classification

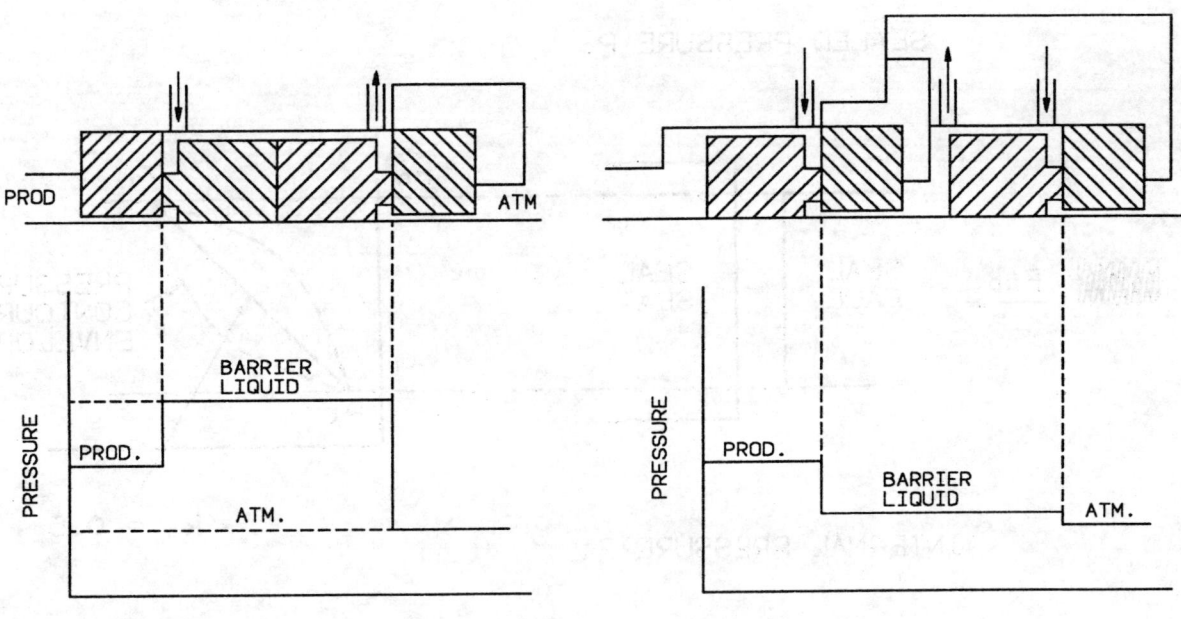

Figure 2. Double seal

Figure 3. Tandem Seal

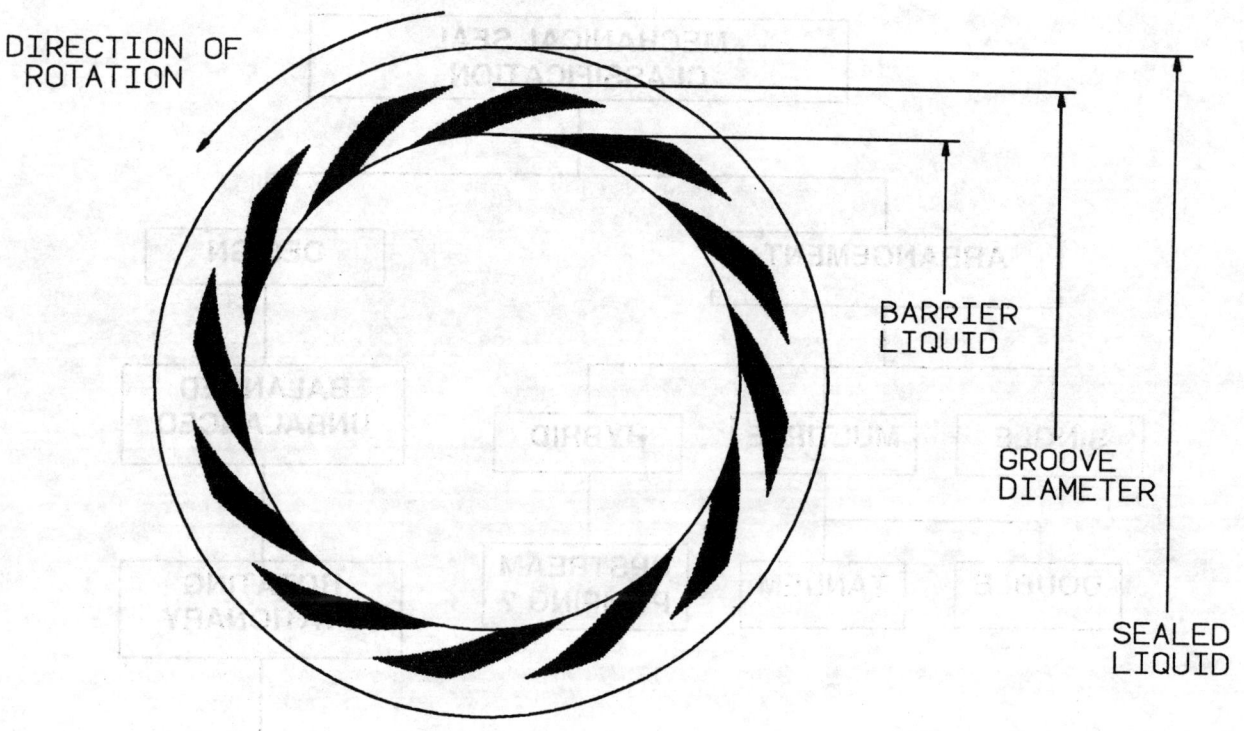

Figure 4. Spiral groove pattern

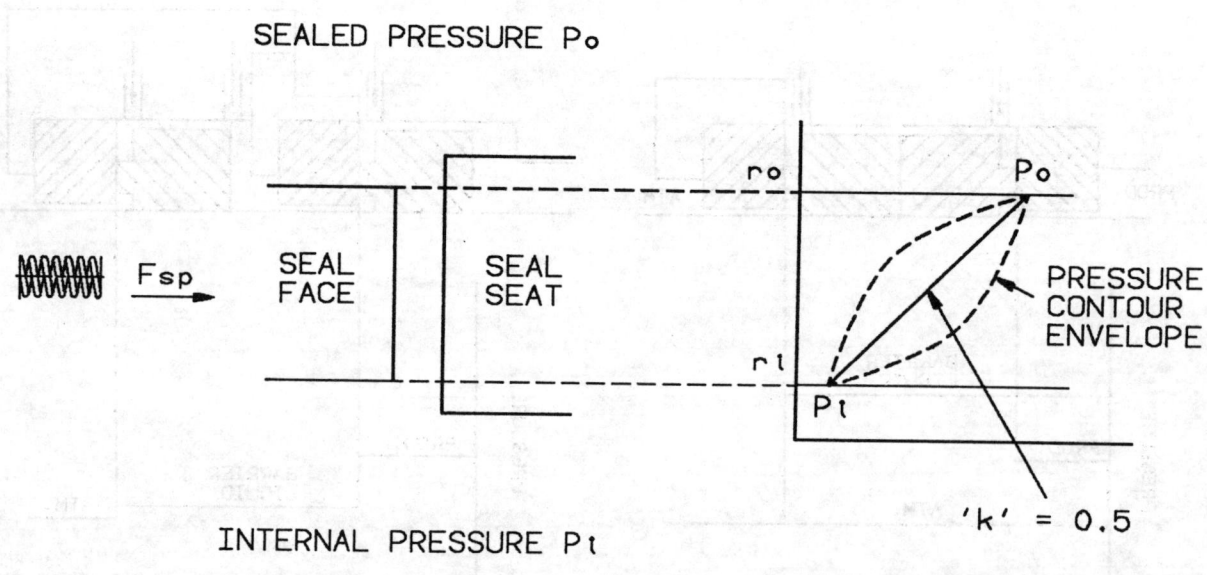

Figure 5. Conventional flat face seal pressure decay contour

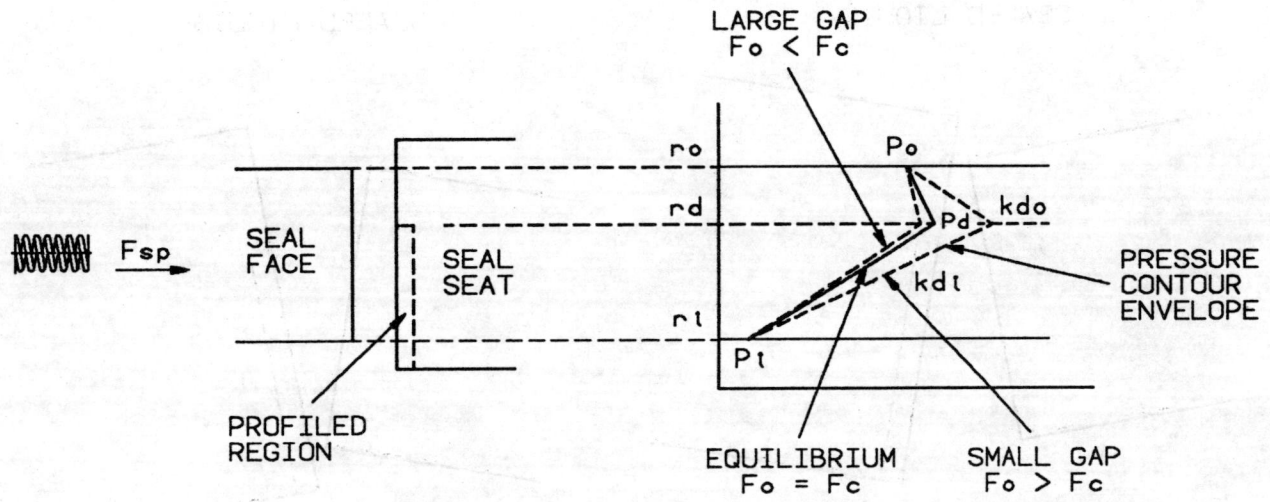

Figure 6. Profiled face pressure contours indicating stability mechanism

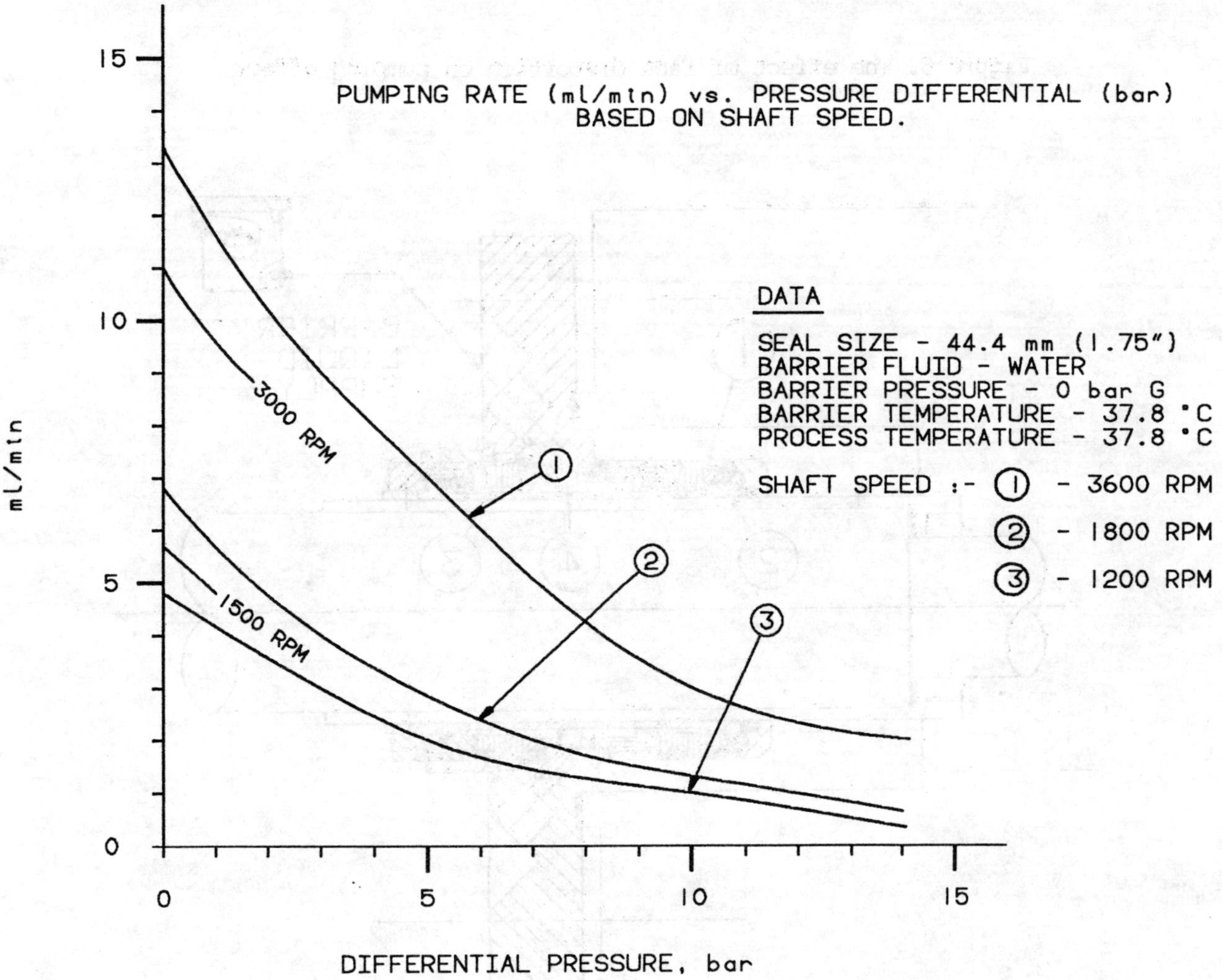

Figure 7. Upstream pumping seal performance

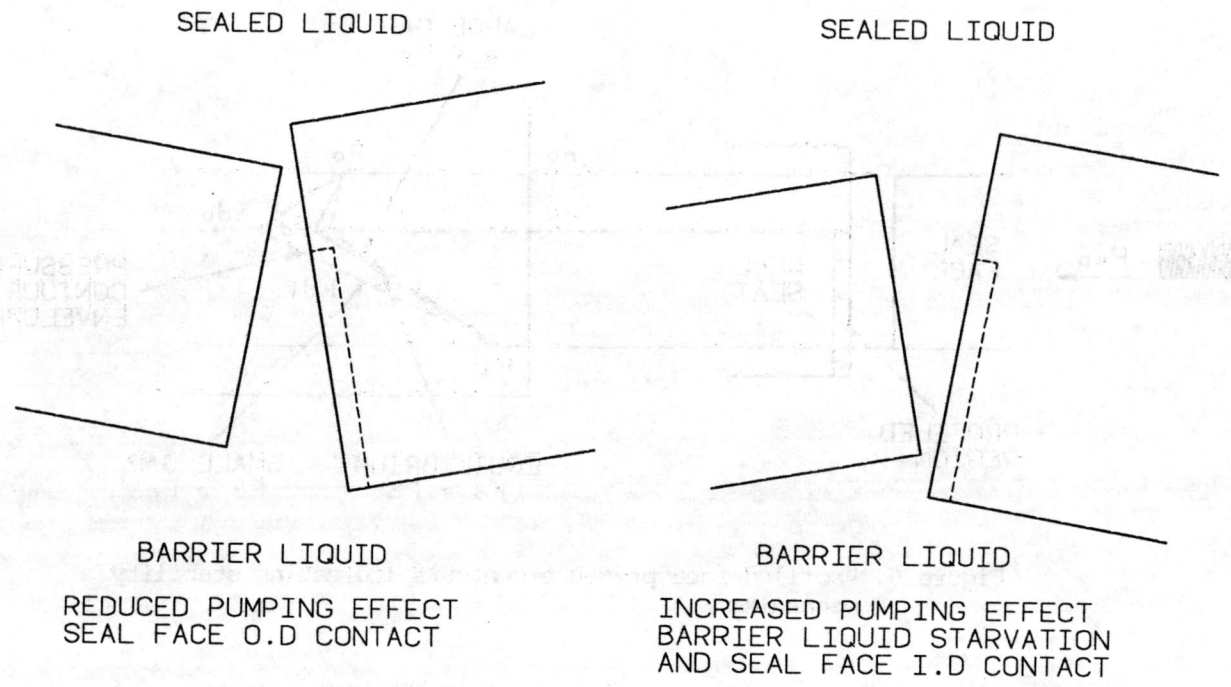

Figure 8. The effect of face distortion on pumping effect

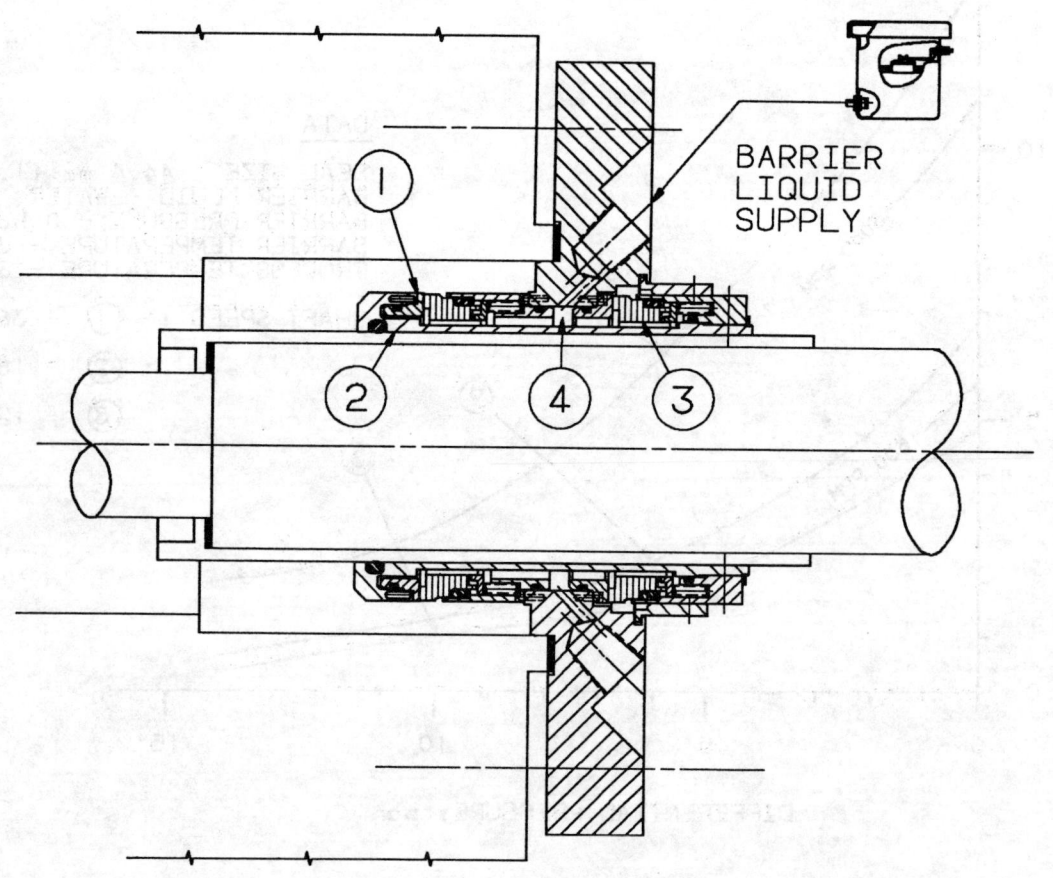

Figure 9. Upstream pumping seal cartridge with header tank

JAN 1 0 1990